civics+
citizenship

history

geography

economics+
business

contents

good humanities

Matilda Education Australia acknowledges the Aboriginal and Torres Strait Islander peoples of this nation. We acknowledge the traditional custodians on whose unceded lands we conduct the business of our company. We pay our respects to ancestors and Elders past, present and emerging. Matilda Education Australia is committed to honouring Aboriginal and Torres Strait Islander peoples' unique cultural and spiritual relationships to the land, waters and seas; and their rich contribution to society.

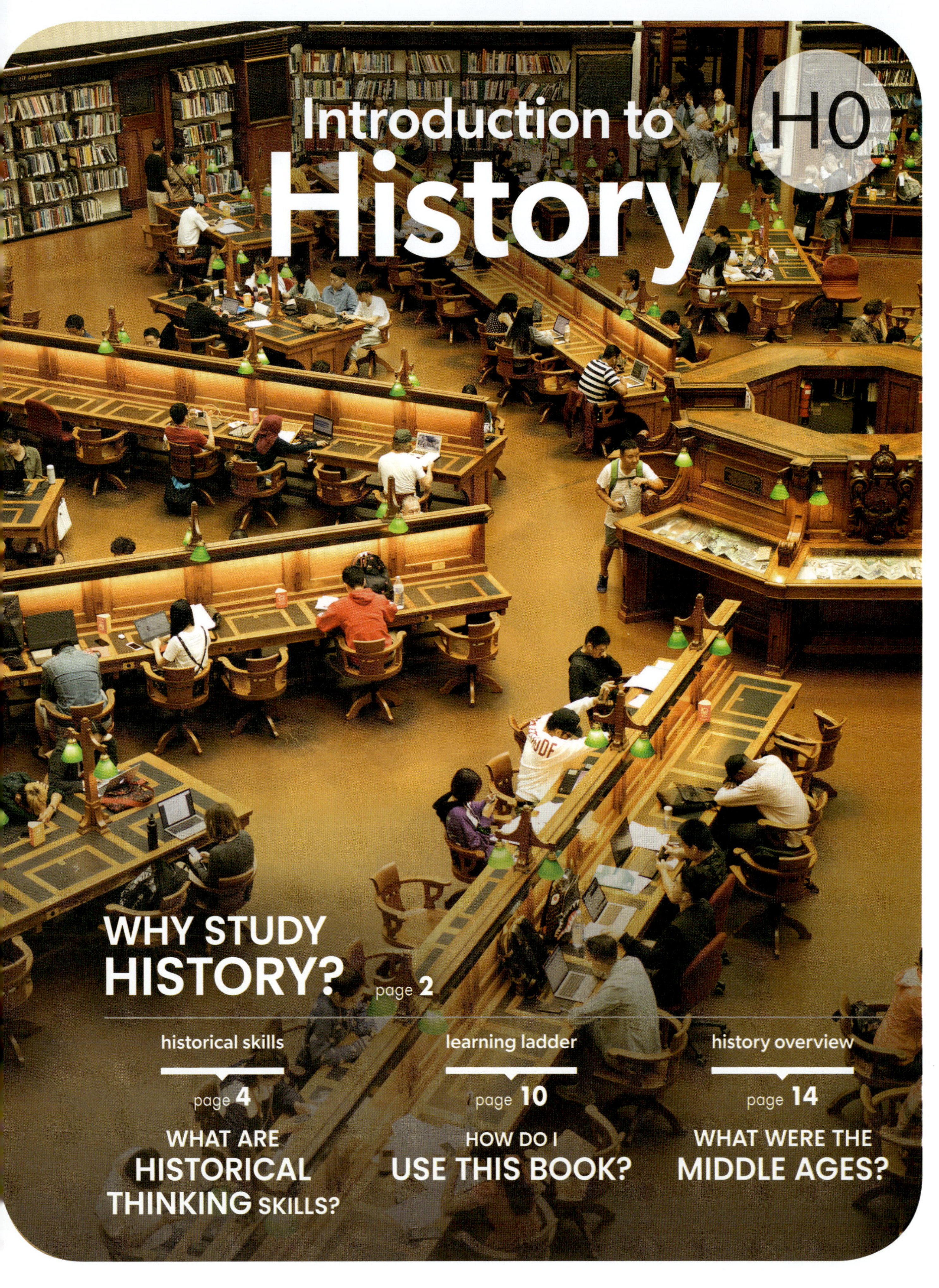

Introduction to History

H0

WHY STUDY HISTORY?

Why study history?

History is not just a set of facts about what happened in the past. Studying history shows you how the world you live in came to be.

The benefits of studying history

It is easy in the busy, noisy modern world to forget that every invention, every nation and every concept has a history that is years or even thousands of years old. The world didn't pop into existence yesterday. Learning about history helps to remove our bias towards the present. It helps us realise there is a lengthy chain of events leading to modern times.

Another benefit of studying history is that it develops important skills that are useful in many areas of life: how to ask intelligent questions, express opinions, think critically and to research and communicate your findings.

Source 1

Stonehenge, in England. Mystery surrounds its construction, as stones were somehow transported hundreds of kilometres before heavy transport was available.

Source 2

Pietro Giannone was an 18th-century historian. By studying sources from this period, historians can travel back to Giannone's time.

Learning about history is also the closest thing you'll ever get to travelling back in time. Do you want to know how Stonehenge was built, or how the Aztecs constructed a pyramid-filled city on a lake? Interested in samurai and ninja? Do you want to know how Vikings explored North America? Or do you wonder how the Mongols built the biggest land empire in the world? Take a trip back in time by studying history.

Perhaps most importantly, studying history can help you to gain a better understanding of yourself as a human being. Human history is a long story about everything people have ever done. There is a lot we can learn from this story – not only what we should and shouldn't do, but also who we are. Learning about history helps shape our identity.

What do historians do?

Many people think history is just a list of information about what happened in the past, but history is not just a catalogue of facts to be memorised. History is a process and a way of thinking.

Those who study history, follow its processes and use its way of thinking are called **historians**. There are three parts to a historian's role:

1. Ask a question, such as 'What happened here?'. 'How did it happen?' or 'Who was behind it?'
2. Examine sources to uncover relevant evidence.
3. Use the evidence to answer the question or to tell a story about the past.

As a history student, you are an apprentice historian. Your goal when learning about history is not to memorise facts, although having content knowledge helps. Instead, you should aim to develop your historical skills.

Learning ladder H0.1

1. List four reasons why we should study history.
2. What do historians do?
3. Why is developing historical skills more important to your studies than memorising facts?
4. Select an event from your family's own history, then ask three historical questions about it. How could you go about answering your questions?
5. As a class, discuss this statement: 'Only by understanding history can we understand ourselves'. Do you agree or disagree?

What are historical thinking skills?

History is a way of thinking, and several important skills support this way of thinking. As an apprentice historian, you will develop and practise your historical thinking skills over the course of your Year 8 studies.

Chronology

Chronology is the process of organising events into the order they happened. This brings structure and order to events, allowing historians to study them more effectively.

Ages and eras

Historians divide time into ages, to make large durations of time easier to understand. There are many ways historians divide up time. Here are a few examples:

History/prehistory

- **Prehistory**: any time before 3500 BCE
- **History**: any time after 3500 BCE

Why 3500 BCE? That is when writing was invented in ancient Sumer (modern-day Iraq). Once writing spread to other civilisations, there were many more sources of information historians could study.

Stone/Bronze/Iron Ages

This classification is based on the most advanced material used in technology at that time.

- Old Stone Age (Palaeolithic): 3 300 000–10 000 BCE (before farming is developed)
- New Stone Age (Neolithic): 10 000–3000 BCE (after farming is developed)
- Bronze Age: 3000–1200 BCE (bronze metal and materials used)
- Iron Age: 1200 BCE–500 BCE (iron materials used)

Curriculum eras

In Victorian schools, students study four eras of history outlined by their curriculum.

- Year 7: Ancient history 60 000 BCE–650 CE
- Year 8: Medieval history 650–1750 CE
- Year 9: Premodern history 1750–1918 CE
- Year 10: Modern history 1918 CE–Present

Source 1

Iron soup pots like these were advanced technology during the Iron Age. This is the Battersea Cauldron found near the River Thames in London, England. It is made of copper alloy and has been dated to the Iron Age, circa 800–700 BCE.

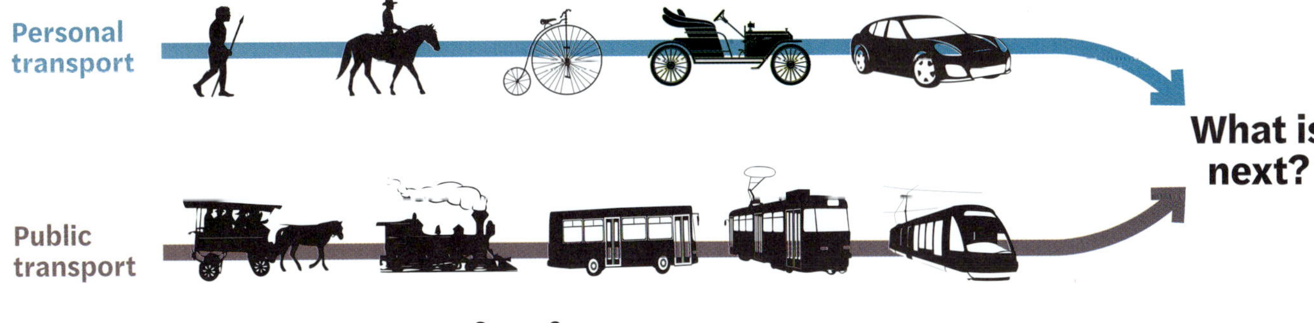

Source 2

Within a few hundred years, land transport moved from walking or horse riding to using cars, trains and bicycles.

The march of progress

As history progresses, the speed of change increases. For thousands of years there was little change in how humans behaved. Now, a century is a very long time. People living in 1900 had very different daily lives than people today.

We also know a lot more about modern history than ancient history. The further back in history we go, the less we know about it. This is because there are fewer sources from the distant past. For some ancient peoples, we might only have a few pieces of buried pottery or skeletons. In the modern world, on the other hand, people upload over 300 hours of video to YouTube every minute!

Chronology terms

Historians use a set of common terms to help them organise dates chronologically. These terms are also important in historical writing. Source 3 lists some major terms.

Numbering years and centuries

Although we live in the 21st century, the numbers for our years all start with '20 ...'. Ever wonder why? It's because the 1st century was from 1–100, not from 101–200. For example:

- 1–100 CE: 1st century
- 101–200 CE: 2nd century
- 1001–1100 CE: 11th century
- 1801–1900 CE: 19th century
- 1901–2000 CE: 20th century
- 2001–2100 CE: 21st century (the one we are in now)

Note: When creating timelines, there is no 'year zero'. Timelines go from 1 BCE straight to 1 CE.

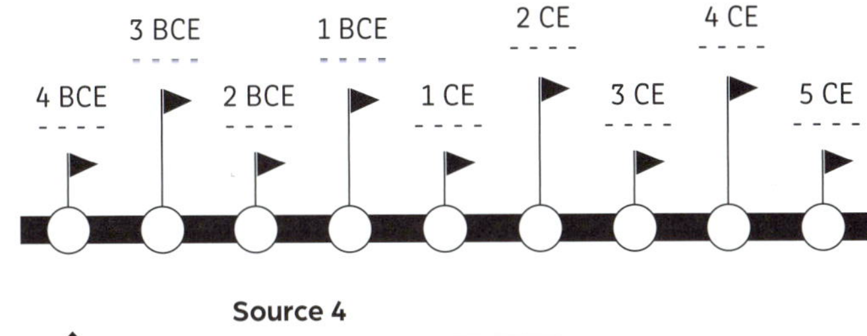

Source 4

This timeline goes from 1 BCE straight to 1 CE.

Source 3

Common chronology terms

Term	Stands for ...	Meaning
BCE	Before Common Era	The period of time before the birth of Christ (before 1 CE); this used to be called 'BC', for 'Before Christ'
CE	Common Era	The period of time after the birth of Christ (after 1 CE); this used to be called 'AD' for the Latin phrase Anno Domini, which means 'in the year of our Lord'
BP	Before Present	Years before the present day (often standardised to mean `years before 1950')
c.	circa	The Latin word for 'approximately'
decade	10 years	
century	100 years	
millennium	1000 years	

Source analysis

Using sources to answer historical questions and build narratives is essential to what historians do. Sources come in two types.

Primary sources were created at the time of a historical event, or created by someone who had first-hand experience of the event. Examples of primary sources include books, diaries, photographs, archives, letters, **artefacts**, buildings and ruins.

Primary sources often show the perspectives of the people who experienced the event itself. They might have unique information about an event, because 'you had to be there'.

Secondary sources were created *after* the time or event being studied. Examples of secondary sources include textbooks, websites and documentaries. These sources are often created by historians combining primary sources together to interpret the past or to tell a narrative story about it.

Bias and reliability

Just because primary sources were created at the time being studied, it doesn't mean they are always reliable. Primary sources can be **biased** (unfair or prejudiced). Perhaps a medieval scribe hated a leader and so wrote down nasty and untrue things about them. Historians might then find this ancient text centuries later and read it. Should they believe what it says?

The way to minimise bias is to look at many different sources. Secondary sources can be helpful for this, as they often bring many primary sources together.

Secondary sources can also be biased, of course. The most common secondary source you will use are websites, and judging the reliability of a website is very important. There are many sites that don't state where their information is from, making it hard to determine their reliability.

Continuity and change

History is a story of continuity and change: some things stay the same, others change.

Some things have changed a lot over the course of history, such as transport or communication technology. Other things have changed very little, such as the bond between family members. Then again, ideas of what makes a 'family' have changed – so you need to consider continuity and change, without making assumptions.

Narratives (historical 'stories') and timelines will help you work out what has changed and what has stayed the same. You will learn how to recognise continuity and change, and how to describe how quickly and to what degree change happened.

Cause and effect

Historians are always trying to figure out *why* things happened. During your studies, you will learn about several different causes of change:

- actors (individuals and groups)
- conditions (social, political, economic, cultural and environmental)
- short-term triggers
- long-term trends.

Causes and effects can be organised using timelines or by writing historical narratives. One thing to remember is that most causes themselves have even earlier causes, and most effects have further effects in the future.

Source 5

Books written during a historical period are primary sources of that period; books written afterwards are secondary sources.

Source 6

Genghis Khan is considered historically significant. These models are wearing historically accurate traditional Mongolian dress.

Historical significance

History is everything that has ever happened. How do we narrow this down and decide what is worth studying? Determining whether something is historically significant requires you to make an *evaluation* or judgement.

Historians have come up with models to help guide us when trying to decide what is important. In this textbook, we use a model developed by Australian historian Geoffrey Partington. To help work out how important something is, ask:

1. How important was it to people at the time?
2. How many people were affected?
3. How deeply were people's lives affected?
4. For how long did these effects last?
5. How relevant is it to modern life?

Learning ladder H0.2

1. Rank the five history skills in order from easiest to hardest, in your opinion.
2. Chronology: Why do historians divide up time?
3. Sources: Which is more reliable, a primary or a secondary source?
4. Continuity and change: This year you will be studying the medieval period: the time of castles, Vikings and *shoguns* (and many other things!). List three things that have stayed the same since the Medieval period.
5. Cause and effect: What different kinds of causes are there?
6. Significance: What is one method to decide what is important in history?

What communication skills do historians need?

While thinking skills are vital for historians, communication skills are also very important. You need good reading and comprehension skills in order to research historical sources. You also need strong writing skills so that other people can understand your work.

How do I write for history?

Literacy is an essential skill for a historian. Being able to communicate is one of the most important skills you will learn. The writing process can be split into six stages:

- Step 1: I can identify the writing purpose
- Step 2: I can gather information
- Step 3: I can organise information
- Step 4: I can structure a piece of writing
- Step 5: I can write a draft
- Step 6: I can edit and proofread

The History How-To section discusses these stages in depth on pages 211–215.

Source 1

Researching is a core skill for historians.

How do I research like a historian?

Another core historical skill is researching. The research process can be divided into six parts:

- Step 1: I can define the problem
- Step 2: I can decide what information to find and where to find it
- Step 3: I can find information
- Step 4: I can extract information
- Step 5: I can organise and present information
- Step 6: I can evaluate information

The History How-To section discusses these stages in depth on pages 216–217.

What is plagiarism and how do I avoid it?

Plagiarism is when you use someone else's words or ideas as your own. Plagiarism is a problem because it is dishonest – you are pretending someone else's writing or research is yours. In addition, when you plagiarise you don't gain new knowledge or skills of your own.

Plagiarism includes:

- copying and pasting from a website, or just changing the words around
- using information from a source but not including it in your source list
- copying work from another student.

Avoid plagiarism by taking your own notes from research sources *in your own words*. When writing or researching, write using *your own notes*.

Source 2

When studying a society like the Vikings, you will learn about their defining characteristics, beliefs and values. [Hermann Vogel, *Viking Raid* (1881), wood engraving.]

What historical knowledge should I learn?

For each historical period you study, there are key questions posed by the Victorian Curriculum:

- How do we know about the ancient past?
- Why and where did the earliest societies develop?
- What emerged as the defining characteristics of ancient societies?
- How did societies change from the end of the ancient period to the beginning of the modern age?
- What key beliefs and values emerged and how did they influence societies?
- What were the causes and effects of contact between societies in this period?
- Which significant people, groups and ideas from this period have influenced the world today?

Throughout your Year 8 studies, ask yourselves these key questions as you learn about different periods and societies.

Learning ladder H0.3

1. How do historians benefit from good communication skills?
2. Why is plagiarism bad for your learning?
3. Is literacy important to historians? Justify your answer.
4. How can historical research be divided into stages?
5. List three of the key historical questions posed by the Victorian Curriculum.

How do I use this book?

Good Humanities has been built to help you thrive as you move through the Level 8 Humanities curriculum and to enable you to demonstrate your progress in every single lesson. The History section of this book includes five chapters of medieval history and a History How-To skills section. The History How-To section is vital – you should refer to it often.

Climb the History Learning Ladder

Each chapter begins with a Learning Ladder. The Learning Ladder is your 'plan of attack' for the skills you will practise in each chapter. It lists the five historical skills you will be learning, and has five levels of progression for each of those skills.

Each skill described in the Learning Ladder is of a higher difficulty than the one below it. To be able to achieve the higher-level skills, you need to be able to master the lower ones. Practising activities at all the levels will help you to master more involved skills, such as evaluating. This approach is called 'developmental learning' – and it puts you in charge of your own learning progression!

Read the ladder from the bottom to the top. As you progress through the chapter, you will climb up the Learning Ladder.

Source 1

The Learning Ladder helps you to take charge of your own learning!

learning ladder

Step	Chronology	Source analysis	Continuity and change	Cause and effect	Historical significance
step 5	I can describe patterns of change	I can use the origin of a source to explain its creator's purpose	I can evaluate patterns of continuity and change	I can evaluate cause and effect	I can evaluate historical significance
step 4	I can distinguish causes, effects, continuity and change from looking at timelines	I can use my outside knowledge to help explain a source	I can analyse patterns of continuity and change	I can analyse cause and effect	I can analyse historical significance
step 3	I can create a timeline using historical conventions	I can find themes in a source	I can explain why something did or did not change	I can explain why something is a cause or an effect	I can apply a theory of significance
step 2	I can place events on a timeline	I can list specific features of a source	I can describe continuity and change	I can determine causes and effects	I can explain historical significance
step 1	I can read a timeline	I can determine the origin of a source	I can recognise continuity and change	I can recognise a cause and an effect	I can recognise historical significance

Check your progress

Each chapter is divided into sections, with each section designed to cover one lesson. Sometimes your teacher might decide to spend more or less time on a particular section. A section is either two or four pages long.

At the end of most sections, you will find a block of questions called 'Learning Ladder', which has two different types of questions or activities:

1. **Show what you know:** These questions ask you to look back at the content you have read and viewed and to show your understanding of it by listing, describing and explaining.
2. **Learning Ladder:** These activities are linked to the Learning Ladder. You can complete one of the questions or more of them. In each chapter you will complete an activity for each level of the Learning Ladder. This will sharpen your historical skills.
The digital version of this book has additional multiple-choice questions for every section to help you test your abilities for each of the five historical skills.

key individuals   key events

Throughout every history chapter, you will discover a variety of features that focus on key individuals and events that have changed the course of history. You will also find 'History mystery' and 'how it works' features to explore. Many of these features link to interactive sources, so you can explore further using your eBook.

Source 3

Discover the people and events that have shaped our world.

Source 2

Check your progress regularly. You can attempt one or more of the Learning Ladder questions.

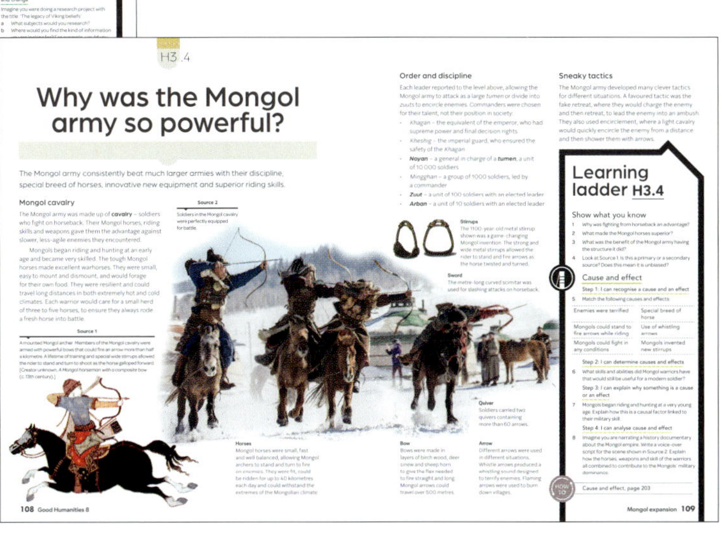

civics+ citizenship

economics+ business

The study of Humanities is more than just History and Geography – it is also the study of Civics and citizenship and Economics and business. In every chapter of this book you will discover either a Civics and citizenship lesson or an Economics and business lesson. School is busy and you have a lot to cover, so designing a textbook where the important Civics and citizenship and Economics and business content is placed meaningfully next to relevant History or Geography lessons makes good sense, and will help you to connect your learning.

As you work through the Civics and citizenship and Economics and business sections in this book, you will also be working your way up a Learning Ladder for these subjects too!

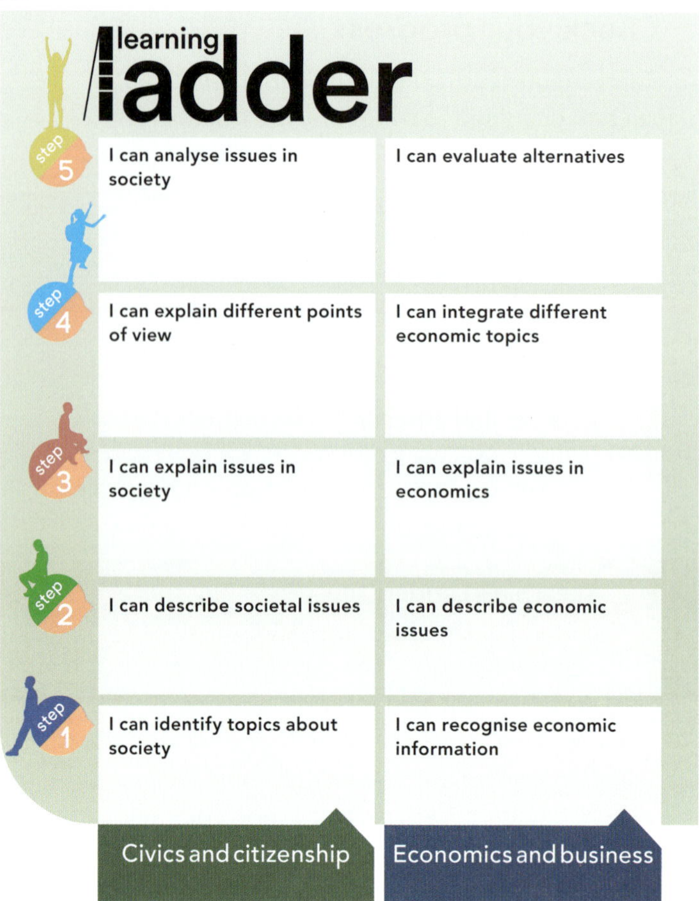

Source 4

Explore Civics and citizenship, and Economics and business, alongside your History course.

History How-To

In the middle of the book, you will find a skills section called 'History How-To'. This section explains how to perform each skill and the steps in writing and research. There are *lots* of worked examples. Refer to it often, especially when answering the Learning Ladder questions and completing the review activities.

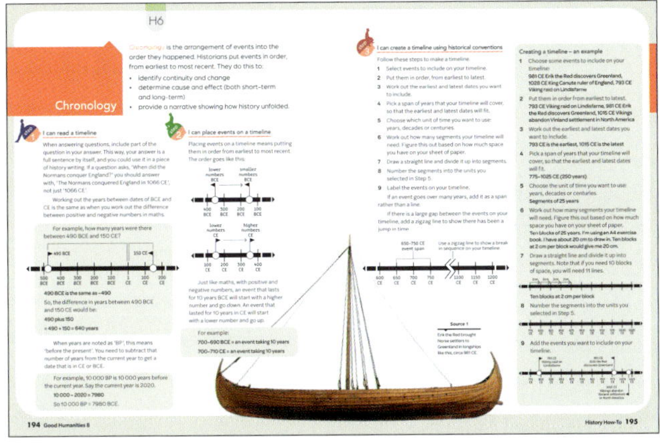

Source 5

The History How-To section is your key to success – refer to it often!

Masterclass

At the end of each chapter is a review section, called the Masterclass. The questions here are organised by the steps on the Learning Ladder. You can complete all of the questions *or* your teacher might direct you to complete just some of them, depending on your progress.

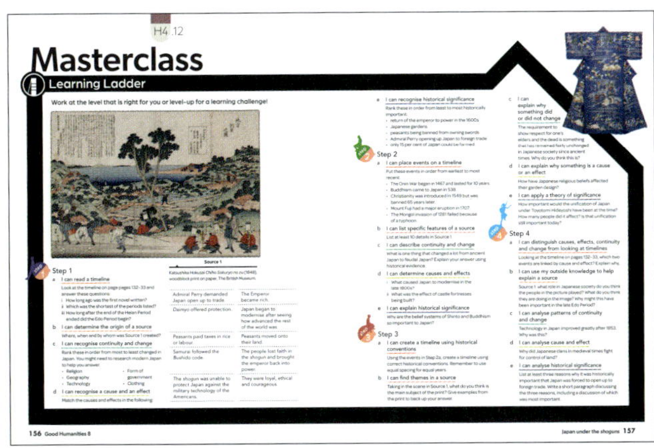

Source 6

You did it! We knew you could! The Masterclass is your opportunity to show your progress. Take charge of your own learning and see if you can extend yourself.

Capstone

After you complete a chapter, it's time to put your new knowledge and understanding together for the capstone project to show what you *know* and what you *think*. In the world of building, a capstone is an element that finishes off an arch, or tops off a building or wall. That is what the capstone project will offer you, too: a chance to top off and bring together your learning in interesting and creative ways. It will ask you to think critically, to use key concepts and to answer 'big picture' questions. The capstone project is accessible online; scan the QR code to find it quickly.

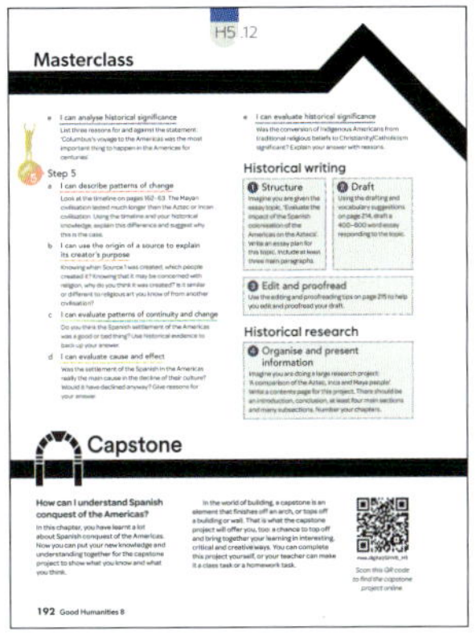

Source 7

The capstone project brings together the learning and understanding of each chapter. It provides an opportunity to engage in creative and critical thinking.

Learning ladder H0.4

1. What are the different types of questions in this textbook? Describe them in your own words.
2. How can you use the Learning Ladder to monitor your progress in Year 8 History?
3. As a class, discuss the idea of 'monitoring your own progress'. Why is this important?
4. Read through the steps of the History Learning Ladder and consider where you might already be up to for each skill, based on your prior learning.

What were the Middle Ages?

The **Medieval Period** is the name given to the time from about 400–1400 CE. The period is also known as the **Middle Ages** because it is in the middle of two other major periods in European history. Before the Middle Ages was the classical period of Greece and Rome, which ended when the Roman Empire fell. The Medieval Period was then followed by the Renaissance, a rise of cultural and intellectual advancement. Its beginning marks the end of the Middle Ages.

Both of these names are Eurocentric ways of dividing up time, meaning the period mostly makes sense when we consider European history. However, there were also many historical developments happening in non-European societies.

A period of change

The Medieval Period was around one thousand years long. A lot can happen in a thousand years, and a lot did; the Middle Ages were a time of significant change.

The Roman Empire began to decline from 200 CE and fell in 476 CE. The end of the Roman Empire made Europe more dangerous. There were more minor wars, and travel and trade were more dangerous. Vikings from northern Europe impacted the continent from about 800–1050 CE as they raided coastal areas. Mongols conquered huge areas of land across central and northern Asia in the 13th century.

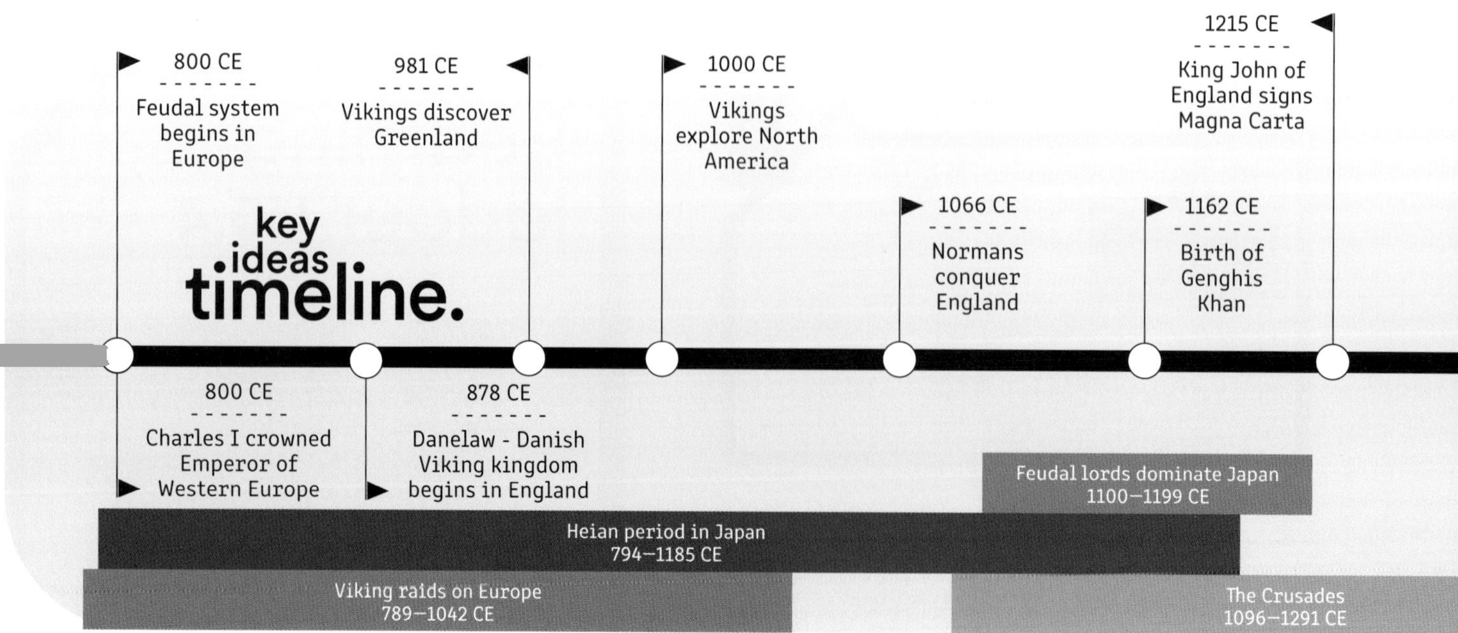

Source 1

This painting from the 1411 Toggenburg Bible (Switzerland) shows a man and woman suffering from the bubonic plague.

Agriculture improved in many ways. Better farming methods and tools meant more food, which meant population growth. Bigger populations meant more towns and larger cities, while better land and sea transport led to a big increase in trade within and between regions. However, the increase in the human population and the number of domesticated farm animals living close to humans also led to an increase in disease. The bubonic plague, or **Black Death**, devastated the continent in the Middle Ages, killing 30 to 50 per cent of Europeans.

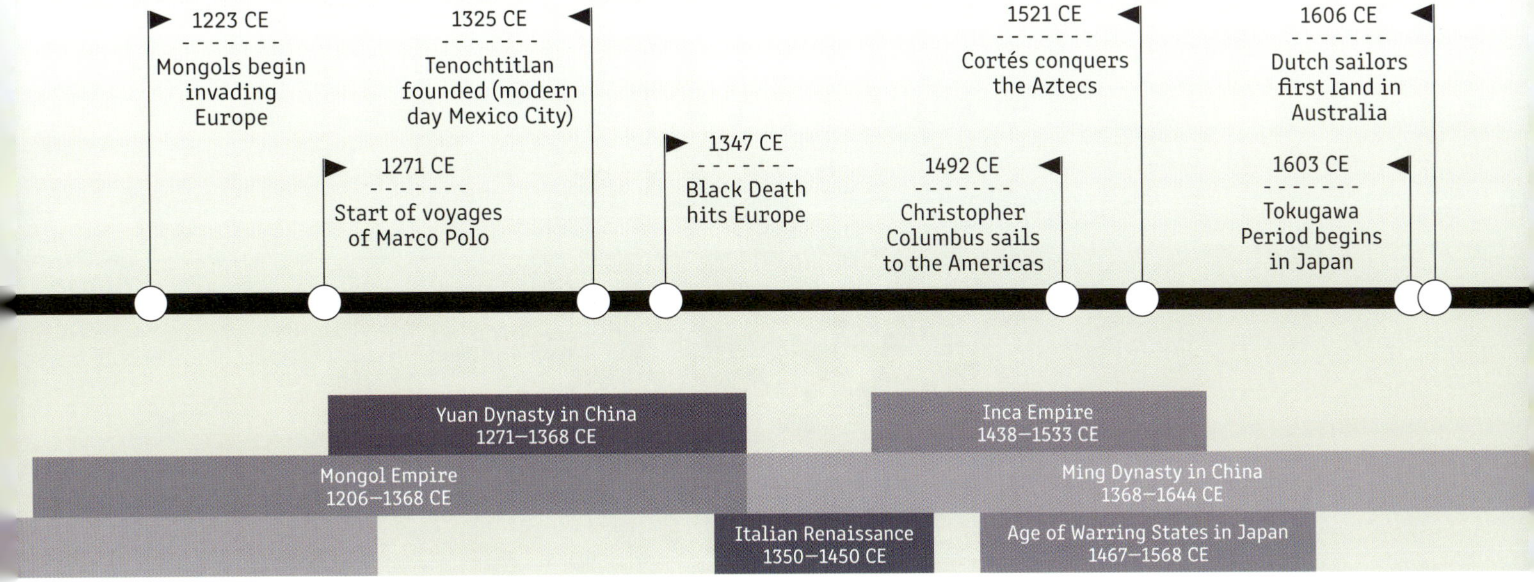

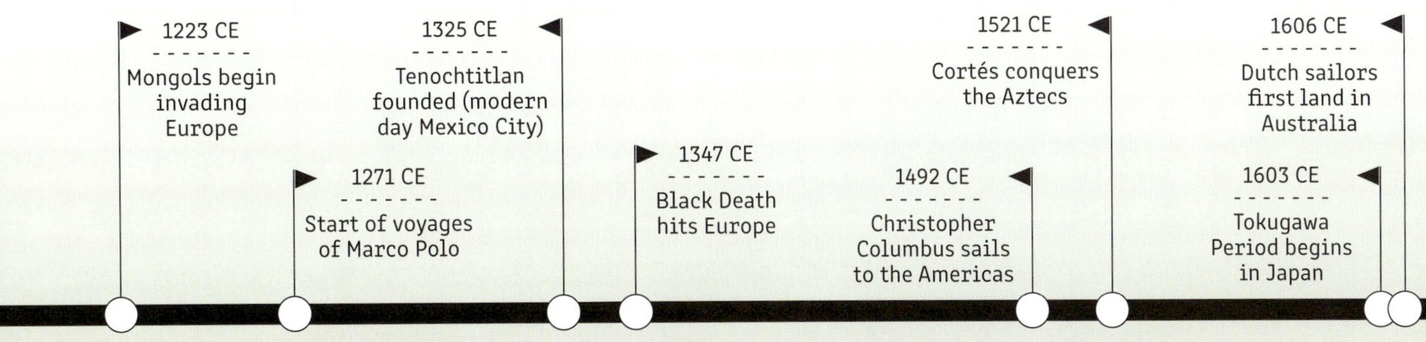

The rise of empires

An empire is a group of nations ruled by a single ruler. While the Roman Empire fell at the start of the Middle Ages, this period also saw the rise of empires, kingdoms and dynasties across the world. Those living outside empires often led nomadic lifestyles.

Americas

In the Americas, three great civilisations prospered. The Mayans developed a complex society based in the jungles of central America. Their civilisation declined from about 900 CE, perhaps because of environmental ruin, though historians still debate why. The Aztec civilisation built its great capital, Tenochtitlan, in the middle of a lake. This artistic, religious society peaked for 200 years before being conquered by the Spanish in 1519–21. The Inca ruled an enormous area of land high in the mountains of South America, with a complex communal economic system. They too were defeated by the Spanish in 1532–33.

Asia

The 13th century was the Mongol era in Asia. These nomadic, warlike peoples created a huge land empire stretching from China to Eastern Europe. The empire fragmented after its greatest leaders, Genghis Khan (1227) and Kublai Khan (1294), died, but Mongol influence lived on through China's Yuan Dynasty.

China itself saw a series of united **dynasties** across the period – the Sui, Tang, Song, Yuan and Ming. During the Middle Ages, the Chinese had more advanced technology and living standards than other peoples.

In India, various dynasties and kingdoms came and went. It was a time of great cultural diversity in the region. Northern parts were invaded by Turkish people who set up the Delhi Sultanate (1210–1526). Later, Muslim settlers established the Mughal Empire (1483–1739), and introduced Islam to the region.

Source 2

Machu Picchu, in modern-day Peru, was a great city of the Incan Empire, and its ruins still stand today.

Africa

Muslim Arabs conquered the northern coast of Africa in the 7th century, established Islam and built numerous trading posts along the eastern coast. A number of kingdoms dominated different areas of Africa at this time, such as in Ghana on the east coast. There was little contact between sub-Saharan African people and the rest of the world until the end of the Medieval period, when European powers begun to explore and colonise the area.

Polynesia

Across the Pacific, Polynesians made heroic long-distance sea voyages. The last large land masses in the world were populated when Polynesians reached Hawaii (about 500 CE) and New Zealand (about 1300 CE).

The spread of world religions

A world religion is one that has many followers spread over a large area. In the Middle Ages, several world religions experienced periods of change.

Buddhism

Buddhism began in northern India around 500 BCE. It is based on the teachings of Siddhartha Gautama, an Indian prince who gave up his riches and found enlightenment through meditation. Buddhism teaches that people should lead a good life by avoiding harm to others, ending suffering and leading a balanced life. During the Middle Ages, Buddhism spread to the rest of India, southeast Asia, east Asia and central Asia.

Judaism

Judaism is based on the Torah, and claims origins to 5781 BP. After a period of slavery in Egypt, the first and second Jewish Commonwealths were partly destroyed by the Babylonian (586 BCE) and Roman (70–136 BCE) Empires. A fraction of Jewish people stayed in the Middle East, while by the Middle Ages, a majority had formed communities in the Christian and Islamic Empires.

Islam

Islam is a religion founded by the Prophet Muhammad early in the 7th century CE. The religion's teachings emphasise faith, regular prayer, charity, fasting and pilgrimage. After Muhammad's death in 632 CE, the religion spread quickly around the Middle East, across northern Africa and briefly into southern Spain.

Christianity

Christianity is based on the teaching of Jesus Christ, a prophet from the Middle East who lived in the 1st century CE. Christians believe in salvation through Christ, love for others, forgiveness and compassion. The Roman emperor Constantine converted to Christianity in 313 CE, and this helped spread Christianity around most of Europe by about 800 CE. In 1517 the Church split into two factions – Catholics and Protestants.

Source 3

Isaac Newton analysing the colours in a ray of light. [19th-century hand-coloured woodcut]

Intellectual developments

The Middle Ages saw many advances in technology, including many tools and devices we now take for granted – horseshoes, paper money, eyeglasses and wheelbarrows were all invented during this time. Towards the end of the era, there were two periods of intellectual discovery that changed the world forever.

The Renaissance

The **Renaissance** (about 1350–1450 CE) saw a rebirth of the classical Roman and Greek focus on philosophy, art and architecture. It also represented a shift towards **humanism**, a view of the world focused on human talents and life on Earth rather than religion and the afterlife.

Scientific Revolution

The Scientific Revolution (roughly 1450–1700 CE) produced huge advances in scientific study and knowledge. The scientific method of experimentation was established by Francis Bacon. Isaac Newton revolutionised physics and theorised the existence of gravity. In astronomy, Copernicus stated that the Sun was the centre of the solar system rather than Earth, and Galileo discovered other planets like our own.

Learning ladder H0.5

1. Why is the Medieval Period also called the 'Middle Ages'?
2. Where did many of the world's largest religions begin?
3. What is the difference between the Renaissance and the Scientific Revolution?
4. What was the result of the improvements in agriculture in the middle ages?
5. Compare the Aztec, Mayan and Inca civilisations. What did they all have in common? What was different about them?
6. How might the daily life of someone living in an empire be different to a nomad living outside an empire?
7. Do you think a Eurocentric approach to history is appropriate? Explain your answer.

The Vikings

H1

WHAT BELIEFS CAME FROM VIKING CULTURE?

page 36

Economics and business

page 42

HOW DID THE VIKINGS INSPIRE MODERN ENTREPRENEURS?

key evidence

page 46

HOW DID VIKINGS DISCOVER NORTH AMERICA?

key individual

page 48

WHY WAS HARALD THE LAST OF THE VIKINGS?

How can I understand the Vikings?

During the Viking Age, Viking settlers spread across Europe, discovering Iceland and Greenland and briefly colonising North America. They held parliaments centuries before other European countries, and had greater gender equality than most other civilisations. Politically, Vikings established kingdoms in England, Scotland, Russia and Scandinavia. Their advanced boat building and seafaring allowed them to travel vast distances, and their artistic achievements included sagas, stories, poems, mythology and intricate metalwork.

learning ladder

Step	Chronology	Source analysis	Continuity and change
step 5	**I can describe patterns of change** I read timelines and see the 'big picture'. I group timeline events about the Vikings and see if they show patterns of change. I know typical historical patterns to check for.	**I can use the origin of a source to explain its creator's purpose** I combine knowledge of when and where a source was created to answer the question, 'Why was it created?'	**I can evaluate patterns of continuity and change** I answer the question, 'So what?' about patterns of continuity and change. I weigh up ideas and debate the importance of a continuity or a change.
step 4	**I can distinguish causes, effects, continuity and change from looking at timelines** I read Viking timelines and find events linked by cause and effect. I also find things that are the same or different from then until later times.	**I can use my outside knowledge to help explain a source** I have enough outside knowledge about the Vikings to help me explain a source.	**I can analyse patterns of continuity and change** I see beyond individual instances of continuity and change in the Viking Age and identify broader historical patterns, and I explain why they exist.
step 3	**I can create a timeline using historical conventions** When given a set of events, I construct a historical timeline using the dates, and using correct terminology, spacing and layout.	**I can find themes in a source** I look a bit closer into a source and find more than just features. I find themes or patterns in the source.	**I can explain why something did or did not change** I answer the question, 'Why?' something changed or stayed the same between historical periods.
step 2	**I can place events on a timeline** When given a list of events and developments about the Vikings, I put them in order from earliest to latest, the simplest kind of timeline.	**I can list specific features of a source** I look at a Viking source and list detailed things I can see in it.	**I can describe continuity and change** I have enough content knowledge about two different historical periods to recognise what is similar or different about them, and can describe it.
step 1	**I can read a timeline** I read timelines with Viking events on them and answer questions about them.	**I can determine the origin of a source** I can work out when and where a Viking source was made by looking for clues.	**I can recognise continuity and change** I recognise things that have stayed the same and things that have changed from the Viking Age until now.

Warm up

Source 1

Viking warriors terrorised villages and monasteries in surrounding countries. [Panel painted by Danish artist Lorenz Frolich, *Vikings Plundering a Monastery*, 1883.]

Cause and effect	Historical significance
I can evaluate cause and effect I answer the question, 'So what?' about cause and effect. I weigh up ideas and debate the importance of a cause or an effect.	**I can evaluate historical significance** I answer the question, 'So what?' about things that are supposedly historically important. I weigh up events and cast doubt on how important things are.
I can analyse cause and effect I don't just see a cause or an effect as one thing. I determine the factors that make up causes and effects.	**I can analyse historical significance** I separate out the various factors that make something historically important.
I can explain why something is a cause or an effect I can answer 'How?' or 'Why?' a cause led to an effect in the Viking Age.	**I can apply a theory of significance** I know a theory of significance. I use it to rank the importance of Viking events.
I can determine causes and effects Applying what I have learnt about the Vikings, I can decide what the cause or effect of something was.	**I can explain historical significance** I answer the question, 'Why?' about things that were important in the Viking Age.
I can recognise a cause and an effect From a supplied list, I recognise things that were causes or effects of each other in the Viking Age.	**I can recognise historical significance** When shown a list of things from Viking history, I can work out which are important.

Chronology

1 In what century was Source 1 painted?

Source analysis

2 Source 1 and Source 2: Which source is a 'Viking source'? Explain your answer.

Continuity and change

3 Read the introductory paragraph on page 20. Identify similarities between the Viking Age and 21st-century Australia.

Cause and effect

4 What did the Vikings' advanced boat building and seafaring skills allow them to do?

Historical significance

5 Source 2: The Cuerdale Hoard was discovered in 1840 and is still one of the largest Viking hoards ever found. Do you think this is historically significant? Explain your answer.

Source 2

The Cuerdale Hoard 905 CE, silver and gold, Cuerdale, England

Why learn about the Vikings?

The **Vikings** are a fascinating group of people. Some see them as vicious warriors who pillaged Europe for 250 years, while to others they were settlers, traders and distributors of people and ideas. The period of the Vikings dates from circa 789 CE to 1066 CE. Learning more about these intriguing voyagers, who arguably have given our culture as much as the ancient Greeks or Romans did, requires us to challenge our ideas about historical significance.

Source 1

A recreation of a Viking raider

key ideas timeline.

793 CE – Vikings raid Lindisfarne monastery in England

840 CE – City of Dublin founded by Vikings

874 CE – Vikings settle Iceland

878 CE – Vikings begin settling permanently in England

911 CE – Viking chief Rollo granted land in northern France; founds Normandy

Danelaw – Danish Viking kingdom in England 880–954 CE

Viking raids on Europe 789–1042 CE

Source 3

This painting is a secondary source of evidence painted in the late 1800s. [Hans Dahl, *Leif Eriksson Discovers America*, oil on canvas, 50 x 75 cm]

Source 2

Jarlshof Viking Age ruins in the Shetland Islands, Scotland, make up the largest such site visible anywhere in Britain and include a longhouse (see page 24).

Source 4

Iron helmet from a Viking chief's grave c. 900 CE. Discovered in Eastern Norway in 1943, it is now exhibited in the Historical Museum in Oslo, Norway.

Learning ladder H1.1

Show what you know

1 Describe the two opposing views about the Vikings and their place in history.

2 List three countries that the Vikings invaded.

3 List three places that the Vikings settled.

Chronology

Step 1: I can read a timeline

4 How many years were there between the discovery of Greenland and North America?

Step 2: I can place events on a timeline

5 Place the following events in chronological order:

799 CE Vikings began to raid France

795 CE Vikings first raided Ireland

840 CE Vikings began to raid the south coast of England

794 CE Vikings raid the island of Iona, Scotland

Step 3: I can create a timeline using historical conventions

6 Recreate the beginning of the timeline from 793 CE to 840 CE in your notebook. Add the events from question 5.

Step 4: I can distinguish causes, effects, continuity and change from looking at timelines

7 What does the timeline tell us about the Vikings as explorers? What do you think was the cause of this need to find new areas to raid and settle?

HOW TO

Chronology, page 194

981 CE – Erik the Red discovers Greenland

1000 CE – Leif Eriksson explores North America

1015 CE – Vikings abandon settlement Vinland in North America

1028 – King Cnut is ruler of England, Denmark and Norway

1066 CE – Death of Harald Hadrada, marking decline of Viking era/Norman conquest of England

What was daily life like for the Vikings?

The Viking's homelands were located in the group of countries in Northern Europe now known as Scandinavia. Most Vikings lived in simple homes on small farms grouped together in villages. They lived off their crops, and by hunting and fishing for meat and gathering wild plants. There were only a small number of large towns along the coast, where talented craftspeople made goods to trade locally and to export.

Farm life

Most Vikings were farmers, growing crops and raising animals to sustain the family. They usually lived on small farms, grouped together into villages. Viking families grew vegetables and crops of hay, barley, oats and rye. They raised cattle, pigs, horses, sheep, goats, geese and chickens.

Winters in Scandinavia are long and dark. Villagers grew crops in the summer, harvested them in autumn and stored them to sustain the family through the long winter. Cattle were kept indoors in winter and fed hay, either in the barn or inside the **longhouse** with the family. Longhouses were rectangular buildings up to 25 metres long and 5 metres wide.

Viking men left the farm on fishing or raiding expeditions (see pages 30–31), returning for the autumn harvest and winter. In their absence, Viking women managed farms and therefore held some power in Viking society. Children also worked on the farm as the Vikings did not have schools.

Source 1

The fire was placed in the middle of Viking longhouses for light, warmth and cooking. The longhouses were basically one big room where people cooked, ate, washed and slept. Many also had a loom for weaving cloth.

Craftspeople

Skilled craftspeople were important in Viking society. The most highly skilled artisans created items with elaborate decoration. Metalworkers made everything from simple farm tools to swords, armour and helmets for Viking warriors. Carpenters fashioned wood for ships, produced shields and even made toys. Other craftspeople included leatherworkers, jewellers, weavers and potters.

Viking craftspeople worked with imported materials such as silver, jet (black stone), amber (yellow-orange resin) and glass. Jewellers and metalworkers also used silver that was stolen on raids. They would also melt down coins or candlesticks to make their goods.

Source 2

Ornate dragon head post carved by a craftsperson around 850 CE. It was uncovered at the ship burial site at Oseberg, discussed on page 38.

Housing

Nearly all homes and workshops in Scandinavia were constructed of timber. Where wood was scarce, homes could be built of stone or **sod** (turf or grass).

Source 3

Internal view of a sod-covered Viking longhouse. The living quarters and the fire were in the middle and a barn was at one end.

Buildings were long, with thatched roofs made of reeds or straw. Only the blacksmith had a chimney, with other buildings having openings for the smoke from the fire to escape. There were no windows, so oil lamps and candles provided the only light.

The largest building was a Viking longhouse. Viking families or groups of people lived in the central part of the building. Large fireplaces in the middle of the longhouse would have been used for heating, lighting and cooking. One end of the longhouse could be used as a barn to shelter livestock in the winter and the other end could be used as a workshop.

The longhouses of the chiefs of the communities, the *konungr*, or nobles, *jarls*, were the largest and most highly decorated. Feasts that lasted for days were held here following successful raids, with plenty of ale or **mead**, an alcoholic drink made from honey, as well as music, dancing and poetry readings.

Most villages also had a smaller building which was a workshop for a **blacksmith** containing a forge for heating iron.

1. Roman fort
2. Cross
3. Assembly Hall
4. Boat builders
5. Bakery
6. Blacksmith
7. Leather workers
8. Silks
9. Roman wall
10. Well
11. Splitting wood

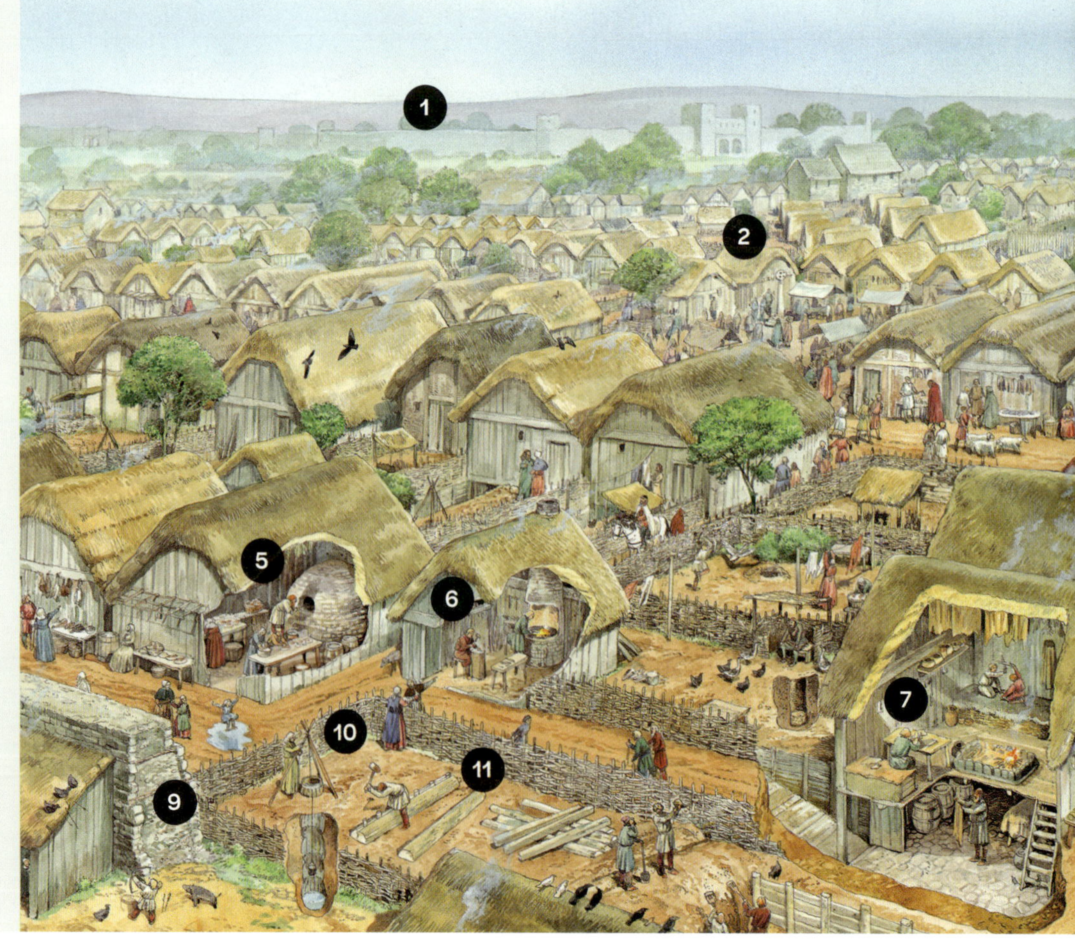

Source 4

This illustrated Viking town includes evidence of old Roman buildings. Activity was centred around the harbour, where boats were built, loaded and unloaded. Markets were held near the docks while craftspeople built their workshops nearby.

Food

Vikings ate the food they produced on their farms or caught fishing. They collected fruits and berries from the forests and hunted wild boar, elks and hares. Salt was bought from travelling merchants to preserve meat and fish to eat when fresh food was scarce during winter.

The barley and rye were used to make flour for bread. This bread was baked on flat stones or iron plates over the open fire pit in the middle of the home. Vegetables and meat were often stewed in a big cauldron of water over the fire.

Communication

The Vikings used a system of writing called **runes**. Letters from the runic alphabet were carved into wood, bone or stone. At first it was thought that runes were only used for formal purposes, but a discovery at Bryggen in Norway in 1955 found runes also had an everyday use.

Not all Vikings could read and write, so many stories including **sagas** were passed on through word of mouth.

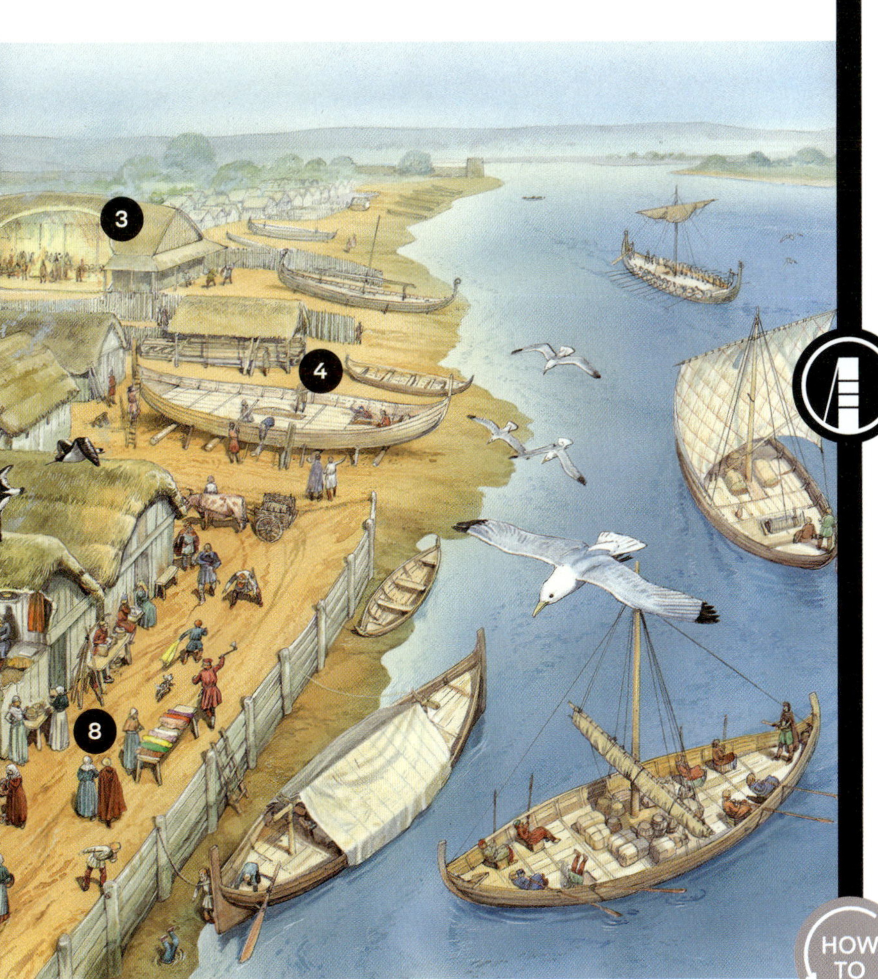

Source 5

Runestone from 1070 CE found near the ruins of a king's dwelling on Adelso Island in Sweden

Learning ladder H1.2

Show what you know

1 What power did women have in Viking society?

2 Source 1: Describe three activities taking place in this Viking house.

3 Describe what Vikings might have eaten for breakfast, lunch and dinner.

4 List at least four things from Viking life that are still common in everyday life today.

Source analysis

Step 1: I can determine the origin of a source

5 From where and when is Source 2?

Step 2: I can list specific features of a source

6 Describe in detail the markings on the stone in Source 5. Only describe what you see, not what you think the markings might mean.

Step 3: I can find themes in a source

7 What are all the different types of work you can see in Source 4?

Step 4: I can use my outside knowledge to help explain a source

8 Suggest three different reasons the carvings in Source 5 might have been made.

HOW TO

Source analysis, page 197

Were Vikings bloodthirsty pillagers?

Vikings have a fearsome reputation as vicious killers who would raid and steal from innocent villagers. But they were also cunning **entrepreneurs** who developed powerful trade networks.

Viking raiders

'The great white sail cracked as the vicious Atlantic wind lashed against it, but still the ship sailed on. Long and sleek, the warship, crafted from mighty oak, crashed through the waves, sending a sharp spray of water across the deck. The men inside rowed as one, their mighty muscles straining as they plunged the oars deep into the water and drove the ship forward through the turbulent waves. Their strength alone brought the ship to land and they poured out onto the beach. Dressed in thick woollen tunics, the warriors were armed with an array of weapons, from long sharpened spears to hefty battle-axes. With a booming voice one man yelled to the others, thrusting his sword into the air, and the rest bellowed in response. Then onward he ran, as the united force thundered uphill against the billowing wind. Their destination? A coastal monastery bursting full of gold, gems and hefty food supplies ripe for the taking, and only a collection of quiet, unassuming monks to protect it.'

Source 1

Frances White, 2015, *All About History* 22

The attack on Lindisfarne

The first recorded Viking attack was on the holy island of Lindisfarne off the northeast coast of England. In 793 CE, Viking raiders destroyed the monastery of St Cuthbert, killed the monks and other inhabitants and stole crucifixes and other religious items made of gold, silver and ivory. This marked the start of 300 years of brutal Viking raids on Britain and Ireland.

News of the raid reached the scholar Alcuin of York. He wrote a letter to Ethelred, King of Northumbria, extracted in Source 3.

'Lo, it is nearly 350 years that we and our fathers have inhabited this most lovely land, and never before has such terror appeared in Britain as we have now suffered from a **pagan** race (those following lesser religions) other than, nor was it thought that such an inroad from the sea could be made. Behold, the church of St Cuthbert spattered with the blood of the priests of God, despoiled of all its ornaments; a place more venerable than all in Britain is given as a prey to pagan peoples.'

Source 3

Letter from Alcuin of York describing the Viking attack on Lindisfarne in 793 CE.

Source 2

This image of vicious raiders is a **stereotype** that people have of the Vikings: they terrorised their victims as they attacked villages with no warning. They had the fastest ships in the world and could quickly and quietly reach their destination. Armed with superior weapons, the Viking raiders launched furious attacks on their unsuspecting victims; killing them, **pillaging** their possessions and burning their churches and homes to the ground.

Planning a Viking raid

1 Careful preparation

The Vikings planned their raids carefully. First, they identified a target they thought would have precious goods, such as a monastery. Then they allocated ships to the attack – their ships were fast and were designed to be rowed up rivers as well as sailed on the open ocean. Viking raiders could therefore attack both coastal and inland villages. Raids on small villages may have involved just a few Viking ships, but attacks on larger cities involved fleets of 300–400 ships (see pages 38–39).

Sometimes Vikings would raid the villages near their ships to steal horses. They would then ride to more distant villages and carry out a raid on horseback. The horses were then used to transfer the loot from the raid back to their ships.

2 Sudden attack

Vikings would quietly sneak into the unsuspecting village and unleash a sudden and furious attack on the population, killing many and taking some prisoners to sell as slaves.

3 Loot and burn

When the townspeople were no longer a threat, the Vikings stole valuable items and any food they could get their hands on. Once they had what they wanted, they burned the villages to the ground.

4 Making a profit

The Vikings loaded their ships with the **loot** from their raid. They would sell the gold, jewels and other valuable items, as well as their captured prisoners.

Vicious Vikings

Some of the most murderous Vikings include:

1 Ragnar Lodbrok (c. 765–c. 852 CE) – Ragnar was a Viking king who was known as the scourge of France and Britain for his many raids and conquests. Leading 120 Viking ships, he boldly sailed up the River Seine in 845 to raid Paris. After occupying and looting the city, Ragnar finally agreed to leave when paid a ransom of more than 2500 kilograms of gold and silver. He was eventually captured by an English king and thrown into a pit of snakes to die.
2 Bjorn Ironside (9th century CE) – Bjorn Ironside was the legendary king of Sweden and son of Ragnar Lodbrok. Ironside raided the Italian town of Luni by pretending to be dead. He was carried in a coffin through the city gates, then jumped out of the coffin and killed anyone in his way. Ironside then opened the city gates to allow the Viking army outside the gates to raid the city.
3 Eric Bloodaxe (885–954 CE) – Eric Haraldsson was a Norwegian ruler. He earned the nickname 'bloodaxe' by killing anyone who stood in his way. He murdered all but one of his brothers.

Source 4

Bloody Viking raids like this painting of one on the monastery of Clonmacnoise in Ireland spread fear throughout Europe. As time went on, more and more Vikings settled in the areas they had raided. [Tom Lovell, *Norse Marauders Wreak Mayhem at Clonmacnoise*, Ireland.]

Source 5

Viking armour and weapons as worn by nobles and professional warriors

Were Vikings more than bloodthirsty pillagers?

It is undeniable that Vikings preyed on the weak and stole their possessions to increase their own wealth. There is also much evidence to suggest that Viking raids were violent, leading to the deaths and capture of countless victims and the destruction of their homes.

But there was another side to the Vikings. For most of the year, the Viking raiders farmed the land or worked as craftspeople, creating ships, weapons, tools and even jewellery (as discussed on pages 24–27).

During the 300 years of the Viking Age from the late 8th century to the late 11th century, the Vikings built a powerful civilisation. The Vikings didn't just raid, pillage and leave. Many Vikings stayed as migrants and set up new colonies (see pages 44–47), embracing Christianity and marrying local people.

In the lands they invaded, the Vikings created important trade routes to help their civilisation become one of the richest in the world. While the Viking raids provided a quick intake of wealth, the specially designed Viking trading ships helped develop a prosperous and powerful trading network (see pages 40–41).

Viking warrior armour

Wealthy Vikings would have carried a spear, a wooden shield and an axe or a sword on a raid. Only the very richest would have a helmet made from leather or metal. The helmets did not have horns, despite the stereotype. Ordinary Vikings fought wearing their everyday woollen clothes.

Learning ladder H1.3

Show what you know

1 Why did Vikings raid?

2 Why did Vikings frequently attack monasteries?

3 Source 3: What language suggests bias?

4 Source 1: What is Frances White's opinion of Vikings? Provide evidence from the text for your answer.

5 Do you think Vikings were *just* bloodthirsty raiders? Give evidence for your answer.

Chronology

Step 1: I can read a timeline

6 Look at the timeline on page 22. When and where was the first recorded Viking raid?

Step 2: I can place events on a timeline

7 List six events or spans of time mentioned in this section in order from oldest to most recent.

Step 3: I can create a timeline using historical conventions

8 Create a timeline that features the Vikings as settlers using the events from the timeline on page 22.

Step 4: I can distinguish causes, effects, continuity and change from looking at timelines

9 What was special about Viking boats that made raiding and exploration so successful?

Chronology, page 194

How did Viking raids force change?

The Vikings' quests for riches spread fear throughout Europe. They started taking territory, sending large armies to grab land by force. The Vikings were successful in controlling new areas, such as Normandy in France and Danelaw in England.

Fearsome raiders

From 793 CE, Viking raiders began a reign of terror over the people of Europe that lasted around 300 years (see pages 28–31). The poem in Source 1 shows how the Irish knew the Vikings were travelling across the sea to attack.

> 'Bitter is the wind tonight.
> It tosses the ocean's white hair.
> Tonight I fear not the fierce
> warriors of Norway
> Coursing on the Irish Sea.'

Source 1

The Viking Terror, an ancient Irish poem

From raiders to settlers

Gradually, Viking raiders stayed for longer periods in seized lands, first in winter camps, then settling permanently. Norwegian Vikings settled in Scotland, Ireland, Orkney, Shetland and the Faroe Islands. Danish Vikings settled in England, the Netherlands and in Normandy in France. Swedish Vikings settled in Russia and Ukraine.

Vikings expand into England

Viking forces grew from small bands of raiders **pillaging** villages, to armies focused on taking large territories. The Great Heathen Army of Danish Vikings captured York in 866 and that city became their English capital. The Vikings then pushed south and west. By 874, the Great Heathen Army had conquered the English Kingdom of Mercia.

Danelaw

In May of 878, the army of Alfred the Great, King of Wessex, defeated the Vikings at the Battle of Edington. Alfred made peace with the Viking leader, Guthrum. Guthrum agreed to convert to Christianity and King Alfred allowed the Vikings to rule an area of England across the middle of the country that became known as Danelaw. Alfred remained king of the rest of England. In Danelaw, people lived as they had in Scandinavia, using Viking languages and laws.

Alfred the Great's grandson, Aethelstan, defeated the Vikings at the Battle of Brunanburh in 937 to become the first true king of all of England.

King Cnut

Viking attacks on England did not stop after Aethelstan's victory. There were ongoing battles, and short-term rules from both sides of the conflict until King Cnut of Denmark defeated Edmund Ironside at the Battle of Ashingdon in 1016. Soon after, Cnut became ruler of both Denmark and England.

The reign of King Cnut brought stability to the region after years of raids and battles. Cnut was a Christian and he did not force the English to obey Danish laws and customs.

Norway and parts of Sweden also came under his control and he established a large empire, known as the North Sea empire. However, he died at the age of 39 and his sons could only manage short reigns after him.

Viking influence and settlement, 8th – 11th century

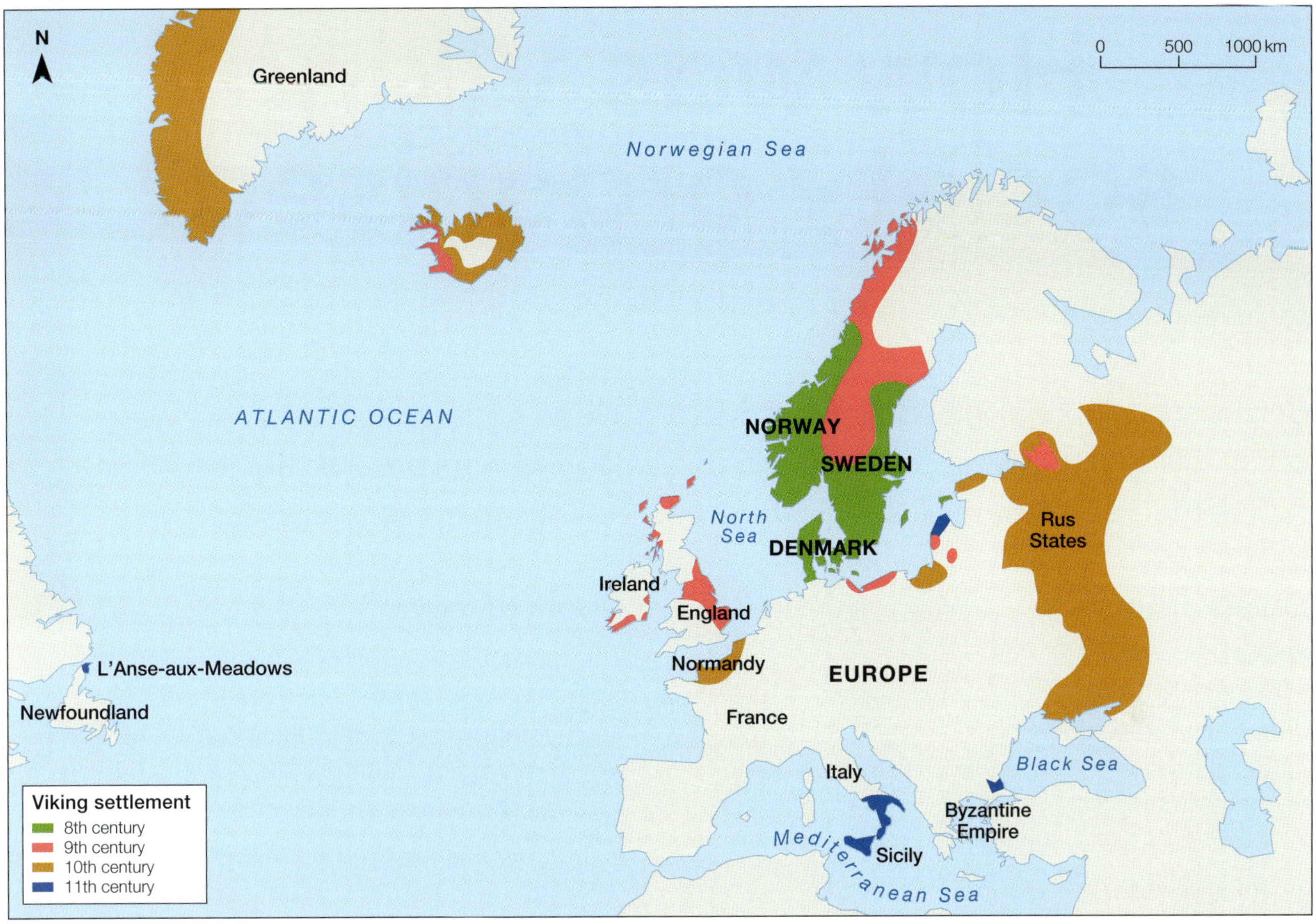

Viking settlement
- 8th century
- 9th century
- 10th century
- 11th century

Source: Matilda Education Australia

Source 2

Scandinavian settlement

Learning ladder H1.4

Show what you know

1 What changed about Viking raiding that enabled them to capture York?

2 Why do you think Alfred gave the Vikings a large piece of land?

3 Is Source 2 a primary or a secondary source? How do you know? What sources do you think were used to create this source?

Historical significance

Step 1: I can recognise historical significance

4 Put these in order from most to least historically important:

King Sweyn Forkbeard died in 1014

Vikings attacked the monastery of Lindisfarne

Vikings settled permanently in England

Step 2: I can explain historical significance

5 Why was the establishment of Danelaw historically important?

Step 3: I can apply a theory of significance

6 Using Partington's model of significance to explain the importance of Viking raids and settlements (see page 7).

Step 4: I can analyse historical significance

7 Using a map of Europe, which modern countries did Vikings settle in from the 8th to the 11th centuries? Analyse what effect you think Viking settlements had on modern Europe.

Historical significance, page 207

How was Viking society organised?

Behind the wild and undisciplined image of savage raiders lies a different view of the Vikings. They were men and women who lived in an ordered society that had democratic assemblies and a desire for honour, fairness and justice.

Levels of society

Viking society was hierarchical, like other societies in medieval Europe. Its people belonged to different groups, with each level of society more influential than the one below it. Women usually had the same social rank as their fathers, brothers or husbands. There were four social classes in Viking society:

1 Kings or ***konungr*** – each village or town had a chief or king as there was no central government.
2 ***Jarls*** – the wealthy nobles employed others to work for them and kept slaves. *Jarls* could become rich and powerful enough to eventually become a king.
3 ***Karls*** – everyday farmers, craftspeople, warriors, sailors and their families.
4 ***Thralls*** – **slaves** who had been abducted in Viking raids, purchased overseas or sold themselves into slavery to pay a debt. Slaves did the dirtiest and hardest jobs.

Room to move

The social hierarchy of the Vikings was not as rigid as many other societies of the time. People could move up or down in social rank. Kings were not simply given their title. They earned the role through showing qualities such as strength, bravery, generosity and leadership.

Jarls could become rich and powerful enough to become a chief or king. Alternatively, they could lose wealth and power and move down a social rank to become a *karl*. A *thrall* was allowed to sell crafts they had made to earn money to buy their freedom and become a *karl*.

Code of honour

It was expected that members of Viking society would live by an unwritten code of honour. They were expected to:

- support friends and groups
- be loyal to leaders
- keep their word
- take action against enemies
- not disgrace their community.

Source 1 Viking social hierarchy

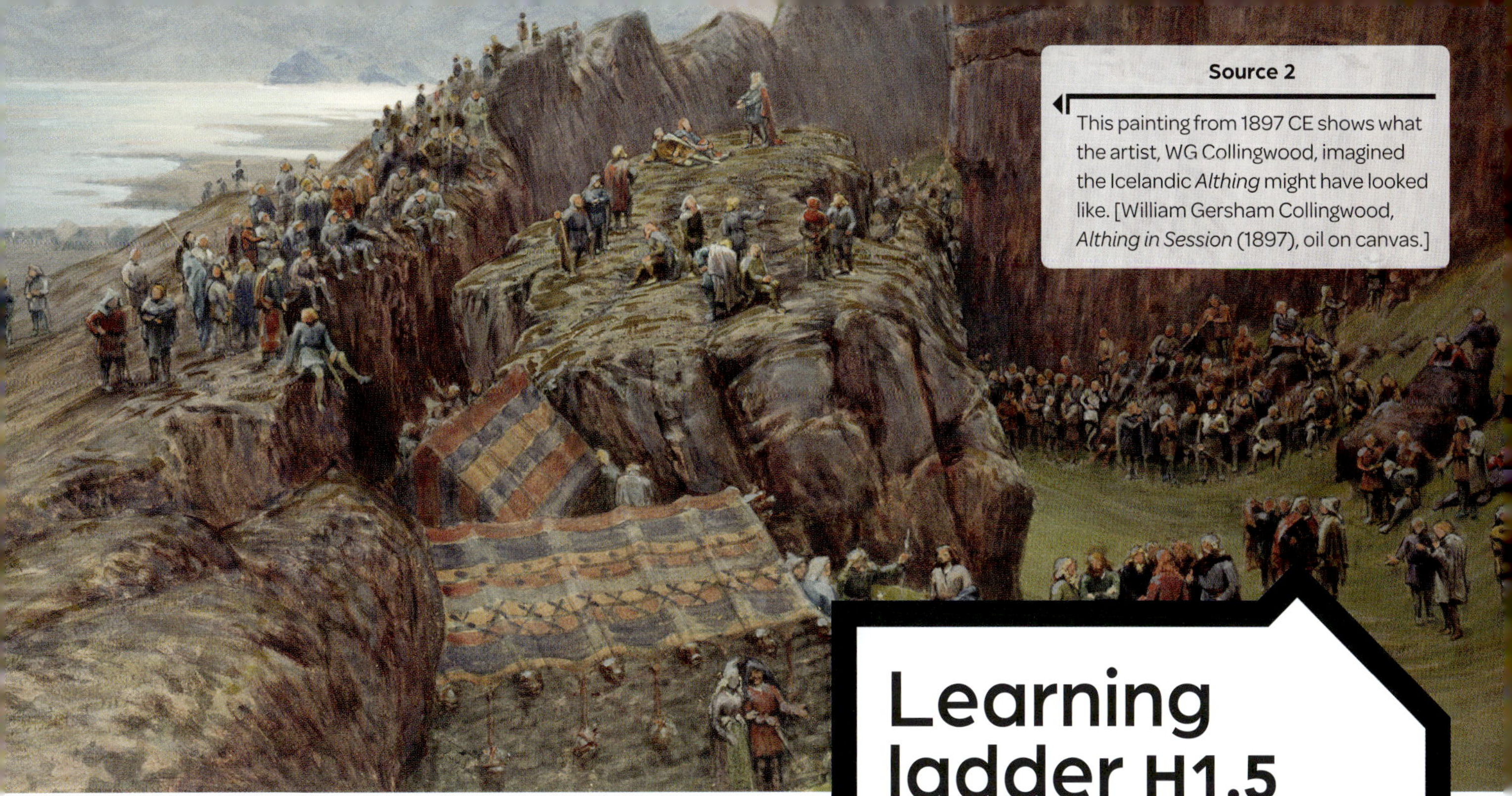

Source 2

This painting from 1897 CE shows what the artist, WG Collingwood, imagined the Icelandic *Althing* might have looked like. [William Gersham Collingwood, *Althing in Session* (1897), oil on canvas.]

People often swore promises in an oath to Thor, the god of thunder, storms and strength (see pages 36–37). Those who broke their promises were usually not forgiven.

Among Vikings, honour was of the highest importance. To settle an issue, one party could challenge another to a duel known as ***holmgang***. Not showing up for a *holmgang* led to a person being labelled a *niðingr*, or coward, and being cast out from society.

A Viking *Thing*

A society comprising farmers and warriors spread out over vast distances required something to bring people together to develop rules to live by. *Jarls* and *karls* had a duty to attend a local meeting called the ***Thing*** to discuss local issues and decide what to do about them. The *Thing* took place over a few days every autumn and spring to decide guidelines for how the chief or king was to rule.

New laws were made at an annual event known as the ***Althing***. Opinions of all of the population could be heard on important topics such as taxes, treaties and even selecting a king. The *Althing* also hosted a religious festival and gave scattered people a chance to trade. The *Althing* in Iceland began as an outdoor assembly in 930 CE and continues to this day – the longest running parliament in the world.

Learning ladder H1.5

Show what you know

1 Describe the Vikings' code of honour.

2 How did someone become a Viking slave, and how could a slave become free?

3 Source 2: What kinds of things were discussed at the *Althing*?

4 Looking at Source 1, describe the different kinds of clothes that *jarls* wore, compared to *karls* and *thralls*.

Continuity and change

Step 1: I can recognise continuity and change

5 Which of these has changed the most from Viking times to the present day?
   a Parliament
   b Clothing
   c Levels of society

Step 2: I can describe continuity and change

6 Describe the difference between Viking kings and other European kings, such as English kings.

Step 3: I can explain why something did or did not change

7 Explain the similarities between the Viking code of honour and your own beliefs.

Step 4: I can analyse patterns of continuity and change

8 What are the similarities and differences between the Viking *Althing* and the modern Australian parliament?

Continuity and change, page 200

What beliefs came from Viking culture?

The Vikings worshipped a group of warrior gods who lived in a land in the sky known as Asgard. They sacrificed animals in their honour and told great tales of the gods' battles with giants and other monsters. These gods are called the Norse gods, as they were the gods worshipped by the people of the medieval Nordic (Scandinavian) region.

Heroic Norse gods

Vikings are known through history for their courage; be it bravery on the battlefield or sailing expeditions into the unknown. Bravery in the face of danger is one of the key themes of Norse mythology.

The Norse gods were heroic and showed reckless courage fighting against giants and other monsters who lived in the human world, called Midgard. The monsters were invisible to humans but not to the Norse gods.

The Norse gods were like humans – they looked like humans, they could be killed and had human faults such as anger and jealousy. However, the Norse gods had magical powers and lived in Asgard in palaces of gold and silver. A rainbow bridge called Bifrost connected Asgard to Midgard.

Impact of Norse gods today

Norse mythology is still popular today. Nearly all the days of the week are named after the Norse gods.

Odin (Woden): The one-eyed ruler of the gods. Odin traded one of his eyes for all the wisdom in the world and became the god of knowledge. He had two ravens that sat on each of his shoulders. Each day they flew around spying on humans, monsters and gods, and returned to Odin each evening to report on what they had seen.

Thor: Odin's son and the god of thunder, storms and strength. He could level mountains with his mighty hammer. Thor caused thunder and lightning as he rode across the sky. Thor was the most popular of the Norse gods, with a cult following throughout Scandinavia.

Loki: Odin's adopted son. He was a mischief-maker and a shape-shifter who could turn himself into any shape – a fly, a horse or an old woman.

Heimdall: Heimdall guarded the rainbow bridge so that no uninvited guest could enter Asgard. He could see for over 150 kilometres in the night or day and could hear grass grow.

Freyja: The goddess of love and war. She would ride into battle on a chariot pulled by two cats, and take half of the souls of fallen Viking warriors. The other half were taken by Valkyries (Odin's handmaidens) to the afterlife hall of the slain in Valhalla, the home of Odin.

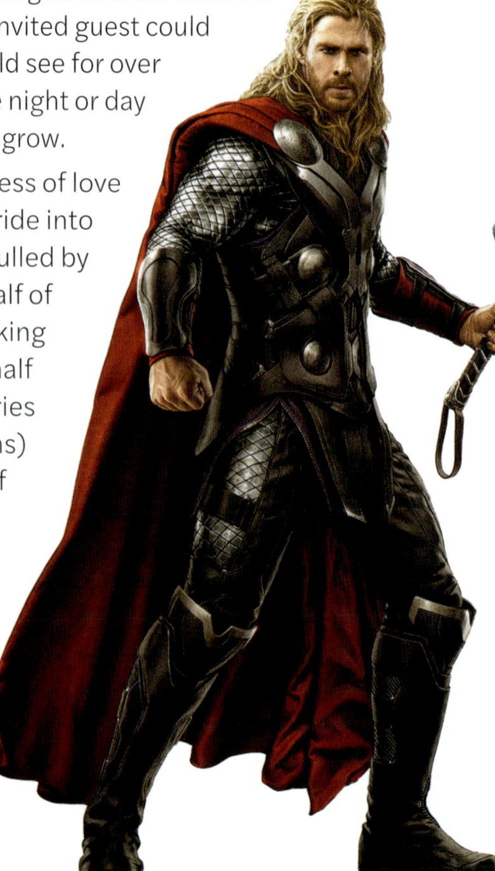

Source 1

Australian actor Chris Hemsworth portrays Thor, the god of thunder, in the Marvel superhero movies.

Source 2

This painting shows Thor riding across the sky in a chariot pulled by two goats called Toothgnasher and Toothgrinder to battle against giants. As Thor travelled across the sky, he created thunder and lightning and brought rain to water the crops. [Mårten Eskil Winge, *Thor's Fight with the Giants* (1872), Nationalmuseum, Stockholm, Sweden.]

How did the Vikings worship their gods?

Unlike other religions, the Vikings had no religious leaders or churches. To stay on good terms with the gods, the Vikings made sacrifices to them at an event called a ***blót***. Pigs and horses were usually sacrificed. The meat was boiled in large cooking pits and shared with the participants. The blood was sprinkled on statues of the gods and on the participants themselves. People drank in honour of the dead, sang songs and read poems to honour the gods.

Day	God	Original name
Sunday	Sol, goddess of the Sun	Sol's day
Monday	Mani, goddess of the moon	Mani's day
Tuesday	Tyr, god of war	Tyr's day
Wednesday	Odin, also called Woden, King of the Gods	Woden's day
Thursday	Thor, god of strength and storms	Thor's day
Friday	Frigg, goddess of marriage	Frigg's day
Saturday	From the ancient Roman god, Saturn	Saturn's day

Source 3

The Vikings wore miniature silver hammer pendants to symbolise their respect for Thor. This pendant was made in Sweden in the 10th century.

Learning ladder H1.6

Show what you know

1 What made Viking gods similar to humans?

2 How did Vikings stay on good terms with their gods?

3 What was the difference between Asgard and Midgard?

4 If you had to worship a Viking god, which one would it be and why?

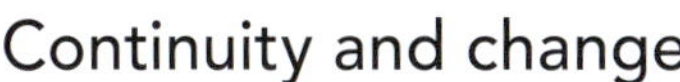

Continuity and change

Step 1: I can recognise continuity and change

5 From which two civilisations do we get the names for the days of the week?

Step 2: I can describe continuity and change

6 What technology do you think you would need to create the pendant in Source 3?

Step 3: I can explain why something did or did not change

7 Create a Venn diagram showing the similarities and differences between Viking religion and a modern religion that you know about.

8 Source 2: What natural phenomena were explained by Thor travelling across the sky? How do we explain these events today? Why has this changed?

Step 4: I can analyse patterns of continuity and change

9 Imagine you were doing a research project with the title: 'The legacy of Viking beliefs'.

a What subjects would you research?

b Where would you find the kind of information you are looking for? For example, would you use databases, dictionaries, encyclopedias, archives, collections, museums, journals, graphs, tables or letters?

c Note down six dot points you have learned from this section under the heading 'The legacy of Viking beliefs'.

Continuity and change, page 200

Why did the Vikings become seafarers?

The Vikings depended greatly on their ships and navigation technology to build a powerful civilisation. Viking ships were bigger, lighter and faster than those of other groups, offering huge advantages in raiding, trading and settling in lands far from home.

A history of seafaring

The sea and sailing were always important to the Vikings. Fishing was an important source of food but, more importantly, sailing boats were required for travel. Scandinavia was very sparsely populated with long distances between settlements. Nearly all major settlements were on the coast or a major river or **fjord**.

The need for sailing boats led to a wave of Viking inventions for shipbuilding and navigation. Their most famous invention was the **longship** – a fast, versatile ship with a very shallow draft, able to travel in shallow water. Some other innovations were:

- the half wheel and the **sunstone**, which used the position of the Sun to help navigation
- building ships with overlapping planks of wood, forcing air bubbles under the hull of the ship and thereby reducing drag from the water
- placing the mast in the middle of the boat and strengthening it to a solid oak keel. This meant more wind energy could be used to move the ship.

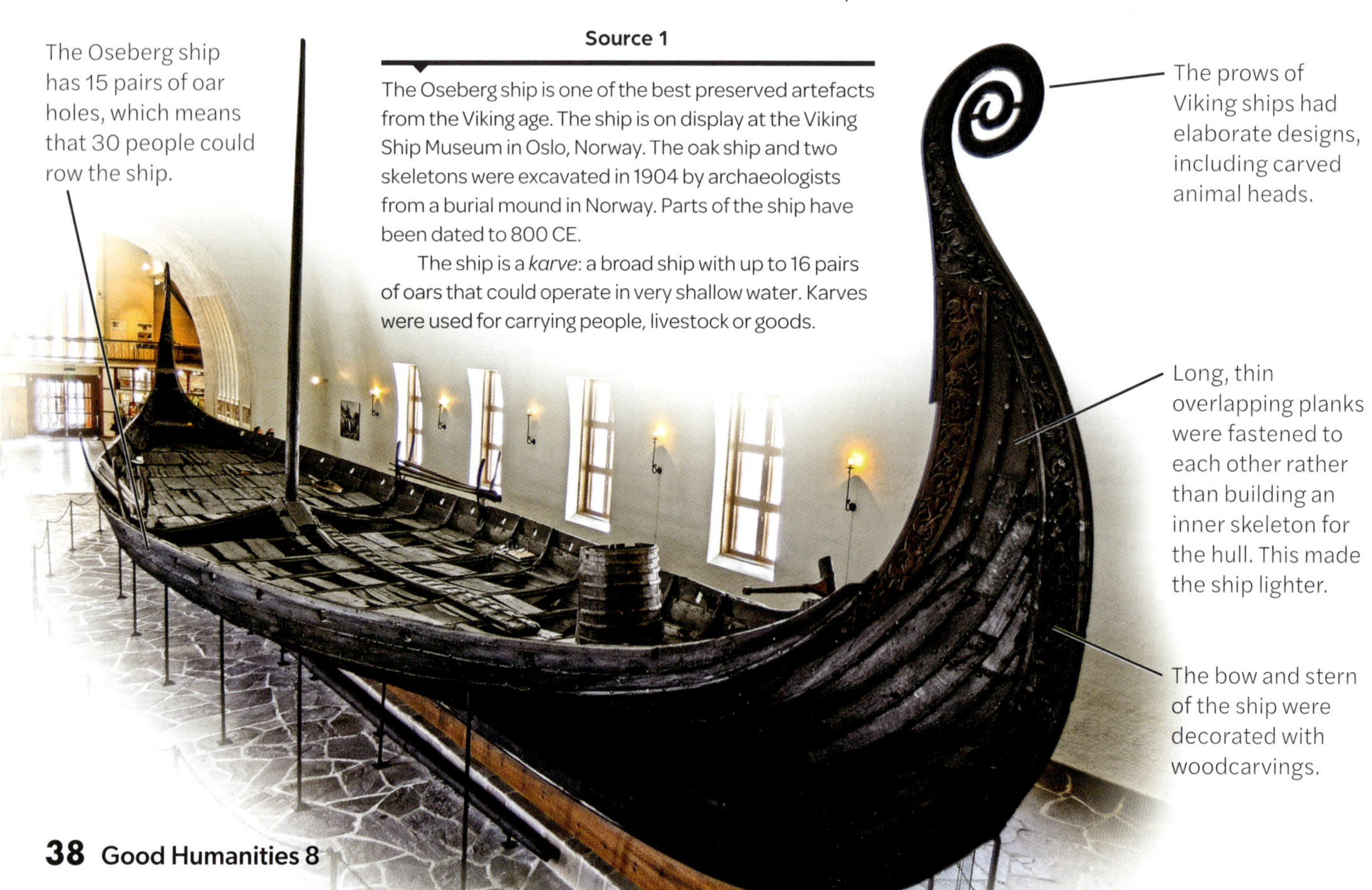

Source 1

The Oseberg ship is one of the best preserved artefacts from the Viking age. The ship is on display at the Viking Ship Museum in Oslo, Norway. The oak ship and two skeletons were excavated in 1904 by archaeologists from a burial mound in Norway. Parts of the ship have been dated to 800 CE.

The ship is a *karve*: a broad ship with up to 16 pairs of oars that could operate in very shallow water. Karves were used for carrying people, livestock or goods.

The Oseberg ship has 15 pairs of oar holes, which means that 30 people could row the ship.

The prows of Viking ships had elaborate designs, including carved animal heads.

Long, thin overlapping planks were fastened to each other rather than building an inner skeleton for the hull. This made the ship lighter.

The bow and stern of the ship were decorated with woodcarvings.

Source 2

A *knarr*

A competitive advantage

Other European nations used large ships based on Roman galleys. They were poorly designed for open-sea or shallow-water travel. As well as being the fastest, Viking longships were narrow enough to travel down rivers and light enough to beach on the riverbank or shore.

Longships were designed for open-sea travel with a deep keel board for stability and the ability to take waves and drain effectively. They were powered by wind hitting the large sails, or up to 80 Viking warriors to row the boat.

The dangers of sea travel

It took determination and bravery to undertake the risks of long travel through rough seas. With no cover in the longships, the sailors sheltered under wet blankets and animal skins. They ate salted meat and drank water, beer or sour milk. Many ships sunk on these long journeys. When Erik the Red led an expedition to Greenland, just 14 of the original 25 ships arrived safely.

What different ships did the Vikings have?

Archaeologists have uncovered various Viking ships built for different purposes. There were broad cargo ships, known as ***knarrs***, which were designed for overseas trade. ***Karves*** such as the Oseberg ship in Source 1 were used to transport heavy cargo around coastlines and up rivers. For warfare, the Vikings used their longships. Nobles were sometimes laid to rest in special burial ships.

Learning ladder H1.7

Show what you know

1 Why could Viking boats go up rivers? How would this have helped raiding?

2 How do we know about the different types of boats Vikings had?

3 What made Viking boats go so fast? Discuss at least two things.

4 Suggest what the ship in Source 2 was used for, using evidence from the picture.

5 Write a diary entry from the point of view of a Viking sailor on a difficult long-distance sea voyage.

Cause and effect

Step 1: I can recognise a cause and an effect

6 Match the following causes and effects:

Longboats had no shelter.	Erik the Red lost 11 ships travelling to Greenland.
Sea voyages were dangerous.	Sailors sheltered under blankets and animal skins.
Vikings lived along the coast.	They invented a boat with a shallow draft.
Vikings wanted to travel up shallow rivers.	They needed to travel by sea.

Step 2: I can determine causes and effects

7 What advantage did Vikings gain by building ships with overlapping planks of wood?

Step 3: I can explain why something is a cause or an effect

8 Explain why the design of longships gave the Vikings a competitive advantage over other European nations.

Step 4: I can analyse cause and effect

9 What three problems did the Vikings overcome when they developed shallow drafts, sunstones and overlapping pieces of wood under boat hulls?

Cause and effect, page 203

Were the Vikings raiders or traders?

The Vikings earned a reputation as vicious raiders, but it was their activities as traders that supported the Viking **economy**. The Vikings' expertise in ship design gave them an advantage. They could travel vast distances to sell their wares and to import goods that were needed at home, such as silver and salt.

Raiders or traders?

Raiding and pillaging provided a quick route to profit, but were not long-term ways to build a prosperous civilisation. The Vikings dedicated much more of their time to establishing a powerful trading network where merchants would travel far and wide to sell and buy goods.

The loot the Vikings stole on their raids did provide opportunities for trade. Viking metalworkers melted down stolen silver and gold to make jewellery or weapons that were traded overseas. People who were captured in raids were sold as slaves.

Source 1

A Viking trading town

Viking trading and exploration network

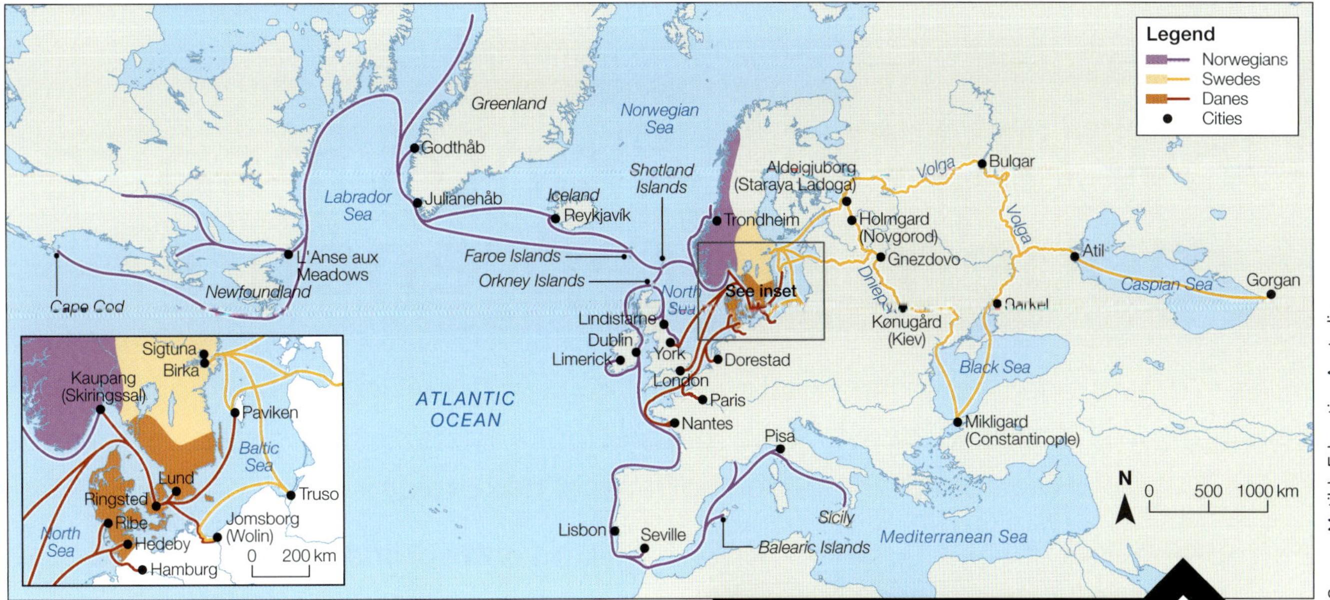

Source 2

Viking routes for trading, raiding and exploration

Where did they go?

The superior Viking trading ships enabled them to travel through the rough Atlantic Ocean and across the rapids of European rivers. The Viking cargo ships, or *knarrs*, could carry up to 35 tonnes of silver, timber and livestock (see page 39).

Trading markets appeared on the west coast of the Baltic Sea in the middle of the 8th century. As markets grew, trading towns such as Birka in Sweden, Kaupang in Norway and Ribe in Denmark developed. People living in these towns were either merchants or craftspeople.

Viking merchants sailed further afield in search of trading opportunities. Valuable trading networks emerged with York in England, Dublin in Ireland, Kiev in Ukraine and Novgorod in Russia. Vikings made even longer treks to Istanbul, Jerusalem and Baghdad.

What did they trade?

Vikings **exported** goods such as furs, skins, timber, smoked fish and jewellery. *Thralls* (see page 34) were valuable, and these slaves were both bought and sold. One female *thrall* could buy a cow and an ox. The Vikings **imported** items such as silver, silk, salt, wine, spices and pottery.

By the 9th century, silver coins were the most common form of payment. Merchants weighed the coins to determine their value. Many coins were melted down and crafted into jewellery for trade.

Learning ladder H1.8

Show what you know

1 Source 2: Who travelled the furthest: Danish, Norwegian or Swedish Vikings?

2 Why wasn't raiding and pillaging a long-term solution for Viking prosperity?

3 What do the trading towns listed have in common?

4 Why do you think Vikings had to import silver, spices and salt?

5 What evidence would historians have found that allowed them to create the map in Source 2?

6 Describe at least three things you can say about daily life for Vikings from looking at Source 1.

Historical significance

Step 1: I can recognise historical significance

7 How highly were *thralls* valued in the Viking Age?

Step 2: I can explain historical significance

8 How did boat technology help the Vikings' ability to trade?

Step 3: I can apply a theory of significance

9 How important was trading to Vikings?

Step 4: I can analyse historical significance

10 Which is more important historically, Viking raids or Viking trading? Give historical evidence for your answer.

Historical significance, page 207

How did the Vikings inspire modern entrepreneurs?

Modern entrepreneurs are influenced by Viking culture. The Vikings were great innovators in shipbuilding, helping them raid and loot targets far from their home. They also developed a huge trading network and explored new lands.

Entrepreneurs

An **entrepreneur** is a person who sets up a business, taking on risks to explore an opportunity that will earn a profit or satisfy a personal goal. Successful entrepreneurs often have these characteristics:

- a passion for what they do
- a willingness to take risks
- resilience to recover from failures
- the ability to work hard
- sound financial judgement.

Viking entrepreneurs

The Vikings displayed enterprising behaviours such as:

- taking risks to travel long distances to foreign lands to find ways build wealth
- building innovative ships that enabled them to travel further and in shallower water than anyone else
- building trading networks to sell their natural resources and produce
- using their trading networks to obtain resources for their craftspeople to work with
- on-selling the artisan-created items, which again built wealth for their communities.

Source 1

The Viking Laws compared with entrepreneurs

Vikings	Modern entrepreneurs
1 Be brave and aggressive	
Be direct	Be clear about what you want
Take all opportunities	Be open to opportunities
Use different methods of attack	Use more than one approach to marketing and sales
Attack one target at a time	Stay focused and don't spread resources
2 Be prepared	
Choose one chief	Select one leader
Find good battle colleagues	Choose the right people for your team
Agree on important points	Ensure you are pulling in the one direction
Stay in shape	Look after your mental and physical health
3 Be a good merchant	
Find out what people need	Respond to real market needs
Keep your promises	Deliver what you promise to
Don't demand over-payment	Never overcharge
Make sure you can return	Ensure a good chance of return business
4 Keep the camp in order	
Keep things organised	Keep accurate data
Arrange fun group activities	Create good bonding opportunities
Ensure people do useful work	People need to understand where they fit in
Ask people for advice	Involve staff in decisions

Source 2

Erik the Red was a Viking entrepreneur.

Source 3

Richard Branson is a modern entrepreneur. He dropped out of school at the age of 16 to start a youth culture magazine in 1966. He then began a mail-order record company that grew to become the huge Virgin Records business. Using the Virgin name, Branson launched Virgin Airlines, and more recently the Virgin Galactic venture, which aims to take tourists into space.

Erik the Red

Banished from Iceland for killing two men, a Viking named Erik the Red made a perilous journey through the icy north Atlantic waters in search of new lands. After three years, he returned to Iceland with tales of a 'green land', perfect for growing crops and raising cattle. Erik the Red gave it the name Greenland (even though it was almost completely covered in ice) to tempt others to join him in settling there. His **marketing** was a success and in 985 he set sail for Greenland leading a fleet of 25 ships with 500 men and women, livestock and supplies to build a settlement in a new country.

The Viking Laws

Modern entrepreneurs are influenced by the way Vikings lived their lives as warriors and traders. These are expressed as the Viking Laws (Source 1).

Learning ladder H1.9

Economics and business

Step 1: I can recognise economic information

1 Why are both Erik the Red and Richard Branson considered entrepreneurs?

Step 2: I can describe economic issues

2 Marketing activities are undertaken by businesses to promote a service or a product. How did Erik the Red market Greenland to customers?

3 List the Viking Laws that relate to marketing for modern entrepreneurs.

Step 3: I can explain issues in economics

4 The Viking Laws were made about 1000 years ago. What can modern business people still learn from them?

5 How could the Viking Laws help you to be a better student? Discuss at least three points.

Step 4: I can integrate different economic topics

6 What advantage did shipbuilding innovations give to the Vikings?

7 Use a search engine to look up British engineer James Dyson. What innovations did he introduce and how has he used these innovations to build a successful business?

Step 5: I can evaluate alternatives

8 Which of the four Viking Laws do you think is the most important for modern entrepreneurs and why?

Why did the Vikings settle in other lands?

The Vikings are among the greatest explorers the world has known. They sailed further than anyone had gone before. They made the first known voyages to Iceland, Greenland and North America and set up colonies in each of them.

Where did they go?

Early Vikings did not travel far from their Scandinavian homes along the coasts of Norway, Denmark and Sweden. Around 789 CE, Viking warriors began raiding and **looting** villages in Europe, developed a trading network throughout Europe and then established settlements in their newly conquered lands in eastern Russia, Britain and northern France.

In the 9th century, the Vikings colonised Iceland. Free land was given to the new settlers, with controls on how much could be claimed. For example, a woman could claim the land she could walk a cow around in a day. By 900 CE, 50 000 Vikings lived in Iceland.

Why did they go?

There are various theories to explain why the Vikings moved from Scandinavia to Europe and more distant lands, including:

- their endless search for further wealth and riches
- better opportunities and locations for trade
- usable land in Scandinavia was scarce and they searched for better farmland
- their fast-growing population put pressure on available land
- land was left only to the oldest sons, so other family members had to find their own land.

Fearless explorers

The Vikings travelled thousands of kilometres in search of new lands. Many ships and their crews were lost in the icy waters of the north Atlantic, but the Vikings discovered and set up colonies in Iceland, Greenland and eventually North America.

Viking exploration and settlement, 8th–11th century

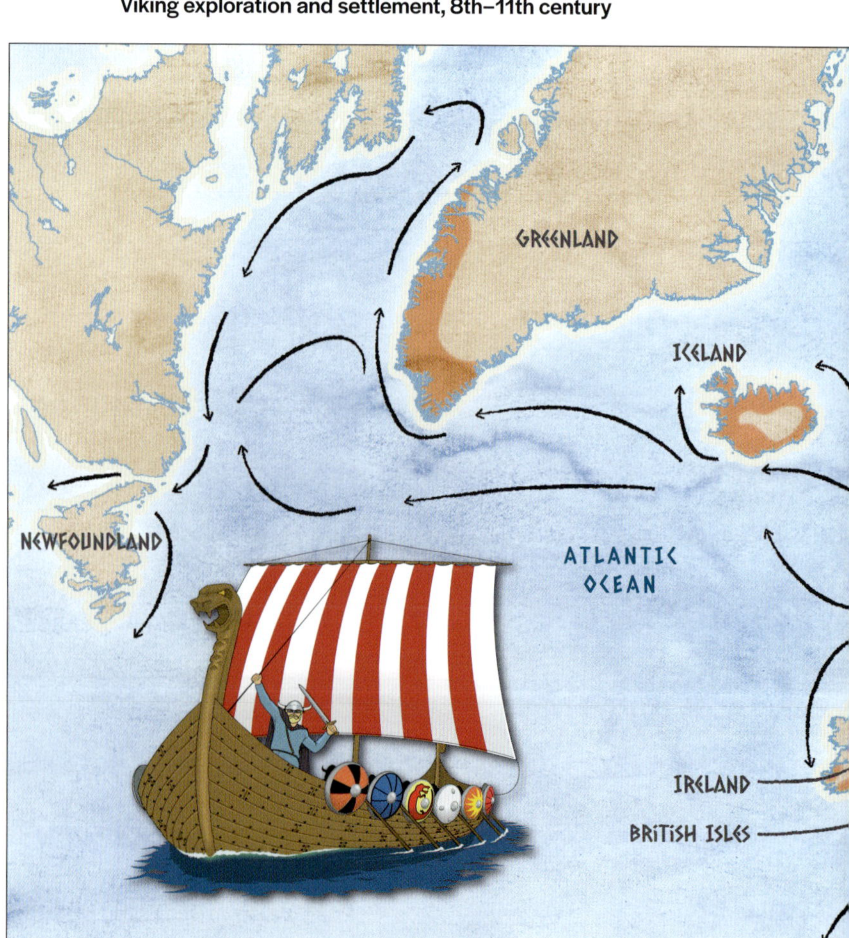

Source 1

Viking exploration and settlements

Source 2

In this modern painting, Leif Eriksson, son of Erik the Red, sails past icebergs in the north Atlantic Ocean. Eriksson was on his way to become the first explorer to land on the North American continent (see pages 46–47). [Unknown artist, *Leif Eriksson sails past icebergs on his voyage to Newfoundland in c. 1000* (20th century), colour lithograph.]

As discussed on page 43, after being exiled from Iceland for murder, Erik the Red discovered and settled Greenland. The island was covered in snow and ice, but he called it Greenland to attract settlers. On an ice-free region in the south of Greenland settlers could grow barley and raise cattle.

Source: Matilda Education Australia

Learning ladder H1.10

Show what you know

1 Why are Vikings considered fearless?

2 How many years ago did the Vikings begin raiding?

3 How did Greenland get its name?

4 What would be the difference between settling an area with people already living there compared to settling an uninhabited land?

Cause and effect

Step 1: I can recognise a cause and an effect

5 Why might their growing population have made the Vikings explore new lands?

Step 2: I can determine causes and effects

6 Why would farmers have wanted to voyage to new lands?

Step 3: I can explain why something is a cause or an effect

7 Source 1: Why were Viking settlers attracted to Greenland when it was covered in ice and snow?

Step 4: I can analyse cause and effect

8 Why do you think there are many different theories to explain why the Vikings settled new areas?

Cause and effect, page 203

key evidence

How did Vikings discover North America?

We have learned about Viking exploration from stories, known as sagas, and archaeological finds such as L'Anse aux Meadows.

Discovery of North America

In 1000 CE, Leif Eriksson, son of Erik the Red, landed on the North American continent and explored a region he called Vinland. He came into contact with the inhabitants of North America who the Vikings referred to as *skræling*. Leif Eriksson reached the North American continent nearly 500 years before Christopher Columbus arrived in 1492. A translation from 'The Saga of the Greenlanders' (Source 2) tells of this discovery.

'Nature was so generous here that it seemed to them no cattle would need any winter fodder, but could graze outdoors. There was no frost in winter, and the grass hardly withered. The days and nights were more nearly equal than in Greenland or Iceland ... Leif told his crew, "From now on we have two jobs on our hands. On one day we shall gather grapes, and on the next we shall cut grape vines and chop down the trees to make a cargo for my ship ..." Leif gave this country a name to suit its resources: he called it Vinland [wine land].'

Source 2

This translation from 'The Saga of the Greenlanders' tells of the enthusiasm that the Viking explorers had for the newly discovered land, named Vinland after the wild grapes that grew there. [Translation by Professor Einar Haugen, 1942.]

Source 1

Reconstruction of the L'Anse aux Meadows site uncovered by archaeologists between 1961 and 1968 in Newfoundland, Canada. So far, it is the only primary evidence of a Viking settlement in North America.

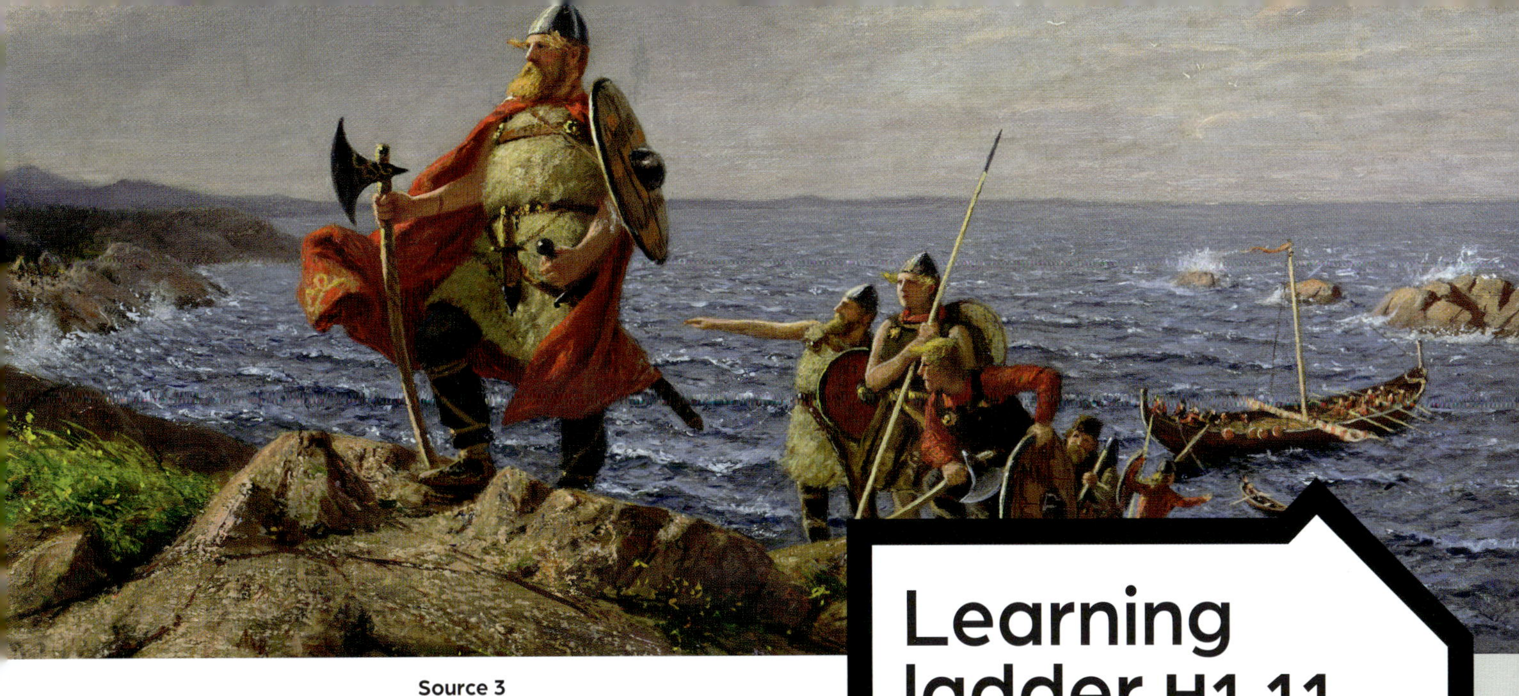

Source 3

Artist Hans Dahl's *Leif Eriksson Discovers America* is a secondary source painted in the late 1800s. Some historians have suggested that Eriksson sailed off course on his way back to Greenland and landed on the North American continent. Others say he may have heard about the land from information passed on by an Icelandic trader.

The search for primary evidence

The **sagas** provided clues that modern **archaeologists** could use to find physical evidence of Viking settlements. In 1960, the Norwegian explorer Helge Ingstad came upon strange bumps and ridges in the land at L'Anse aux Meadows in Newfoundland, Canada. They uncovered definite evidence that people of Viking origin had occupied the site for anywhere between three and ten years.

L'Anse aux Meadows

The ridges discovered by Helge Instad were the remains of eight buildings. They are believed to have been made of **sod** (soil and grass) built up over a wooden frame.

The largest building was a Viking **longhouse** where Viking families lived and kept livestock. One of the smaller buildings was a workshop for a blacksmith containing a forge for heating iron. **Carbon dating** of the evidence at L'Anse aux Meadows suggests the settlement was active between 990 and 1050 CE.

Hundreds of Viking **artefacts** were also found, including a bronze pin used to fasten a cloak; a large number of iron boat nails, used by Viking shipbuilders; and knitting needles and a spindle whorl used for spinning thread, suggesting that women were part of the settlement.

Learning ladder H1.11

Show what you know

1 What does Source 2 suggest about what the Vikings expected when they came ashore in North America?

2 Why did archaeologists start digging at L'Anse aux Meadows in Canada?

3 Looking at the list of artefacts found at L'Anse aux Meadows, what would these items have been used for in Viking life?

4 Why might Einar Haugen's translation of 'The Saga of the Greenlanders' not be an accurate description of Leif Eriksson's voyages in North America?

Source analysis

Step 1: I can determine the origin of a source

5 What evidence suggested to archaeologists that the buildings in L'Anse aux Meadows were Viking?

Step 2: I can list specific features of a source

6 What does Leif think of North America, according to Einar Haugen's translation of 'The Saga of the Greenlanders' in Source 2? Provide quotes from the translation to back up your answer.

Step 3: I can find themes in a source

7 Explain what the Vikings in Source 3 were thinking and feeling. How is this shown in the painting? Why do you think the Vikings felt this way?

Step 4: I can use my outside knowledge to help explain a source

8 What would the Vikings have used to build the structures in Source 1? Why are the buildings shaped that way? Why would there have been fences around the buildings? Why were they built near the coast?

Source analysis, page 197

key individual

Why was Harald the last of the Vikings?

The introduction of **Christianity**, professional enemy soldiers and the migration of Vikings to other lands had all weakened the Viking civilisation. The death of the Norwegian King Harald Hardrada in 1066 was the final blow that brought the age of the Vikings to a close.

The Viking Age

The brave and fearless Vikings were admired and feared by Europeans for almost 300 years. However, their powerful reign that had begun with their savage attacks on Europe, beginning in 793 CE, eventually came to an end with the death of King Harald Hardrada at the Battle of Stamford Bridge in 1066.

Harald Hardrada

Mighty King Harald Hardrada was the last great Viking and the most feared warrior of his time. Brendan Manley describes King Harald as he and his invasion force of 10 000 Viking warriors were preparing to attack York in England in September 1066:

'The Viking commander alone was enough to strike terror in the hearts of English defenders: King Harald III Sigurdsson of Norway, aka Harald Hardrada ("the Hard Ruler"), was a career warlord, a broad-shouldered giant of a man who stood well over 6 feet and who had spent the preceding 35 years

continued >>

Source 1

King Harald Hardrada was the last great Viking and the most feared warrior of his time. His death at the Battle of Stamford Bridge in 1066 marked the end of the 300-year Viking Age, as well as the beginning of the Norman Conquest of England.

honing his martial skills in a variety of conflicts, taking him from the royal court in Kiev to the palaces of Byzantium. Soon after assuming the throne of Norway in 1047, Hardrada – who was as flamboyant as he was fierce and a prolific composer of heroic sagas – launched into a protracted war with Denmark, not tasting victory until 1064. By 1066 the ever-ambitious warrior – who, like Duke William of Normandy, was a potential claimant to the English throne – hungered for a new conquest. At the urging of a future ally, Hardrada set his sights on England.'

Source 2

Brendan Manley, HistoryNet, 2008

The end of the Viking Age

In 1066, Harald Hardrada tried to conquer England, but was defeated by the **Anglo-Saxon** King, Harold II, and his army at the Battle of Stamford Bridge. King Harald's final battle is described in *The Saga of Harald Hardrada*:

'... when they had broken their [the Vikings] shield-rampart, the Englishmen rode up from all sides, and threw arrows and spears on them. Now when King Harald saw this, he went into the fray where the greatest crash of weapons was, and there was a sharp conflict, in which many people fell on both sides. King Harald then was in a rage, and ran out in front of the array, and hewed down with both hands; so that neither helmet nor armour could withstand him, and all who were nearest gave way before him ... King Harald was hit by an arrow in the windpipe, and that was his death-wound. He fell, and all who had advanced with him, except those who retired with the banner.'

Source 3

From 'The Saga of Harald Hardrada' by Snorri Sturlason (c.1179–1241)

In his final battle, Harald Hardrada is said to have entered the trance-like state used by ferocious Viking warriors, known as **berserkers**. He went berserk by racing forward and hacking at his foes until he was felled by an arrow in the throat.

Why did the Viking Age end?

The death of Harald Hardrada and the Viking defeat by Harold II at Stamford Bridge spelt the end of the Viking Age. Less than a month later, Harold II was killed at the Battle of Hastings by the Duke of Normandy (commonly referred to as William the Conqueror) who began the Norman rule of England. However, the power of the Vikings had been declining for more than 50 years. Some of the reasons for this were:

- the Vikings were frequently defeated in battle
- the feudal system (see pages 66–67) involved lords granting men land in exchange for military service. The Vikings were beaten more often by these well-trained soldiers
- many Vikings had moved to other regions of Europe and had become loyal to their new leaders
- by 1066, most Scandinavians had converted to Christianity and no longer believed in the Norse gods and the need to be warlike.

Source 4

Harald Hardrada is fatally wounded when shot by an arrow in the neck at the Battle of Stamford Bridge on 5 September 1066. Anglo-Saxon victory under King Harold Godwinson over Hardrada's invading Norwegian army led to the end of the Viking Age. [Peter Nicolai Arbo, *Battle of Stamford Bridge* (1870), oil on canvas, 143 x 212 cm.]

Source 5

This stained-glass window of Harald Hardrada is from Lerwick Town Hall in Scotland's Shetland Islands. The early history of the islands was greatly influenced by the Vikings.

Learning ladder H1.12

Show what you know

1. Where does the modern phrase 'going berserk' come from?
2. How did the feudal system and religion contribute to the decline of Viking dominance?
3. Consider Source 5. Why do you think there is a picture of a Viking king in a Scottish Church? What can we learn about Vikings from looking at this window?
4. In Source 5, what suggests that the figure in it is a king? Describe *all* the evidence.
5. Source 2: What can we learn about Harald's personality from reading Brendan Manley's excerpt?

Chronology

Step 1: I can read a timeline

6. How is 1066 remembered on the timeline on page 22? What other events occurred in this year?

Step 2: I can place events on a timeline

7. Which events mark the beginning and the end of the Viking Age? How long did it last?

Step 3: I can create a timeline using historical conventions

8. Place four events from this section on a timeline using correct historical conventions, which are listed on page 195.

Step 4: I can distinguish causes, effects, continuity and change from looking at timelines

9. William the Conqueror had two main rivals who stood in the way of him sitting on the English throne – King Harald Hardrada, and the Anglo-Saxon King Harold II. Looking at the timeline you created in question 8, explain the events that led up to William the Conqueror's victory in 1066. Refer to your timeline and dates in your response.

HOW TO

Chronology, page 194

Masterclass

Learning Ladder

Work at the level that is right for you or level-up for a learning challenge!

Source 1

Gravestone, 11th century CE, Bibury, Gloucestershire, England

Source 2

Dwelling made from peat bricks and a sod-covered roof at L'Anse aux Meadows.

Step 1

a **I can read a timeline**

Look at the timeline on pages 22–23 and answer the following questions:

i How long did Vikings raid Europe?

ii How long did the Danelaw last?

iii How long ago did Normans conquer England?

b **I can determine the origin of a source**

From when and where is Source 1?

c **I can recognise continuity and change**

Place the following in order from the least changed to the most changed from the Viking Age to the present.

- Clothing
- Religion
- Food
- Farming

d **I can recognise a cause and an effect**

Copy and complete the following diagram, putting the text in the right place.

- Vikings lived on the coast
- Vikings were able to settle in far-away lands
- Vikings were able to trade with distant people
- Vikings were great boatbuilders

e **I can recognise historical significance**

Which of the sections in this chapter do you think was the most important?

Step 2

a **I can place events on a timeline**

List the following events in the correct order, from earliest to most recent. Note that there are five events mentioned.

Vikings began raiding Ireland in about 795 CE.

The Vikings raided Spain from about 844–855 CE.

Iceland is discovered by Vikings in 860 CE. They begin settling it around 14 years later.

The Vikings settled in Greenland, c. 986 CE.

b **I can list specific features of a source**

Source 2: Identify and describe the key features of this source.

c **I can describe continuity and change**

How has transport changed from the Viking Age to the present?

d **I can determine causes and effects**

What caused Viking raids to be so successful?

e **I can explain historical significance**

Which of the sections in this chapter do you think was the least important? Why?

Step 3

a **I can create a timeline using historical conventions**

Put the events from Step 2a on a timeline, using correct historical conventions (page 195). Make the timeline artistic by drawing the line as an image from the Viking world such as a sword or a longship.

b **I can find themes in a source**

What do you think is the artistic message in Source 1? Why did the artist include the figures in the way they did? What are they trying to evoke in the viewer?

c **I can explain why something did or did not change**

The Vikings settled in many new lands. Why do people still move to new countries and live there?

d **I can explain why something is a cause or an effect**

What caused the Vikings to occasionally build houses out of stone?

e **I can apply a theory of significance**

Looking at Source 2, consider the following.

i How important was the Viking settlement in North America at the time?
ii How deeply would those people have been affected?
iii How many people would have been affected and for how long?
iv Is the Viking settlement that existed in North America still important today?

Step 4

a **I can distinguish causes, effects, continuity and change from looking at timelines**

Look at the timeline on pages 22–23. What is one thing that continued throughout the Viking Age? Give examples from the timeline to justify your response.

b **I can use my outside knowledge to help explain a source**

Which civilisation do you think created Source 1? Justify your answer.

c **I can analyse patterns of continuity and change**

The Vikings raided and settled around Europe over a 300-year period, known as the Viking Age. Why is it just this period that is called the Viking Age?

d **I can analyse cause and effect**

List some different effects of the Viking attacks on the rest of Europe, using the following headings.

- Effect on trade
- Effect on sharing of technology
- Effect on population and spread of people
- Effect on language

e **I can analyse historical significance**

Give a one-sentence-long reason for the importance of Viking trading networks under these headings:

- food for ordinary people
- enriching Vikings
- spreading technology
- leaving artefacts as a record of Viking lifestyles.

Masterclass

Step 5

a **I can describe patterns of change**

Look at the list of possible patterns of change on page 196 in the History How-to section. Which of these do you think most applies to the Viking Age? Back up your answer with dates.

b **I can use the origin of a source to explain its creator's purpose**

Think about what you know of the religious views of the civilisation that made Source 1. Why do you think the gravestone was designed in this way?

c **I can evaluate patterns of continuity and change**

Raiding and settling were two major activities Vikings took part in. Which is more common these days? Why isn't the other one still common? Is this a good thing?

d **I can evaluate cause and effect**

Vikings merged with other Europeans after settling in their regions. Many of them changed their religion from worshipping the Norse gods to Christianity. Why do you think this happened? Does this teach us anything about how groups change? Was this change important?

e **I can evaluate historical significance**

Should we bother learning about the Vikings? Was the Viking Age important enough for us to teach it to young people today? Justify your answer.

Historical writing

1 Structure

Imagine you are given the essay topic, 'Explain the Vikings' impact on modern Europe'. Write an essay plan for this topic. Include at least three main paragraphs.

2 Draft

Using the drafting and vocabulary suggestions on page 214, draft a 400–600 word essay responding to the topic.

3 Edit and proofread

Use the editing and proofreading tips on page 215 to help edit and proofread your draft.

Historical research

4 Organise and present information

Imagine you are doing a large research project: 'Viking discoveries'. Write a contents page for this project. There should be an introduction, conclusion, at least four main sections and many subsections. Number your chapters.

Capstone

How can I understand the Vikings?

In this chapter, you have learnt a lot about Vikings. Now you can put your new knowledge and understanding together for the capstone project to show what you know and what you think.

In the world of building, a capstone is an element that finishes off an arch, or tops off a building or wall. That is what the capstone project will offer you, too: a chance to top off and bring together your learning in interesting, critical and creative ways. You can complete this project yourself, or your teacher can make it a class task or a homework task.

mea.digital/GHV8_H1

Scan this QR code to find the capstone project online.

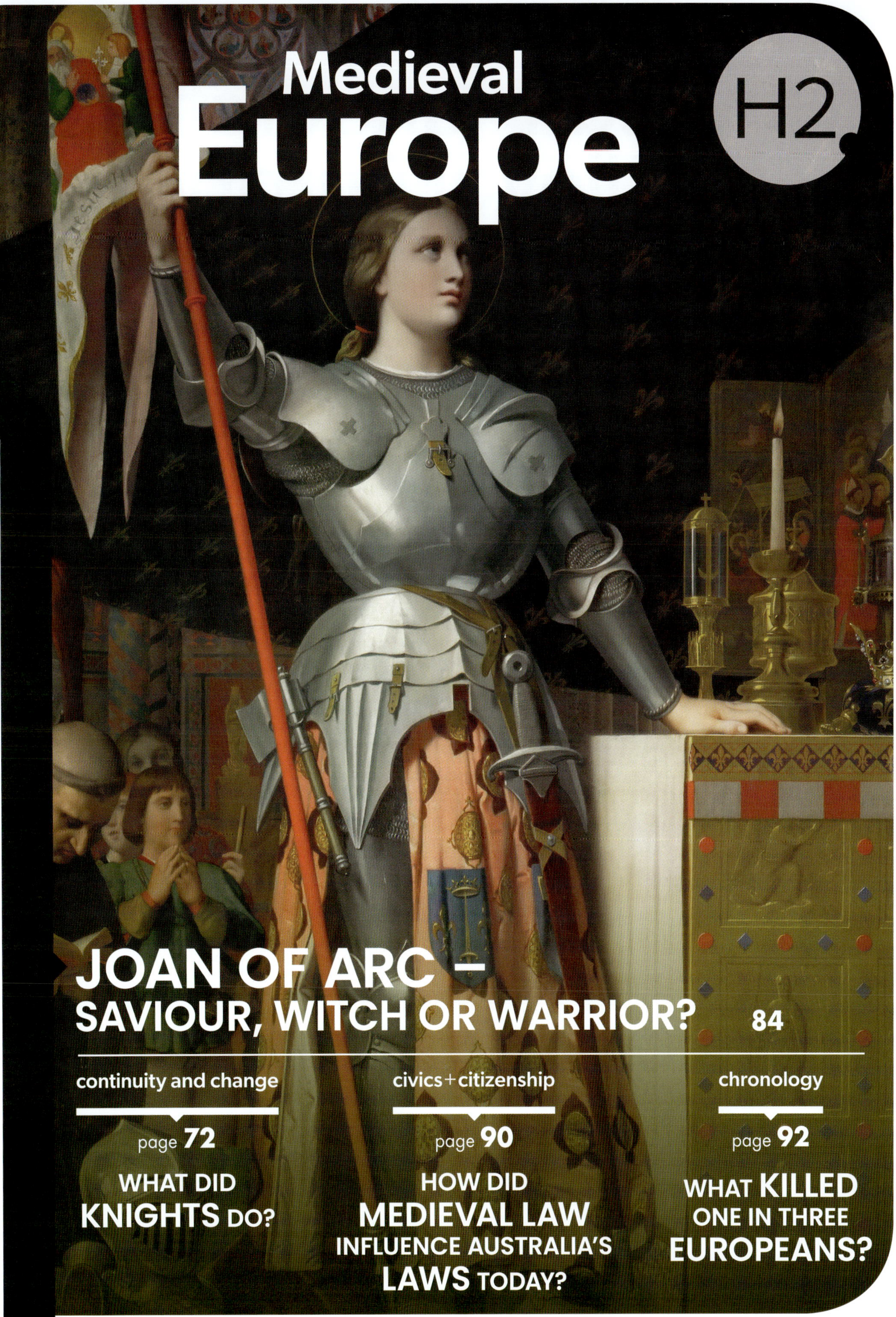

Medieval Europe

JOAN OF ARC – SAVIOUR, WITCH OR WARRIOR? 84

continuity and change

page **72**

WHAT DID KNIGHTS DO?

civics+citizenship

page **90**

HOW DID MEDIEVAL LAW INFLUENCE AUSTRALIA'S LAWS TODAY?

chronology

page **92**

WHAT KILLED ONE IN THREE EUROPEANS?

How can I understand medieval Europe?

With the growth of cities and trade and huge advances in technology, the Medieval Period in Europe gave birth to the modern world. This ever-changing period saw the development of universities, banking, parliament and legal protections like equality before the law and trial by jury.

Source 1

A chalice from central Europe, mid-15th century.

learning ladder

Step	Chronology	Source analysis	Continuity and change
step 5	**I can describe patterns of change** I read timelines and see the 'big picture'. I group timeline events about medieval Europe and see if they show patterns of change. I know typical historical patterns to check for.	**I can use the origin of a source to explain its creator's purpose** I combine knowledge of when and where a source was created to answer the question, 'Why was it created?'	**I can evaluate patterns of continuity and change** I answer the question, 'So what?' about patterns of continuity and change. I weigh up ideas and debate the importance of a continuity or a change.
step 4	**I can distinguish causes, effects, continuity and change from looking at timelines** I read medieval timelines and find events linked by cause and effect. I also find things that are the same or different from then until later times.	**I can use my outside knowledge to help explain a source** I have enough outside knowledge about medieval Europe to help me explain a source.	**I can analyse patterns of continuity and change** I see beyond individual instances of continuity and change in medieval Europe and identify broader historical patterns, and I explain why they exist.
step 3	**I can create a timeline using historical conventions** When given a set of events, I construct a historical timeline using the dates, and using correct terminology, spacing and layout.	**I can find themes in a source** I look a bit closer into a source and find more than just features. I find themes or patterns in the source.	**I can explain why something did or did not change** I answer the question, 'Why?' something changed or stayed the same between historical periods.
step 2	**I can place events on a timeline** When given a list of medieval European events and developments, I put them in order from earliest to latest, the simplest kind of timeline.	**I can list specific features of a source** I look at a medieval European source and list detailed things I can see in it.	**I can describe continuity and change** I have enough content knowledge about two different historical periods to recognise what is similar or different about them, and can describe it.
step 1	**I can read a timeline** I read timelines with medieval European events on them and answer questions about them.	**I can determine the origin of a source** I can work out when and where a medieval European source was made by looking for clues.	**I can recognise continuity and change** I recognise things that have stayed the same and things that have changed from medieval Europe until now.

Source 2

Scenes from the Old Testament of the Bible, from a wall painting in St Stephen's Chapel, London, England, 1355–63 CE. There are 17 fragments from the wall, showing scenes and inscriptions from the Book of Job.

Cause and effect	Historical significance
I can evaluate cause and effect I answer the question, 'So what?' about cause and effect. I weigh up ideas and debate the importance of a cause or an effect.	**I can evaluate historical significance** I answer the question, 'So what?' about things that are supposedly historically important. I weigh up events and cast doubt on how important things are.
I can analyse cause and effect I don't just see a cause or an effect as one thing. I determine the factors that make up causes and effects.	**I can analyse historical significance** I separate out the various factors that make something historically important.
I can explain why something is a cause or an effect I can answer 'How?' or 'Why?' a cause led to an effect in medieval Europe.	**I can apply a theory of significance** I know a theory of significance. I use it to rank the importance of events in medieval Europe.
I can determine causes and effects Applying what I have learnt about medieval Europe, I can decide what the cause or effect of something was.	**I can explain historical significance** I answer the question, 'Why?' about things that were important in medieval Europe.
I can recognise a cause and an effect From a supplied list, I recognise things that were causes or effects of each other in medieval Europe.	**I can recognise historical significance** When shown a list of things from medieval European history, I can work out which are important.

Warm up

Chronology

1 Compare the wall painting in Source 2 with the Bayeux Tapestry (Source 3 on pages 64–65). Which of these is older?

Source analysis

2 Look at Source 2. When and where was the image painted?

Continuity and change

3 Identify three examples from the text on this page that represent continuity between medieval Europe and modern-day Australia.

Cause and effect

4 Looking at the timeline on pages 58–59, the Peasant Revolt occurred in 1381. What event happened before then that may have helped cause the unrest?

Historical significance

5 What was the historical significance of the Black Death? (See the timeline on pages 58–59.)

Why learn about medieval Europe?

During the **Medieval Period**, Europe went from being a backwards, dangerous place after the fall of the Roman Empire to the world's dominant power. In this age, European culture, art, literature, language, sport and religion flourished. Medieval Europe was the time of noteworthy historical figures like Charlemagne (Charles the Great), William the Conqueror, Marco Polo and Joan of Arc. The end of this period saw the birth of Leonardo da Vinci and, later, William Shakespeare.

Medieval Europe 1278 CE

Source: Matilda Education Australia

Source 1

The boundaries of medieval Europe changed often. This map shows the boundaries at 1278 CE.

key ideas timeline.

476 CE — Collapse of the Roman Empire

732 CE — Battle of Tours: Franks defeat Spanish **Moors**

800 CE — Feudal system begins in Europe

800 CE — Charlemagne crowned Emperor of western Europe

Viking raids on Europe 789–1042 CE

Danelaw – Viking kingdom in England 880–954 CE

Source 2

Godfrey of Bouillon, who led the First Crusade, giving thanks to God in the presence of the priest Peter the Hermit after the capture of Jerusalem in 1099. [Emile Signol, *Taking of Jerusalem by the Crusaders, 15th July 1099* (1847), oil on canvas, 324 x 557 cm, Palace of Versailles, France.]

Source 3

The signing of the Magna Carta

Learning ladder H2.1

Show what you know

1 How did Europe change during the Medieval Period?

2 Source 2: Why are the people celebrating?

3 Source 1: List three names of countries or regions that are no longer used today.

Chronology

Step 1: I can read a timeline

4 For how many years did the Crusades last?

Step 2: I can place events on a timeline

5 Place the following events in chronological order:

1187 CE Muslims take back Jerusalem

1099 CE Crusaders capture Jerusalem

1050 CE Muslims and Selijuks control Jerusalem

1096 CE First Crusade leaves Constantinople

Step 3: I can create a timeline using historical conventions

6 Create your own timeline showing the events from question 5.

Step 4: I can distinguish causes, effects, continuity and change from looking at timelines

7 What was the cause and effect of the events in 800 CE?

HOW TO

Chronology, page 194

1066 CE
Normans conquer England

1215 CE
King John of England signs Magna Carta

1271 CE
Start of voyages of Marco Polo

1347 CE
Black Death hits Europe

1381 CE
The Peasant Revolt

1431 CE
Joan of Arc is killed

Italian Renaissance
Approx. 1350–1450 CE

Hundred Years War between England and France
1337–1453 CE

The Crusades
1096–1291 CE

Feudalism
800–1400 CE

What were the Dark Ages?

People sometimes use the term 'the Dark Ages' to describe the uncertain times after the fall of the Roman Empire, from 476 to about 1000 CE. The Romans were no longer able to maintain law and order and ward off attacks from invaders.

Barbarian raids

Historians now prefer to describe the period of the **Dark Ages** as the Early **Middle Ages**. Barbarian tribes took over the land formerly controlled by the Romans. The term **barbarian** was used to describe different non-Latin and Greek speaking groups such as the Celts, Franks, Saxons, Angles, Visigoths and Huns.

The Roman army had kept peace between different parts of Europe, but with the Roman army gone there was much fighting and many kings and leaders tried to create their own empires. In the 9th and 10th centuries, Europe was invaded from every direction – the Vikings from Scandinavia, the Arabs from the Middle East and North Africa, and the Magyars from Eastern Europe.

Source 1

Europe, 814 CE

Europe at the death of Charlemagne in 814 CE

Source: Matilda Education Australia

Emperor Charlemagne

The **Frankish Empire** was the most powerful kingdom in the 8th and 9th centuries. During the reign of Charlemagne (also known as Charles the Great or Charles I) from 768 to 814 CE, the Franks expanded their territory into Germany. Charlemagne was the first ruler in over three centuries to have the military power to create a stable **empire** in Europe.

Charlemagne was helped by armies of loyal supporters who he repaid with grants of land. This exchange of favours laid the foundations for the **feudal system** that spread across Europe in the Middle Ages (see pages 66–67).

Charlemagne supported the Catholic Church and forced conquered peoples to convert to Christianity. In 800, the Pope crowned Charlemagne as **emperor** of the Holy Roman Empire (modern-day western Europe). The empire did not survive long after Charlemagne's death in 814.

Source 2

The Siege of Paris of 885–86 was a Viking raid on the Frankish Empire. Thousands of Vikings sailed up the Seine River and only withdrew when Charles the Fat (great-grandson of Charlemagne) promised to pay them 257 kilograms of silver.

The Catholic Church

The Catholic Church and its leader, the Pope, had great influence over how people behaved (see pages 80–83). **Catholicism** became the Roman Empire's official religion and it spread throughout Europe. Towards the end of the period it then divided into the Eastern Orthodox Church and much later into the Protestant Church and the Church of England.

With the fall of the Roman Empire in 476 CE, the power of the Pope increased. The Catholic Church became a unifying force during a time of great change in the Early Middle Ages. The Pope was a powerful ally and gave kings and other leaders the moral authority to lead their people.

Viking invasions

In response to Charlemagne's conquests, the Vikings retaliated by raiding. After repeated Viking raids, Frankish king Charles the Simple (a descendant of Charlemagne) reached a treaty with Rollo, leader of the Viking raids, in 911. He gave the Vikings land in the north-west of the empire in return for their loyalty and conversion to Christianity. Large numbers of Vikings settled the region and adopted the French language, customs and religion. The region became known as Normandy and the Viking settlers became known as **Normans**.

Learning ladder H2.2

Show what you know

1 What is a barbarian?

2 Who and from where did groups invade Europe after the fall of the Roman Empire?

3 What did Charlemagne do?

4 Why was the Pope powerful?

Cause and effect

Step 1: I can recognise a cause and an effect

5 Which of these was a cause of the Early Middle Ages being called the Dark Ages?
   - a Charlemagne died in 814 CE.
   - b Barbarian raiders took possession of the land
   - c The Roman Empire wasn't around to maintain stability.

Step 2: I can determine causes and effects

6 What was the effect of Vikings raiding northern France?

Step 3: I can explain why something is a cause or an effect

7 Explain the causes behind Charlemagne's ability to create a stable empire in Europe following the Dark Ages.

Step 4: I can analyse cause and effect

8 Analyse the causes and effects driving Europe's period of instability in the Early Middle Ages.

Cause and effect, page 203

H2.3

key individual

Who was William the Conqueror?

In 1066, the Duke of Normandy, William, defeated the English at the Battle of Hastings and became king. The feudal system he introduced changed the way people lived for centuries.

A Viking heritage

Duke William was a **Norman** who spoke French and grew up in Normandy, a territory loyal to the French kingdom. The Normans descended from **Viking** invaders who raided France in the 9th and 10th centuries. In a peace settlement with the French (see page 61), the Vikings accepted Normandy as their own territory.

Fight for the throne

In January 1066, King Edward the Confessor of England died without an heir to the throne. Four people, including William, believed they should be the next King of England. William believed Edward had promised him the throne for helping to put down a rebellion by Harold Godwinson in 1051.

On his deathbed, King Edward anointed the Earl of Wessex, Harold Godwinson, as his successor. Harold was crowned king two days after Edward the Confessor died. William believed Harold had stolen what was promised to him and undertook to take the English throne by force.

Source 1

William the Conqueror leads his Norman cavalry at the Battle of Hastings, 1066. [Illustration from Histoire de Frances, c. 1902.]

The Battle of Hastings

William prepared his troops in France and landed at Pevensey in southern England. King Harold quickly marched his **Saxon** troops (mostly untrained farmers) south to meet the invading French army near Hastings.

The armies were equally matched in numbers; however, William's troops were better trained and many were on horseback. The Saxon soldiers held off the invading Normans who repeatedly charged them. William ordered his archers to fire their arrows high over the wall of Saxon shields. One arrow is thought to have hit King Harold in the eye, killing him. By the end of the day, the Normans had won the battle.

William moved to London and was crowned King of England on December 25, 1066.

Source 2

The Battle of Hastings

1 Norman archers open fire on Saxons;

2 & 3 Normans attack shield wall, but are pushed back;

4 Saxons chase retreating Norman cavalry and Duke William counter-attacks;

5 Saxons again chase retreating Normans and are counter-attacked again. Harold is killed and the English resistance crumbles.

The Battle of Hastings, 1066

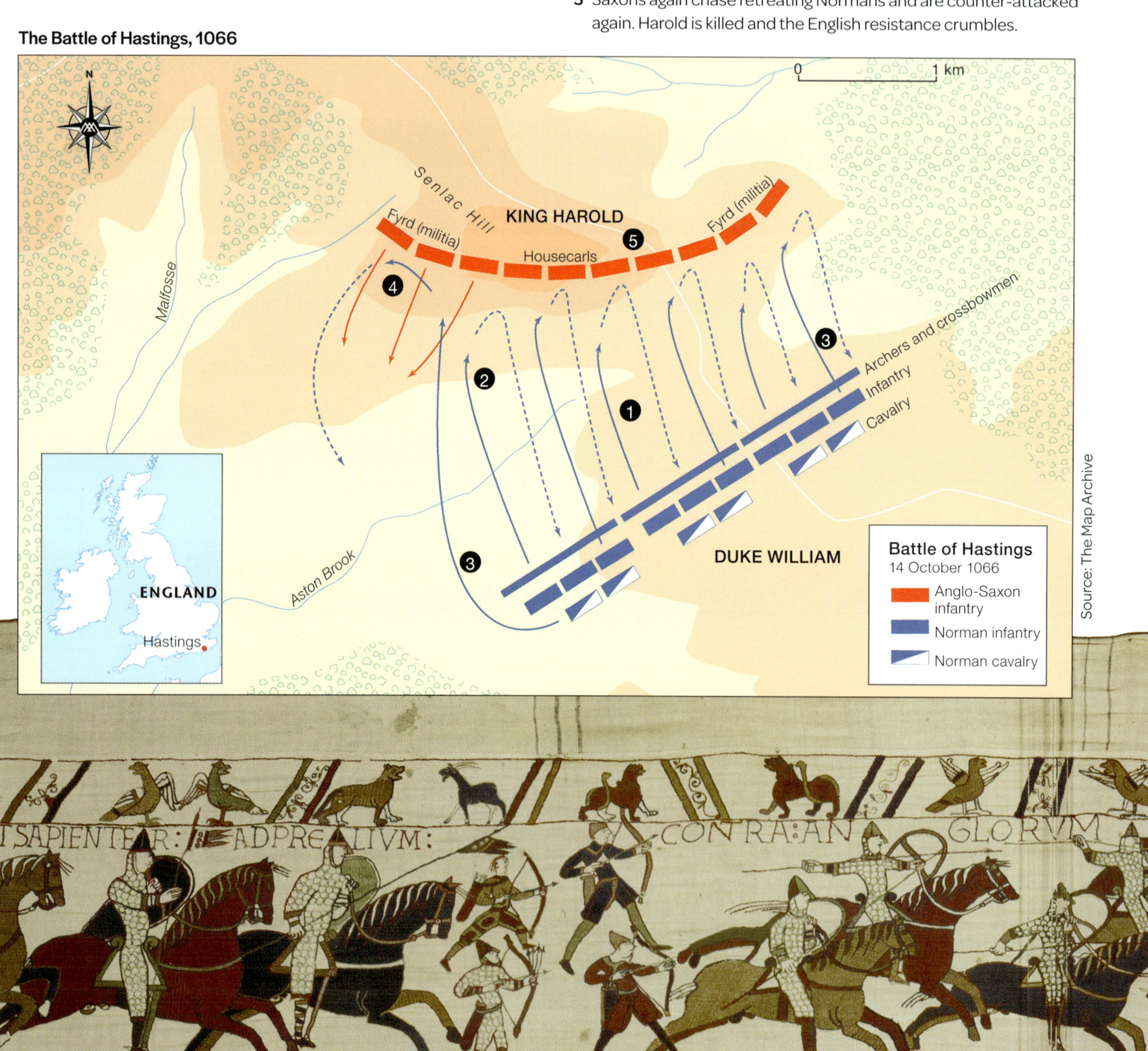

Source: The Map Archive

Norman feudalism

When William the Conqueror became King of England in 1066, he introduced his own **feudal system** into Britain. He confiscated land from Saxon lords and distributed it to members of his family and the Norman lords who had supported his invasion.

However, William still owned the land – the lords became tenants-in-chief of plots known as **manors**. The tenants-in-chief rented the land from the king in exchange for services such as providing knights to fight for the king when necessary. If the lord did not adequately provide this service, he would be removed as the tenant-in-chief.

In 1086, William the Conqueror sent surveyors to every part of England, with orders to list how much land was there, what it was used for, who owned it before 1066, who lived there and how much it was worth. The information was compiled into the **Domesday Book**. This book clearly established what was owed to the King in money, goods or soldiers.

Learning ladder H2.3

Show what you know

1 Why did William think he should be King of England?

2 How did William win the Battle of Hastings?

3 How did William change England after he became king?

4 Why was the Domesday Book written?

Source analysis

Step 1: I can determine the origin of a source

5 Source 3: When was the Bayeux Tapestry created?

Step 2: I can list specific features of a source

6 Describe in detail all the action you can see in the Bayeux tapestry in Source 3.

Step 3: I can find themes in a source

7 Source 1: What techniques has the artist used to show William the Conqueror as a heroic leader?

Step 4: I can use my outside knowledge to help explain a source

8 Source 2: Can you think of any tactics or strategies Harold could have used to win the Battle of Hastings?

Source analysis, page 197

Source 3

The Bayeux Tapestry is a 70-metre-long embroidered cloth that tells of the Norman Conquest of England and the Battle of Hastings. It was created around 1068.

How was society organised?

In medieval Europe, society was organised in a strict hierarchy under the feudal system. All people had rights and obligations, from the King and his lords, down to the poorest peasants.

The origins of feudalism

After the collapse of the Roman Empire, people no longer had the Roman army to protect them against raids from barbarians.

Feudalism, or the feudal system, offered a new way for people to protect and provide for themselves. People moved onto the land run by the wealthy and powerful lords. Here they could be protected by the lord's own army and take shelter in his castle during attacks. As payment, these people provided military service and worked the lord's land for him. Feudalism operated in Europe between the 9th and 15th centuries.

How feudalism worked

Feudalism was a system of rights and obligations throughout society. Beneath the king were lords and knights who acted as **vassals** – someone who received land in return for providing military service.

All the land in a kingdom was owned by the king. The lord was the king's vassal. He was obliged to obey the king and provide him with soldiers for his army in return for a parcel of land known as a **manor**.

Source 1

How the feudal system worked in medieval Europe

The king owned the land, but gave plots of land called manors to the lords.

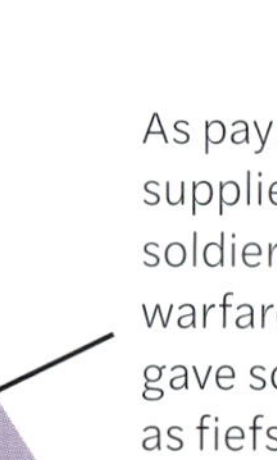

King

As payment, the lords supplied the king with soldiers and horses for warfare. The nobles gave some land, known as fiefs, to the knights.

Lords (vassals to king)

The knights fought for their lords and the king.

Knights (vassals to king/lords)

Peasants farmed the land for their lords and knights. In return, the peasants were given protection.

Peasants

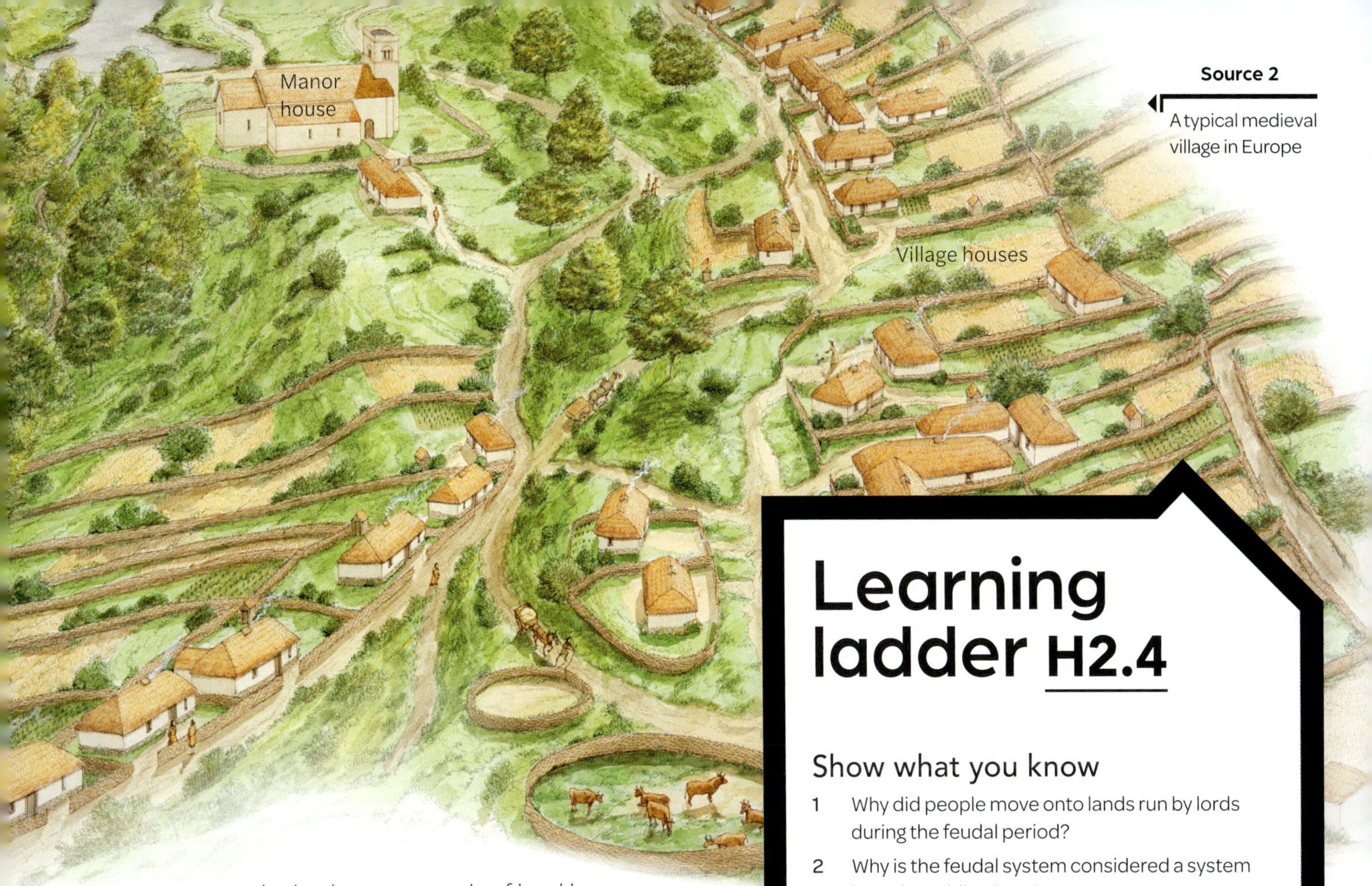

Manor house

Village houses

Source 2

A typical medieval village in Europe

In turn, the lords gave parcels of land known as **fiefs** to their vassals – the knights. In return, the knights were obliged to fight as soldiers for the lord and demonstrate **fealty**. The agreement was made in an official ceremony known as **homage**. To pay homage, the vassal would kneel before the lord and swear the Oath of Fealty: 'I promise on my faith that I will in the future be faithful to the lord, never cause him harm and will observe my homage to him completely against all persons in good faith and without deceit.'

Peasants

The **peasant** class received food, shelter and protection from the lords and knights in return for their work on the land and rent payments. The lord provided peasants on his land with a place to live and the opportunity to grow their own food. In return, the peasants provided the lord with their labour as farmers, or as skilled workers such as blacksmiths or millers. They also also gave a portion of the food they produced as a **tax** to the lord.

The poorest of the peasants were the **serfs**. They worked for the lord and received protection but had very few rights. Most serfs were not free to leave the manor and needed the lord's permission to marry. If you were born a serf, you and your children would always be serfs.

Learning ladder H2.4

Show what you know

1 Why did people move onto lands run by lords during the feudal period?

2 Why is the feudal system considered a system based on obligations?

3 For roughly how long did the feudal system operate in Europe?

Continuity and change

Step 1: I can recognise continuity and change

4 How much have these items changed between the Middle Ages and today? Rank them in order from most changed to least changed.

- The food
- The houses
- The ruler
- The size of towns
- Farming

Step 2: I can describe continuity and change

5 How is the feudal system different to modern society? What parts of it are the same?

Step 3: I can explain why something did or did not change

6 Why was it hard for serfs to improve their position in society?

Step 4: I can analyse patterns of continuity and change

7 During the Medieval Period, the First Nations Peoples of Australia were living and working the land in different ways. Do some research to find out how some of the First Nations worked the land. Compare how they produced food and materials, and how their hierarchies were different.

Continuity and change, page 200

How did peasants live?

Peasants made up 90 per cent of the population in medieval times. They were bound to obey their local lord and the Church, working a six-day week from dawn to dusk.

Peasants at work

Peasants were the poorest people in medieval times and **serfs** were the poorest of the peasant class. Most serfs in the Middle Ages were little more than slaves. Serfs could have property, but they and their property were legally part of the land – if the land was sold, the serfs and their possessions were sold with it. Serfs had to obey their local lord, to whom they swore an oath of obedience on the Bible.

Source 1

Peasant farmers ploughing and sowing in March, from an illustrated manuscript from 1477

In exchange for their loyalty, peasants and serfs were protected by their lord and leased some land to farm. However, peasants and serfs were not given much time to farm their land. They spent three days of every week clearing and farming fields, repairing roads and buildings, and chopping timber on the lord's land. It was back-breaking work, but the workers could not refuse as they faced harsh punishments for not doing as they were told. Serfs were also forced to work church land for fear that God would punish them.

In 1395, medieval author and historian Jean Froissart wrote:

'It is the custom in England, as with other countries, for the nobility to have great power over the common people, who are serfs. This means that they are bound by law and custom to plough the field of their masters, harvest the corn, gather it into barns, and thresh and winnow the grain; they must also mow and carry home the hay, cut and collect wood, and perform all manner of tasks of this kind.

'A peasant had to pay rent for his land to his lord and pay a tax known as a **tithe** to the Church. A tithe was 10 per cent of the value of everything he had produced in a year. Tithes could be paid in cash or produce or seeds. The Church collected so much produce from this tax that they had to build huge tithe barns to store it.'

Source 2

Jean Froissart, 'The Lives of Medieval Peasants', *HistoryLearning.com*, 2015

Source 3

Peasants lived in small cruck houses, with dirt floors and brick walls made from mud, straw and manure. The roofs were thatched with straw or other plant material. Cruck houses were cheap and easy to build, as most materials came from a nearby forest.

Cruck houses had no running water, toilets, baths or wash basins. Windows and doors were holes in the walls, sometimes covered with a curtain. Animals were brought inside at night to ensure they didn't get lost, stolen or eaten by predators.

Peasant families

Peasants had few comforts and a life expectancy of little more than 30 years. Peasant women worked with men as farm labourers, but they also had household duties such as fetching water, cooking, weaving and looking after children and livestock. Young children helped women with household chores and, from the age of 10, they too worked on the farm.

The Peasant Revolt

In 1381, rebel peasants demanded reduced taxation and freedom from serfdom. Author Jean Froissart wrote:

> '… they had determined to be free, and if they laboured or did any other works for their lords, they would be paid for it.
>
> 'The rebels entered London and captured the Tower of London, killing the Lord Chancellor and the Lord High Treasurer. King Richard met the rebels and agreed to most of their demands, including the abolition of serfdom. The peasant unrest continued and King Richard took back his promises to the rebels and had the rebel leaders tracked down and executed. At least 1500 rebels were killed.'

Source 4

The Chronicles of Jean Froissart, 1483

Learning ladder H2.5

Show what you know

1. If a medieval manor had 100 people on it, roughly how many of them would have been peasants?
2. Why did peasants put up with the feudal system when it was so unfair to them?
3. What was the outcome of the Peasant Revolt?
4. Source 3: Compare and contrast your home life to that of a peasant in a cruck house.

Historical significance

Step 1: I can recognise historical significance

5. Put these facts in order from most to least historically important:
   - The Peasant Revolt in 1381
   - 90 per cent of people were peasant farmers
   - Peasant children didn't attend school
   - Peasants paid tax to the lord and the Church

Step 2: I can explain historical significance

6. What is the difference in the proportion of people that were farmers in the Middle Ages compared to now? Why is the difference important?

Step 3: I can apply a theory of significance

7. Source 2: How significant was the power of the nobility for the common people at the time?

Step 4: I can analyse historical significance

8. Source 4: How important historically was the Peasant Revolt in 1381? What do you think was the short- and long-term significance of the action?

HOW TO

Historical significance, page 207

H2.6

economics + business

Has working life changed?

People need to work to earn an income and to achieve a sense of purpose. However, in pursuit of profits, employers can sometimes exploit workers. In modern Australia, laws help protect workers' rights.

Work in medieval Europe

The rich and powerful exploited the peasant class, who sat on the bottom rung of society and made up 90 per cent of the population during medieval times. **Serfs** were the poorest of the peasants. The work contract peasants and serfs had with the lord was an oath of allegiance. These workers had to undertake all kinds of difficult manual labour in return for protection and the lease of a small plot of land. In addition to this, they also had to pay a **tithe** to the Church (see page 68). Peasants and serfs faced harsh punishments if they violated their agreement.

In the middle of the 14th century, the bubonic plague, known as the **Black Death**, swept through Europe, killing one-third of the population (see pages 92–93). With so many deaths there were far fewer people able to work the land. Source 1 outlines the Black Death's effect on work. It was written at Rochester Cathedral between 1314 and 1350.

'There was such a shortage of servants, craftsmen, and workmen, and of agricultural workers and labourers...[that] churchmen, knights and other worthies have been forced to thresh their corn, plough the land and perform every other unskilled task if they are to make their own bread.'

Source 1

Peasants used the situation to demand better working conditions and higher wages, leading to the Peasant Revolt in 1381. Jean Froissart wrote in c. 1483:

'... The evil-disposed in these districts began to rise, saying, they were too severely oppressed ... [that their lords] treated them as beasts. This they would not longer [sic] bear, but had determined to be free, and if they laboured or did any other works for their lords, they would be paid for it.'

Source 2

Source 3

A 14th-century illustration of a reeve overseeing serfs labouring in the fields. Reeves were local officials appointed by the lord to make sure that the serfs completed their work.

Source 4

Worker exploitation still occurs today. A modern version of slavery has developed in Madagascar, where a quarter of children aged 5–17 have to work to support their families. Children are forced to work in mines and stone quarries. Here a young Madagascan girl carries bricks at a brickworks.

Work in modern Australia

Employees in modern Australia have far better working conditions than those in medieval Europe and, indeed, than in most other nations around the world. Australia's **economy** produces enough jobs to employ most of the population. It also has laws and agreements to protect worker's rights.

The Australian government has created laws to prevent mistreatment of workers. Australian businesses are required to provide a safe workplace, pay a minimum wage, and not discriminate against a worker because of their sex, race or beliefs.

Workers' unions

Additionally, workers can join together in unions to protect their rights and seek improved working conditions.

Workers' unions meet with employers to negotiate better terms of employment for their members. Sometimes they use the threat of withdrawing their labour through strike action if negotiations break down. This is similar to the **guilds** from medieval times (see page 89).

Finding cheaper labour

The cost of labour is often the most expensive part of running a business. More Australian businesses are **outsourcing** parts of their operations to other areas of the world where they can take advantage of cheaper **labour**. Using overseas call centres to provide customer service is one example.

Outsourcing work to countries with low labour costs raises concerns about the exploitation of poor workers, who may be forced to work long hours in unsafe working conditions. Some employers also use child labour to reduce expenses, with up to 168 million children aged 5–14 around the world being forced to work.

Learning ladder H2.6

Economics and business

Step 1: I can recognise economic information

1 Define the term 'worker exploitation' using examples from Sources 1 and 2.

Step 2: I can describe economic issues

2 What was a positive outcome of the Black Death for medieval workers?

Step 3: I can explain issues in economics

3 Why do employees in modern Australia have better working conditions than most other countries?

4 What is a workers' union and what steps do they take to improve conditions for workers?

Step 4: I can integrate different economic topics

5 Why do companies choose to outsource parts of their businesses to countries with low labour costs, and what are some problems associated with this action?

Step 5: I can evaluate alternatives

6 How can businesses reduce their labour costs yet ensure that workers in cheaper labour countries are not exploited?

What did knights do?

At first, any skilled soldier with a horse could be a knight, but by the 12th century knights increasingly came from wealthy families, as the costs of providing horses and armour rose. Knights became a privileged class in medieval society.

A promise to serve

The **feudal system** depended on sworn loyalty (**fealty**) in return for land and protection (see pages 66–67). For a **knight**, this meant a promise to serve and protect his lord for life. Knights had to provide their own horse and armour, as well as a team of fighting men if he was called on by the lord to fight. The knight became the lord's **vassal** or tenant and in return the knight received a parcel of land known as **fief**.

A knight was expected to act in accordance with the code of chivalry. He had to stand up for what was right and had a duty to defend those in trouble. If a knight broke the code, his spurs and sword were broken and he was disgraced. Some knights did not live up to the code of chivalry and used their privileged position in society to be cruel and greedy.

Source 1

This image shows the evolution of a knight's armour: from chain mail in the 1100s through to full plate armour weighing about 25 kilograms, with iron or steel gloves (gauntlets) and helmets with a hinged visor.

Training to be a knight

At age seven, a boy who was to undertake training as a knight was sent to another lord's family as a **page**. All the page's play and sports were used as training for physical fitness, horse riding and fighting skills. Sword skills were practised using wooden swords and shields. Pages fought on piggyback to improve their balance and skill in fighting on horseback. Starting with small ponies, they would hone their riding skills and care for the horses in the stables.

At the age of 14, a page became a **squire**. Squires were treated as men and their training became far more dangerous. Fighting with swords and weapons and learning to control a **lance** while on horseback led to many injuries. Training also included learning military strategy such as siege warfare (see pages 78–79).

Source 2

Tournaments were fighting games designed to keep knights in excellent condition for medieval warfare. Knights took part in sword fighting and **jousting**, where mounted knights used a lance to try to knock their opponent off their horse.

Different knights and their families could be identified via their **coat of arms**. Creating a coat of arms is known as **heraldry**. Knights would display their family coat of arms and colours on clothing, banners and shields.

Receiving a knighthood

There were two ways that a man could become a knight.

1. *Earning the right on the battlefield*
   If a soldier fought bravely during a battle, he could be awarded **knighthood** by the king, a lord or sometimes another knight.
2. *Training to be a knight*
   He could earn the title through hard work and training as an apprentice to a knight. A knight had to be able to afford weapons, armour and a war horse. These items were expensive, so only the very rich could pay for them.

If a squire had proven his skill and bravery in battle, he would be eligible to become a knight when he turned 21. The official act of becoming a knight was a dubbing ceremony. Before the ceremony, a squire was required to spend the night alone in prayer. At a dubbing ceremony, the squire would kneel before a king, lord or other knight, who would tap their sword on each shoulder of the squire to 'knight' him. The new knight took an oath to protect and honour both his King and the Church. He was given a sword and a pair of riding spurs to mark the occasion.

Source 3

A dubbing ceremony, where a squire became a knight

Learning ladder H2.7

Show what you know

1. Explain how the following sets of terms are connected:
   - page, squire, knight
   - heraldry, shield, coat of arms
   - lord, fealty, vassal.
2. What skills were needed to be a good knight?
3. Why did knights compete in tournaments?
4. Who could not become a knight?
5. Source 1: What items did knights possess that showed they were wealthy?

Continuity and change

Step 1: I can recognise continuity and change

6. What action did lords take to protect their continuity?

Step 2: I can describe continuity and change

7. Describe in detail the changes in a knight's equipment over time, from looking at Source 1.

Step 3: I can explain why something did or did not change

8. Using evidence from Source 2, explain why knights needed to be wealthy.

Step 4: I can analyse patterns of continuity and change

9. Source 3: We still have knighthoods today. Compare and contrast medieval knights and modern knights. What is similar and what is different?

HOW TO

Continuity and change, page 200

What was castle life like?

In the Middle Ages, kings and lords built large buildings protected with high walls and watchtowers to protect against attack.

Castle design

Military defences or fortifications set castles apart from the palaces of kings and lords. In the Medieval Period most of Europe was divided up between lords and princes who ruled the local land and its people. To defend themselves, they built large castles in the middle of their land. Castles enabled the nobility to defend themselves from attack and launch their own attacks on invaders.

The first castles built in England in the 11th and 12th centuries were motte-and-bailey castles. The **motte** is a raised hill with a wooden or stone **keep** (a fortified tower) on the top. An enclosed courtyard with buildings known as the **bailey** was protected by a ditch and a wall of wooden stakes, known as a **palisade**.

Stone castles

Motte-and-bailey castles were replaced by larger stone castles to provide stronger and more fire-resistant defences. Castles were often built on top of hills or next to a source of water that could be used to help with their defence.

Curtain wall and towers
The curtain wall was the outer wall surrounding the castle. The wall was often connected by towers that housed castle defenders. Curtain walls had walkways for defenders to move along and fire arrows down on attackers.

Moat
A moat was a deep ditch filled with water that went around the castle. A drawbridge was used to allow or restrict entry into the castle.

Battlements
Battlements, also called crenulations, were at the top of the castle walls. The gaps (crenels) allowed defenders to shoot arrows or throw spears at the enemy while sheltering behind the higher parts of the wall (the merlon). Merlons often contained arrow slits to allow the archer to stay behind the wall at all times.

Source 1

A motte-and-bailey castle

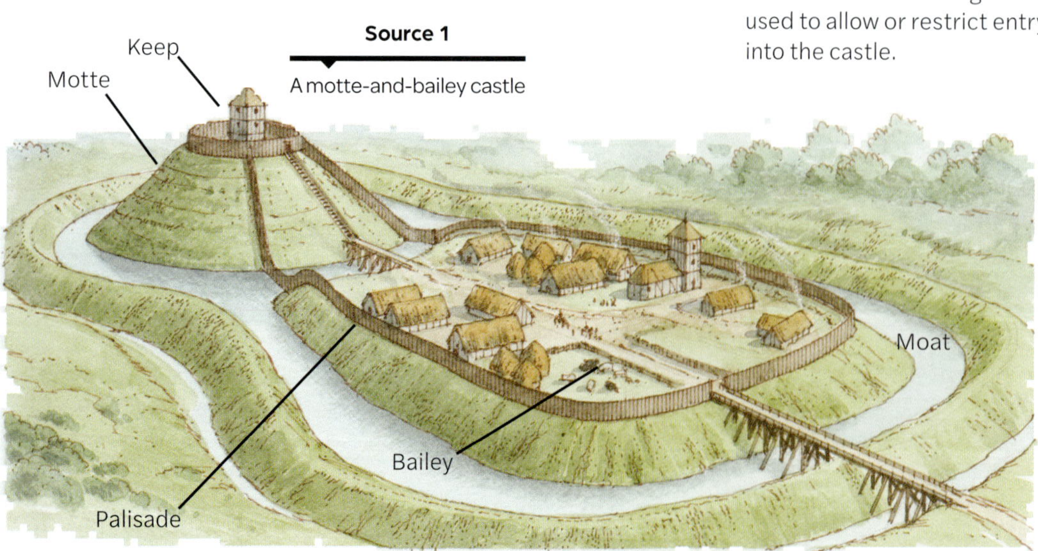

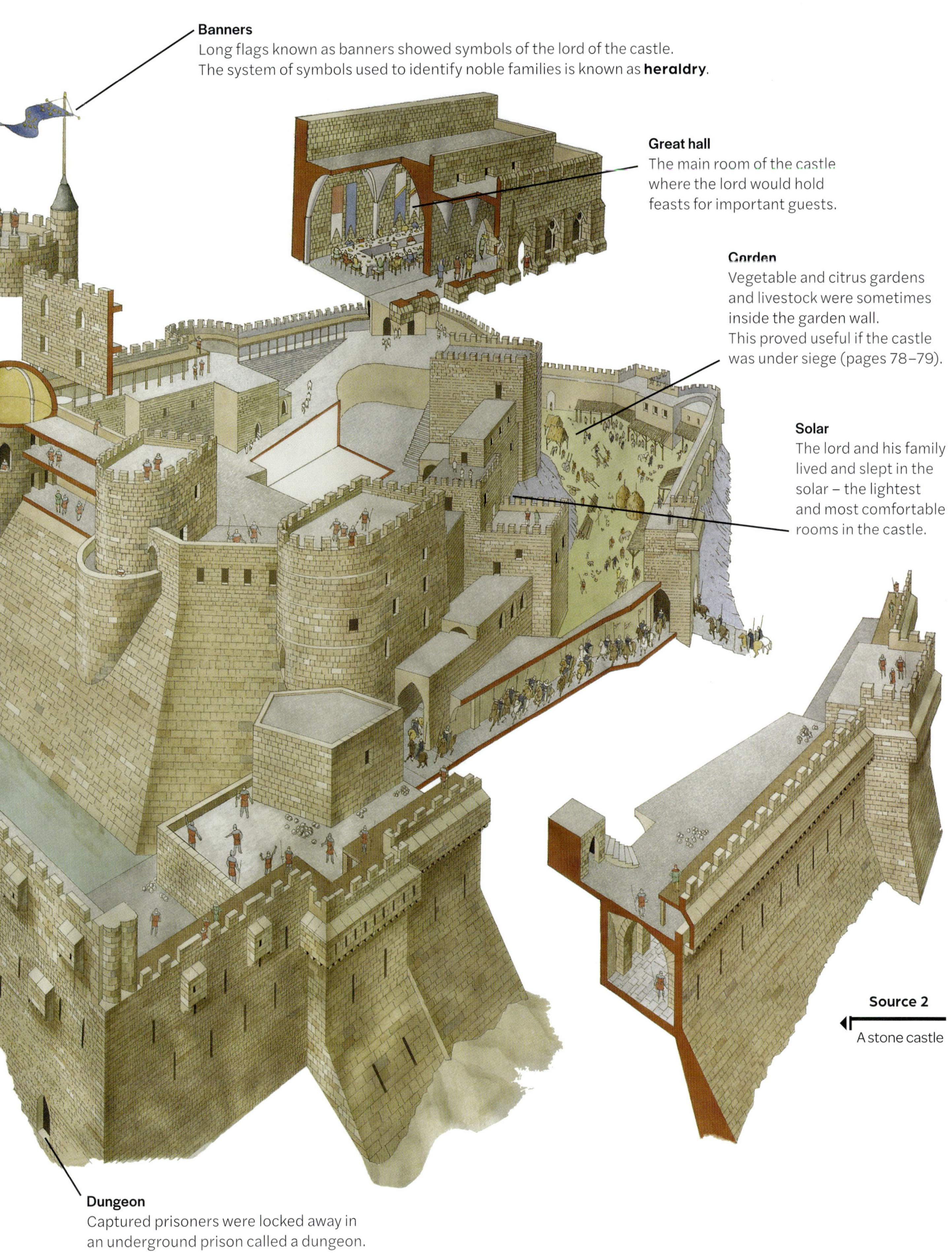

Source 2
A stone castle

Dover Castle

Dover Castle on England's southern coastline was first built in 1066 by William the Conqueror (see pages 62–65) to help prevent anyone repeating his own invasion of England from France. The medieval castle sits atop the white cliffs of Dover, guarding the shortest crossing point between England and France.

Dover Castle is the largest castle in England. What can be seen today is the stone castle rebuilt in the latter half of the 12th century by Henry II. He transformed the castle by adding towered walls and a huge stone **keep**, the strong central tower, to the **bailey**. The keep was surrounded by two rows of towered walls, a perimeter ditch and **barbican** – a towered wall to protect a main gate.

Despite its defences, the castle has been under attack many times in its history. It was besieged by Prince Louis of France in 1216. He cut off supplies to Dover from both land and sea and bombarded the main gate with rocks thrown from the **mangonel** siege engines (see page 78). Archers shot down at the castle defenders from a tall **siege tower**. The attackers took the barbican and marched towards the main gate.

Louis had miners tunnel under the huge stone gatehouse. When part of the gatehouse collapsed, the French poured into the gap only to find a further barrier of boulders and heavy wooden posts. The French attackers were driven out, suffering heavy casualties.

Source 3

Dover Castle today. It is located at the shortest crossing point between England and France. Nearby is a car and rail tunnel that now connects the two countries.

Medieval feasts

The lord of the castle entertained guests with feasts in his great hall. The most important guests sat at the high table with the lord and lady, while soldiers guarded the main entrance.

Feasts involved many courses. Spoons were used for soup and puddings and fingers for everything else. Animals eaten included cows, pigs, deer, boar, swans and peacocks. The lord's dogs waited for leftover food. Performers such as minstrels, jugglers and jesters entertained the guests.

The lady of the castle

The lady of the castle led a privileged life. Her clothes were often trimmed in velvet and fur. The lady was in charge of the domestic duties of the castle such as monitoring food supplies, as well as looking after her children. She was supported in her duties by ladies-in-waiting, who helped the lady to dress and who would read and play music for her.

Source 4

Dover Castle during the siege of 1216. The French invaders have collapsed the barbican and are charging towards the gatehouse, only to be repelled by the strong castle defences.

Source 5

A medieval feast

Source 6

The lady of the castle

Learning ladder H2.8

Show what you know

1 In what important way is a castle different to a palace?

2 Why did stone castles replace motte-and-bailey castles?

3 Why were castles built?

4 Describe the similarities and differences between a feast in a medieval castle and a big feast you have experienced.

Source analysis

Step 1: I can determine the origin of a source

5 Source 3: Why was Dover Castle built where it was?

Step 2: I can list specific features of a source

6 Looking at Sources 1 and 2, list as many ways that castles defended people as you can.

Step 3: I can find themes in a source

7 Sources 5 and 6. Why did the lady of the castle lead a privileged life? Do any modern people live like this?

Step 4: I can use my outside knowledge to help explain a source

8 Look carefully at the illustration in Source 4.

a Which army is trying to attack Dover Castle?

b List the tactics used to try to enter the castle and defeat the enemy.

c What was the result of the attack?

HOW TO

Source analysis, page 197

how it works

How could a castle be defeated?

Conquering a kingdom relied on capturing the enemy's castle. Weapons were designed to climb the outer walls, destroy the thick stone walls and launch fire attacks. Attackers could also adopt siege tactics – where the castle was surrounded to cut off food and water supplies, eventually forcing a surrender.

Siege tactics

To overcome castle fortifications, attacking forces sometimes used siege tactics. A **siege** is a military operation where enemy forces surround a town or building with the aim of cutting off supplies and forcing a surrender. The party conducting the siege holds a strong, stable, defensive position to stop movement in and out of the town. Sieges can be drawn out over many months or years (see the siege of Orléans on page 84). The Siege of Candia in Crete lasted 21 years!

Siege engines

To bring a siege to an end more quickly, the invading forces used a range of siege engines to break down heavy castle doors and thick castle walls. After the Chinese invented gunpowder later in the Middle Ages, attackers began to use cannons.

1 Trebuchets

Trebuchets were tall wooden catapults, up to 30 metres high. The swinging arm of the trebuchet could hurl projectiles high in the air at targets as far as 300 metres away. Trebuchet missiles were usually large rocks weighing up to 100 kilograms, but they could also be fire bombs or even plague corpses.

3 Battering rams

Originally, these were simply a large log that soldiers would carry forward with force, used to break through doors or walls. Later versions were covered in fire- and arrow-resistant canopies and mounted on wheels. The log was swung using ropes to create greater momentum.

Source 1

A castle siege

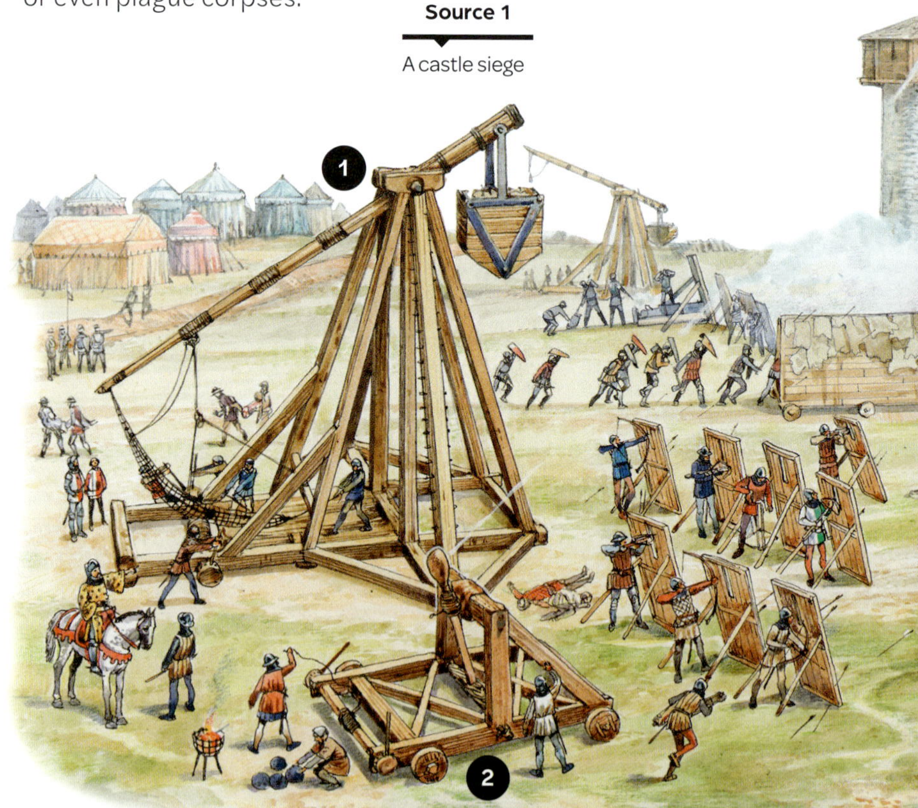

2 Mangonels

The **mangonel** was a weapon from ancient Roman times. It used the power of twisted rope to make a wooden arm spring up and hurl an object. The mangonel threw projectiles lower and faster than the trebuchet, with the intention of destroying walls.

Source 2

The kings, knights, castles and weapons of medieval Europe are a great basis for strategy video games, where heroes take on quests and engage in epic battles with enemies. This scene of a castle siege is from the later medieval period, as cannons were only introduced to Europe in the 14th century, when they quickly replaced other siege weapons.

4 Siege towers

Siege towers were high wooden platforms on wheels that were pushed against castle walls. Soldiers climbed the stairs and let down the gangplank on top of the wall so the attackers could climb into the castle. Siege towers were often covered in animal hides to make them less flammable.

In addition to the siege engines described, attacking forces also used archers to fire arrows at the enemy from behind protective screens.

Show what you know

1 Why would people besiege a castle?

2 What different tactics were involved in a siege?

3 List as many different siege activities as you can from Source 1.

Cause and effect

Step 1: I can recognise a cause and an effect

4 What do the forces besieging a castle need to be successful?

Step 2: I can determine causes and effects

5 In small groups, brainstorm the technology needed to make the siege weapons in Source 1.

Step 3: I can explain why something is a cause or an effect

6 Write a narrative describing how a group of people inside a besieged castle could defend themselves or even fight back.

Step 4: I can analyse cause and effect

7 Which do you think would be the most effective against a castle: a trebuchet, a mangonel or a siege tower? Explain your answer.

HOW TO

Cause and effect, page 203

H2 .10

How powerful was the Church?

The spread of Christianity during the Middle Ages was a lasting change that has greatly influenced the world. The Catholic Church had significant influence over people in medieval society and provided a stable system of beliefs and behaviours.

The Catholic Church

Christianity was a powerful force in Europe during the Medieval Period. Nearly all people in Europe were Christian and the Catholic Church's leader, the Pope, was an influential leader.

The Pope was seen as God's representative on Earth. As head of the Church the Pope had great influence over the kings of Europe and could dictate how people should behave. Following the Pope in the Catholic Church hierarchy were cardinals, bishops, priests, monks and nuns.

The Catholic Church became very wealthy during the Middle Ages. Like knights, bishops could receive **fiefs** from lords in return for their loyalty and services. The Church did not have to pay the king any tax for their land.

People gave the Church one-tenth of their yearly earnings in tithes. **Tithes** could be paid in money or in goods. Peasants had little money, and usually paid their tithes in grain, animals or seed.

The Church's wealth was greater than that of many countries. Many high-ranking Church officials became rich and powerful, and resisted any efforts from kings and nobles to reduce the Church's political influence.

Providing stability

The Roman Catholic Church helped unify a greatly divided society. All classes of people – kings, lords, knights and peasants – were greatly affected by the rulings and teachings of the Church.

The local church or cathedral was the centre of town life. People attended weekly mass and were married and buried at the church. The grounds of the church were used for fairs and sporting events.

A priest represented the Church in each parish. He would run church services every Sunday in Latin (a language that most people did not understand), and listen to people's confessions and advise them on what to do. Priests often taught in parish schools and offered prayers for the sick.

Source 1

An illustration from 1482 in a French book, of a medieval hospital where the four virtues of Prudence, Temperance, Fortitude and Justice teach nuns how to care for the ill.

Source 2

The fear of being sent to hell motivated people to follow the teachings of the Church. This mosaic from the Baptistery of San Giovanni in Florence, Italy, was completed in the 13th century by Coppo di Marcovaldo. The Devil is in hell, devouring those who have not followed the Church's teachings.

Monks and nuns dedicated their lives to worshipping God. Their responsibilities included providing food for the poor, treating the sick, transcribing holy texts, taking care of orphans and helping educate children.

Faith and fear

All people, no matter their social status, were educated in the teachings of the Church. By following Church law, people believed they would be granted entry into heaven. Those who disobeyed the Church's teachings or failed to pay tithes were told they would be sent to hell where they would face horrific torture at the hands of the Devil.

The fear of being sent to hell was a powerful motivator for people to follow the teachings of the Church. Bishops had the authority to **excommunicate** Christians – this meant they could no longer participate in church services and guaranteed they would be sent to hell after they died.

People were also afraid they might be declared a **heretic** or a witch. Heretics were people whose beliefs did not align with those of the Catholic Church. Witches were said to be working with the Devil to cast spells on people.

Witches were blamed for many things – accidents, sickness, death and even crop failure. Old women who lived alone were particularly vulnerable to being labelled a witch. In 1484, Pope Innocent VIII delivered a decree about witchcraft, authorising the imprisonment and punishment of devil worshippers. He claimed,

> '… many persons of both sexes, unmindful of their own salvation and straying from the Catholic faith, have abandoned themselves to devils, incubi and succubi, and by their incantations, spells, conjurations, and other accursed charms and crafts, enormities and horrid offences.'
>
> **Source 3**
>
> Pope Innocent VIII's decree on witchcraft, 1484 CE

As witch hysteria grew, witch-hunts were organised to search for witches or evidence of witchcraft. Between 1450 and 1750, historians estimate there were around 110 000 witch trials, with approximately half found guilty. The punishment was torture, hanging or burning at the stake. (See Joan of Arc, pages 84–87.)

Holy pilgrimages

During the Middle Ages, the Church encouraged people to make spiritual journeys called **pilgrimages** to holy shrines. People believed if they prayed at a shrine they might have a greater chance of going to heaven or being cured from an illness.

Pilgrimages helped Christians prove their devotion to God. Europe had its own shrines, but the most sacred pilgrim shrines were in the Holy City of Jerusalem. Many pilgrims walked up to 5000 kilometres to reach Jerusalem.

The Arab conquest of the Holy lands in 637 CE did not stop the pilgrimages, as the Arabs who first conquered the city tolerated pilgrims. Jerusalem was an important religious site for believers of both Christianity and Islam. However, when a group of Turkish **Muslims**, the Seljuks, took control of Jerusalem in 1050, pilgrimages became more dangerous. The Turks harassed pilgrims and refused them entry into Jerusalem. On Good Friday in 1065, Seljuk Turks massacred 12 000 German pilgrims, leading to calls for action.

A call to arms

Many believed that the new Seljuk Turk Islamic regime would also move to invade Constantinople, the capital of the Byzantine Empire, which was under Christian rule. The Byzantine Emperor asked Pope Urban II for weapons, supplies and skilled troops to help defend his empire against the Turks. The Pope called on Christians to defeat the Turkish invaders, giving a stirring outdoor speech to the common people as well as nobles and church leaders in 1095 at Clermont in France:

> '… a race from the kingdom of the Persians, an accursed race, a race utterly alienated from God … has invaded the lands of those Christians and has depopulated them by the sword, pillage and fire … it has either entirely destroyed the churches of God or appropriated them for the rites of its own religion. They destroy the altars, after having defiled them with their uncleanness … every one that hath forsaken houses, or brethren, or sisters, or father, or mother, or wife, or children, or lands for my name's sake shall receive an hundredfold and shall inherit everlasting life … undertake this journey for the remission of your sins, with the assurance of the imperishable glory of the kingdom of heaven.'

Source 4

Pope Urban II, 1095

The Crusades, 1096–1204

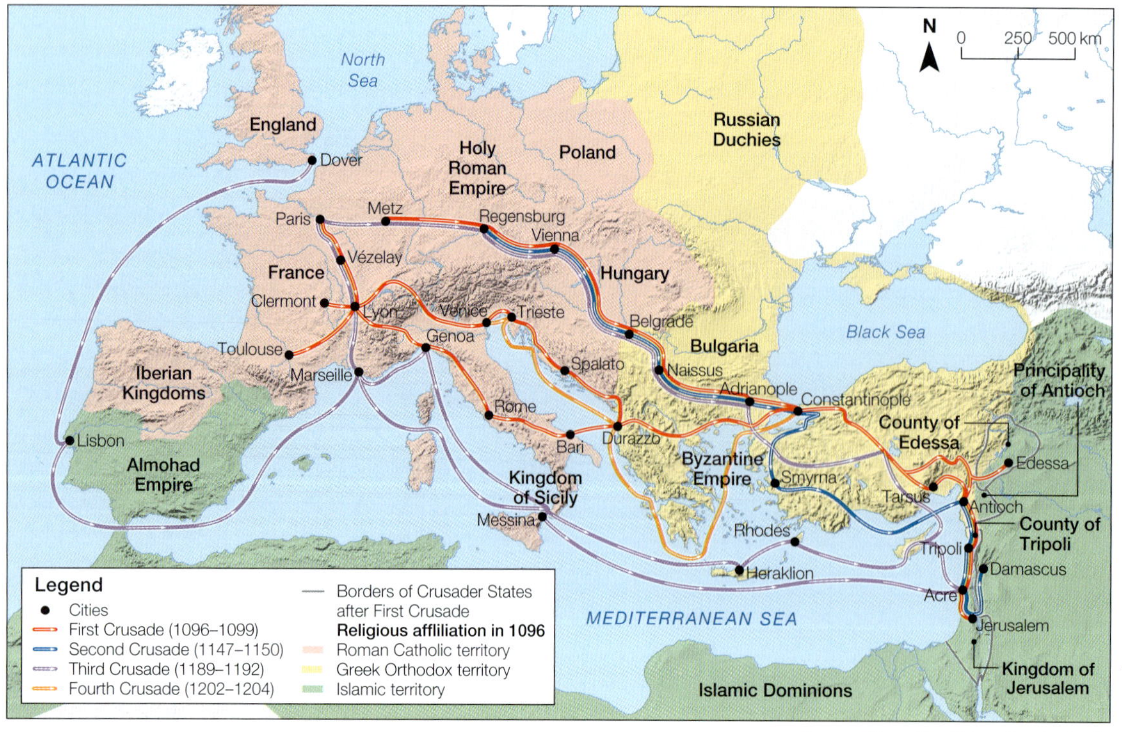

Source: Macmillan Education Australia

Source 5

The Crusades, 1096–1204

Source 6

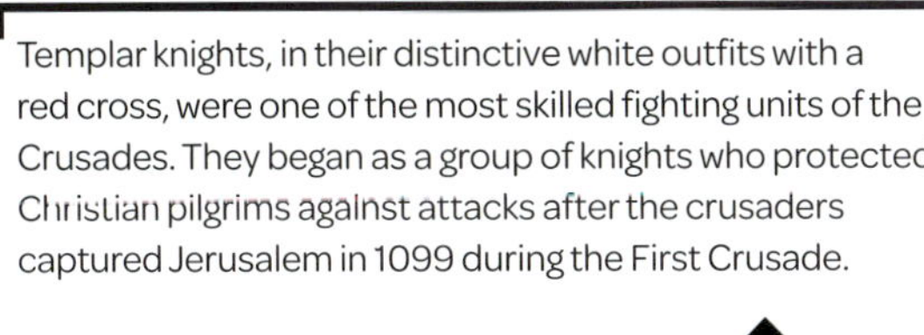

Templar knights, in their distinctive white outfits with a red cross, were one of the most skilled fighting units of the Crusades. They began as a group of knights who protected Christian pilgrims against attacks after the crusaders captured Jerusalem in 1099 during the First Crusade.

Holy Crusades

In 1096, four armies of crusaders gathered in Constantinople to start the First **Crusade** in response to the rise of **Islam**, and loss of Christian holy sites in the Eastern Empire. People from all levels of society joined the Crusades. Many young peasants were encouraged by local parish priests to join the fight.

Most people joined the Crusades to regain control of the Holy Land for Christians and to guarantee entry into heaven when they died. Some were also hoping for adventure, wealth and fame.

By 1099, the Crusaders had captured Jerusalem, but had lost two-thirds of the troops to heat exhaustion and hunger. Many Crusaders returned home, bringing new ideas, customs and inventions. The remaining crusaders settled in the region, building castles and organising orders of knights to protect the pilgrims.

Guarded by Crusader castles, Christians defended the region until around 1130, when Muslim forces began gaining ground with their own ***jihad*** (holy war) against the Christians. Saladin, the King of Egypt, united Muslims to take back Jerusalem in 1187. After three unsuccessful crusades, the Sixth Crusade finally brought Jerusalem back under Christian control, only to lose it again to the Muslims in 1244.

Learning ladder H2.10

Show what you know

1 What were churches used for?

2 Source 1: What duties did monks and nuns undertake?

3 Source 2: Describe what is happening in this image and why it might have encouraged people to follow the teachings of the Catholic Church.

4 What kinds of behaviours might get a person excommunicated?

5 Source 4: Reading the text from Pope Urban II, how did the Pope persuade people to take part in the Crusades? Provide example quotes in your answer.

6 Do you think the Crusades were a good idea? Explain your response.

Historical significance

Step 1: I can recognise historical significance

7 Why was the Pope so influential in medieval times?

Step 2: I can explain historical significance

8 How did the Roman Catholic Church help unify a greatly divided society?

Step 3: I can apply a theory of significance

9 Answer these questions about the Catholic Church in the Middle Ages:

a How important was it to people at the time?
b How deeply were they affected?
c How many people were affected?
d For how long were they affected?
e What role does the Catholic Church play in modern times?

Step 4: I can analyse historical significance

10 Some may say the Catholic Church is not as significant now as it was in the Medieval Period. Make three brief points for or against this view, giving historical examples.

Historical significance, page 207

key individual

Joan of Arc – saviour, witch or warrior?

An uneducated teenage girl leads the French army to an improbable victory at Orléans.

Joan of Arc was a peasant girl from medieval France who believed God had selected her to lead France to victory in the Hundred Years War against England (fought from 1337–1453). With no military training, she convinced Prince Charles of Valois to allow her to lead a French army to the **besieged** city of Orléans where she inspired a key victory over the English and their French allies, the Burgundians.

From humble beginnings to leading an army

Born around 1412, Jeanne d'Arc (Joan of Arc) was the daughter of a farmer in the French village of Domrémy. Joan was not taught to read or write, but her mother encouraged her deep love for the teachings of the Catholic Church. When Joan was 13, she began to hear voices that she believed were sent by God. The voices told her to save France by defeating its enemies and installing Charles as its king.

In May 1428, Joan travelled to Vaucouleurs where she attracted a small group of followers who also believed she was sent by God to save France. Joan cut her hair and dressed in men's clothes to travel across enemy territory to Prince Charles' palace at Chinon.

Joan promised to have Charles crowned king and asked him to give her an army to lead to the besieged city of Orléans, which was under attack from the English. Charles acted against the advice of his generals and advisers and granted her request.

Dressed in white armour and riding a white horse, Joan joined an army convoy trying to bring supplies to Orléans in March 1429. Orléans had been cut off and surrounded by English troops who planned to starve the city to force a surrender.Joan sent a letter to the English commanders at the siege, ordering them in the name of God to 'Begone, or I will make you go'. The convoy reached Orléans and the supplies, Joan and 200 soldiers were smuggled into the city under the cover of darkness.

Joan joined several French assaults against the British forces and also inspired the citizens of Orléans fight the English. The French army was eventually successful in forcing the English and Burgundians to retreat across the Loire River.

After the great victory at Orléans, the French saw Joan of Arc as a heroine. She and her followers escorted Charles across enemy territory to Reims where he was crowned King Charles VII in July 1429.

Source 1

Milla Jovovich plays Joan of Arc in the 1999 film *The Me*

Source 2

Letter by Charles VII, dated 2 June 1429, conferring the title of nobility on Joan of Arc's family.

Capture and execution

A year later the king ordered Joan to protect the town of Compiégne against attack. However, she was pulled from her horse and taken captive by the Burgundians. She was then sold to the English and put on trial to face charges of witchcraft, heresy (anti-religious beliefs) and dressing like a man.

After a year in prison and threatened with immediate execution, Joan of Arc was pressured into signing a confession (that she could not read) denying that she had ever received guidance from God.

On 29 May 1431, Joan of Arc was sentenced to death. The next day Joan was taken to the marketplace at Rouen and burned at the stake in front of a jeering crowd. She asked the priest to hold the cross high and shout prayers so she could hear them above the roar of the flames. Joan of Arc was just 19 years old.

Today, Joan of Arc is a heroine in France, celebrated with her own national holiday. Twenty years after her death, Charles VII ordered a new trial and her name was cleared. In 1920, Pope Benedict XV made her a saint – the highest honour in the Catholic Church. She is the patron saint of France and immortalised in statues, paintings and literature.

Source 3

Joan of Arc, from the c. 1505 illuminated manuscript *Lives of Famous Women* by Antoine Dufour

Source 4

This scene from the 1999 film *The Messenger* depicts Joan riding into battle.

History mystery

Why did Charles put a teenage girl in charge of an army?

In her book *Joan of Arc: A History*, medieval historian Helen Castor looks at evidence surrounding the decision to put a teenage girl in charge of an army. In an interview with Castor, broadcaster Richard Fidler said:

'Imagine yourself into the life of someone living in medieval France ... there would be none of the images and texts that you're flooded with now, instead there are the sky, the forest, the village and the river ... and angels and demons would be as real to you as the house next door.

You would know probably very little about the world beyond your own village, but if you're living in 15th-century France, then you would be aware that a civil war is tearing your country apart and even worse, that the English have invaded.

Living like that and thinking like that, what happens when a humble teenage girl, dressed as a boy, comes into the court of the future King of France with a message? She says she's received a commandment from an angel telling her to drive the English out of France.'

The mystery continues ...

Joan of Arc arrives at Prince Charles' palace at Chinon proclaiming God has sent her to lead an army to defeat the English and make Charles the next king of France. How did she manage to persuade the prince?

She is examined by the prince's religious experts, who conclude that she is a god-fearing girl who leads a pious (strongly religious) life. She says that God has entrusted her with this mission and the experts decide to put her to the test by allowing her to fight at Orléans.

The prince decides he believes in Joan's visions. He orders armourers in the city of Tours to make Joan a very fine and costly suit of armour. She is also given a white silk banner with the name of Jesus on it and a picture of God or Jesus.

She leads her troops dressed in her fine armour, with her silk banner, accompanied by priests singing hymns. The military and religious procession that passed by would have been quite a sight.

Learning ladder H2.11

Show what you know

1 What happened in Joan's early life that made her think she was chosen by God?

2 How did Joan win the battle at Orléans?

3 Why did Charles put a teenage girl in charge of an army?

4 Why was Joan of Arc burnt to death?

Chronology

Step 1: I can read a timeline

5 How old was Joan of Arc when she joined the army convoy trying to bring supplies to Orléans?

Step 2: I can place events on a timeline

6 List these events in order from oldest to most recent.

A year before her death, she was captured by enemy forces.

Joan died aged 19.

Joan was born on 6 January 1412.

She was made a saint about 100 years ago.

When Joan was 10, Charles VI of France died.

Step 3: I can create a timeline using historical conventions

7 Complete a timeline of Joan's life from start to finish based on the information in this section.

Step 4: I can distinguish causes, effects, continuity and change from looking at timelines

8 Joan of Arc is the only person in history to be both condemned and canonised as a saint by the Catholic Church. When did these events occur and why?

Chronology, page 194

Why did towns and trade expand?

By the 11th century, Europe had become a more peaceful place. As barbarian raids stopped and money replaced oaths of allegiance, the feudal system began to break down. People felt free to leave the safety of the manors and move into towns. A network of towns grew along trade routes and surrounding castles.

Growing towns

From 1091 to 1291, the **Crusades** saw people travel to far-away lands and bring back new inventions and ideas (see page 83). As new trade networks opened, the number of towns continued to grow. Many of these towns grew into cities and by 1200 CE there were about 600 cities in Europe.

As towns grew, their communities built large walls around them. The homes of the wealthy were in the town's centre and the poor lived in small wooden dwellings near the walls. The smelly, narrow laneways of the town were filled with rubbish, sewage and animals. The town's marketplace was full of activity with merchants, craftspeople, entertainers and beggars.

Source 2 is a young boy's account of the market-place in the English town of Shrewsbury in 1241.

'Church bells pealed out the hour ... Men wandered the streets shouting 'hot meat pies' and 'good ale' ... itinerant pedlars hawked their goods, offering nails, ribbons, potions to restore health ... People gathered in front of the cramped, unshuttered shops, arguing prices at the tops of their voices. Heavy carts creaked down the streets ... Dogs darted underfoot, and pigs [shuffled] about in the debris dumped in the centre gutter.'

Source 2

quoted in Sharon Penman, *Falls the Shadow*, Penguin Books, 1989

Source 1

Medieval towns developed along trade routes and provided opportunities for skilled workers.

Source 3

A 16th-century painting of merchants selling their wares outside the city walls. [Sebastian Vrancx, *The Fish Market*.]

The merchant class

The earliest medieval merchants were peddlers who travelled from town to town selling small items such as pots and pans, gloves, mirrors and purses. However, by the 12th century, a new social group of middle-class merchants arose from the increase in trade.

European merchants traded with other European cities and with dealers in Asia, the Middle East and northern Africa. By 1300, cargo ships from Venice in Italy were taking precious metals, silks and other luxuries to England, where they picked up wool, timber and coal for the return voyage.

European merchants imported exotic spices, and silk from the Arab world and Asia and **exported** goods such as wool and silver. Merchants also became travel operators by taking pilgrims to holy sites (see pages 82–83). Robbers and pirates also operated along trade routes, increasing the risks for merchants.

By the Late Middle Ages, the new merchant class was a powerful force in medieval society. To grow business, merchants formed partnerships, which eventually led to the development of large trading companies.

New opportunities

The demand for skilled workers was growing and opportunities to produce goods for trade helped people learn skills such as spinning, weaving, dyeing, leather work, brewing, ropemaking and banking.

Guilds were established for each skilled craft to guarantee workmanship, oversee apprentices' training and to settle disputes between employers and skilled workers. Guilds were like modern trade unions (see page 71) except they also included employers.

Learning ladder H2.12

Show what you know

1 What helped bring an end to the feudal system?

2 How did trade influence the growth of towns?

3 What did European merchants import and what did they export?

4 Source 2: Read the young boy's account of Shrewsbury in Source 2. What vocabulary does he use to make the reader feel like they are really in a medieval English town? Explain your answer with examples.

5 How is trade different now compared to the Middle Ages?

Source analysis

Step 1: I can determine the origin of a source

6 What is the origin of Source 3?

Step 2: I can list specific features of a source

7 Look at Source 1. What features and activities in it can you see that would help trade? Explain how for each feature.

Step 3: I can find themes in a source

8 Use five adjectives to describe Source 1. Taken together, what overall feeling do you get from reading the adjectives you have chosen? Do they have something in common?

Step 4: I can use my outside knowledge to help explain a source

9 Looking at Source 3, describe what each person is doing, based on the knowledge you have gained from this chapter.

HOW TO

Source analysis, page 197

How did medieval law influence Australia's laws today?

In the Middle Ages, those found guilty of an offence were harshly punished, even being tortured in front of crowds of cheering onlookers. Eventually, people received protection through the law with the signing of the Magna Carta – the document that underpins the rights of Australian citizens today.

Torture and punishment

Laws and punishments in medieval Europe were extremely harsh, as the rulers believed that peasants would only behave if they feared the consequences of breaking the law. For example, if a woman was found guilty of gossiping or nagging her husband, she would be forced to wear an iron muzzle on her face, called a scold's bridle, to shame her.

Church courts heard charges such as heresy and witchcraft. The King's court heard serious charges such as **treason**, where a person is found to have betrayed their country by, for example, planning to overthrow its rulers. From 1352, the penalty for treason was to be hung, drawn and quartered. This involved hanging a person nearly to death, pulling out his intestines while he watched, and finally tying his hands and legs to four horses who would gallop in different directions to tear him into quarters.

Source 1

A woman wearing a scold's bridle to shame her in public

Trial by ordeal

In the Middle Ages, people were considered guilty until proven innocent. Their guilt or innocence was sometimes determined by a dangerous test where people believed God would help the innocent by performing a miracle on the person's behalf. The accused was innocent if they lived through the ordeal or if their injuries healed. There were two types of trial by ordeal:

1. ordeal by fire – the accused was burned by fire, hot coals or by holding a red-hot iron. If their burns healed after three days, they were judged to be innocent.
2. ordeal by water – the accused was tied up and thrown into a river. If the body sank, they were innocent and they were dragged out of the river.

Duelling was another way to let God decide on the outcome of a dispute. Rich men fought with swords; poorer men fought each other with lesser weapons such as staffs and fists. The modern sport of boxing may have its origins in these public displays of justice. Watching such displays was a popular pastime as spectators could bet on the winner.

Source 2

King John reluctantly signs the Magna Carta in 1215.

Reforming the law

Even though peasants made up 90 per cent of the population, they had no say in how they were governed. When Henry II (William the Conqueror's great-grandson) became King of England in 1154, he introduced **trial by jury**, where groups of 12 or more people were invited to hear the facts of the case.

In 1215, rebel lords in England, backed by the Church, called for the protection of Church rights, legal rights for all and reduced feudal payments to the king. The lords forced the unpopular King John to sign a document called the **Magna Carta**. The document outlined basic rights for all people with the principle that no-one was above the law, including the king. The Magna Carta is seen as one of the first steps in the development of legal and political rights for common people and the start of modern **democracy**. The Magna Carta also abolished trial by ordeal. People could no longer be tortured or killed on the grounds of suspicion.

The Magna Carta and Australia

The legal and political rights outlined in the Magna Carta underpin the **rule of law** in Australia, whereby both the government and the citizens understand the law and obey it. The rule of law has its roots in Chapter 29 of the Magna Carta, which states:

> 'No free man shall be taken or imprisoned, or be disseised [dispossessed] of his freehold, or liberties, or free customs, or be outlawed, or exiled, or in any other wise destroyed; nor will we pass upon him, nor condemn him, but by lawful judgement of his peers, or by the law of the land. We will sell to no man, we will not deny or defer to any man either justice or right.'

Source 3

The Magna Carta, Chapter 29

Learning ladder H2.13

Civics and citizenship

Step 1: I can identify topics about society

1 Describe the punishment in Source 1. What was her crime?

2 Source 2: What changes did the Magna Carta make to English society in the Middle Ages?

Step 2: I can describe societal issues

3 Why do you think 'trial by jury' was introduced by King Henry II?

4 List some of the different types of punishments people could be given in the Medieval Europe.

5 In Australia today, a person is considered innocent until proven guilty. What was the experience in medieval society, and what impact did it have on punishments?

Step 3: I can explain issues in society

6 Explain in your own words what Chapter 29 of the Magna Carta means for modern Australians.

Step 4: I can explain different points of view

7 Find three things that are different about medieval justice and modern Australian justice. Describe them and state whether you think the changes are an improvement or not.

Step 5: I can analyse issues in society

8 In Australia, torture is illegal, punishments don't involve pain, and people believe we should try to reform an offender. Why do you think our justice system has changed so much since the Middle Ages?

What killed one in three Europeans?

The Black Death, also called the plague, was spread from fleas and ticks that transmitted harmful bacteria to humans. The deadly disease, carried around the world by rats and humans, caused a pandemic, killing over one third of the European population.

The disease

The name **Black Death** comes from the swollen glands (called buboes) in the victim's neck, armpits and inner thighs that turned black as they swelled with blood. The Black Death began in Central Asia in the 1200s and was transported via rats and people on trading ships to Europe between 1347 and 1350, as well as via the **Silk Road**.

Spread of the Black Death through Europe

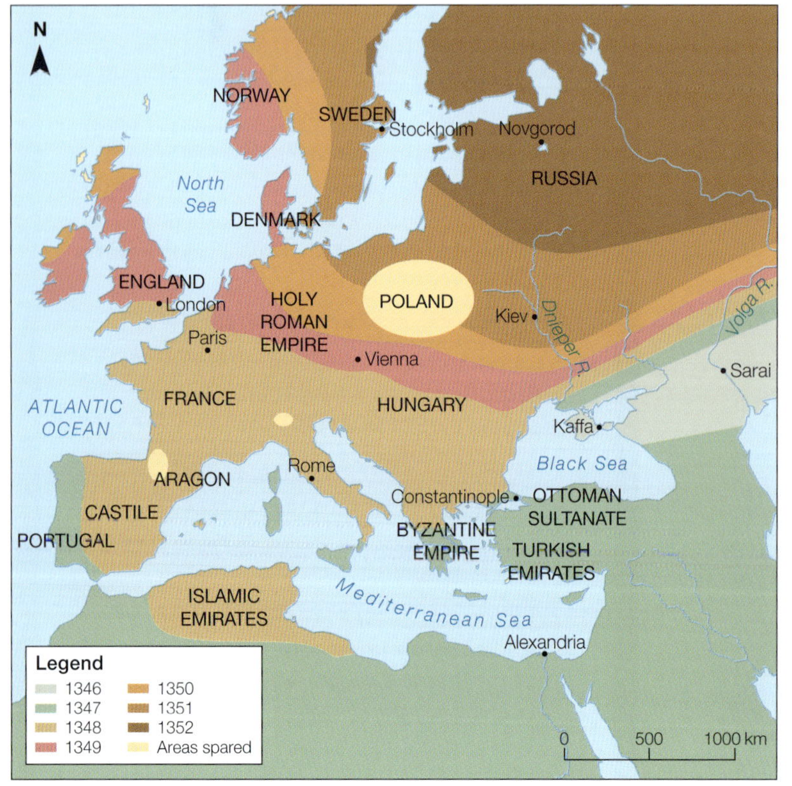

Source: Matilda Education Australia

Source 3

How the Black Death spread to Europe

Source 1

A doctor dressed from head to toe to treat patients during the Black Death. The beaks on their masks were filled with pleasant-smelling oils to overcome the bad smells they thought caused the plague. The plague mask has become synonymous with plague times in medieval Europe but actually wasn't invented until the early 1600s.

In his book *The Decameron*, author Giovanni Boccaccio wrote:

'... [The] deadly pestilence ... showed its first signs in men and women alike by means of swellings either in the groin or under the armpits, some of which grew to the size of an ordinary apple and others to the size of an egg (more or less), and the people called them [buboes]. And from the two parts of the body already mentioned, in very little time, the ... deadly [buboes] began to spread indiscriminately over every part of the body; then, after this, the symptoms of the illness changed to black or livid [bluish] spots appearing on the arms and thighs, and on every part of the body – sometimes there were large ones and other times a number of little ones scattered all around ... [Almost] all died after the third day of the appearance of the previously described symptoms (some sooner, others later), and most of them died without fever or any other side effects.'

Source 2

The Decameron, a collection of stories by Giovanni Boccaccio, probably written between 1349 and 1353.

Source 4

This Italian painting shows a Catholic priest and nuns caring for victims of the plague. Priests and other caregivers were faced with the task of tending to the sick, while knowing the disease was highly contagious. [Miniature from *La Franceschina*, c. 1474, codex by Jacopo Oddi (15th century).]

Treating the Black Death

Medieval doctors didn't understand the cause of the plague or how to treat it. White crosses were drawn on the doors of houses containing plague victims. No-one was allowed to enter or leave the house. Doctors believed that foul smells caused the disease. People wore masks over their faces to stop them breathing in the germs of plague victims. Doctors offered patients sweet-smelling flowers to smell and wore beaked masks filled with pleasant-smelling oils. Other doctors lanced the pus-filled sores to draw out the 'bad blood', but this only helped to spread the disease.

Many believed that the sickness was God's punishment for their sins. Groups of 50 to 500 men called **flagellants** travelled from town to town publicly whipping themselves in the hope that God would take pity on them and protect them from the plague. No-one connected the plague to the rats, which carried the deadly fleas, nor the dumps of waste in villages and towns that attracted them.

The impact of the Black Death

Within four years, one-third of the people in Europe had died. Villages were deserted and workers were in short supply. Peasants began to demand higher wages and lower rents on their land. Farmers increasingly raised livestock, as tending animals required fewer workers than growing crops.

The Church also lost many priests, monks and nuns who fell ill after working so hard to tend to the sick.

The single-roomed houses of the poor were easily kept rat- and lice-free from the plague. Cloisters and castles had a bigger challenge keeping free from rats and hence people in these more lavish dwellings died in greater numbers.

Learning ladder H2.14

Show what you know

1 How did the Black Death spread to Europe?

2 Source 1: Why is the doctor dressed like this?

3 Source 2: How does Giovanni Bocaccio describe the Black Death? Give examples from the text.

4 What can Source 4 tell us about the role of the Church in the Middle Ages?

5 What was the impact of the Black Death?

Chronology

Step 1: I can read a timeline

6 Use the map in Source 3 as a visual timeline. Roughly how long did the disease take to spread from Sarai to London?

Step 2: I can place events on a timeline

7 Use the map in Source 3 as a visual timeline. Place these cities in chronological order based on when the Black Death arrived: Stockholm, Novgorod, Alexandria, Vienna, Paris.

Step 3: I can create a timeline using historical conventions

8 Place these events in a timeline, obeying all conventions.

The Great Plague of London happened in 1665.

The plague arrived in Venice in 1348, and the city-state established maritime quarantine in 1403.

A year before arriving in Venice, the plague arrived in other parts of Italy via ships.

An outbreak of the plague occurred in Russia in 1345.

Step 4: I can distinguish causes, effects, continuity and change from looking at timelines

9 Using the dates above and those mentioned in this section, including the map in Source 3, do you see a causal chain that might help describe the spread of the plague? Write a short paragraph explaining this spread, using dates.

Chronology, page 194

Masterclass

Learning Ladder

Work at the level that is right for you or level-up for a learning challenge!

Source 1

Engraved hunting horn, 1325–50, England

Step 1

a **I can read a timeline**

From looking at the timeline on pages 58–59, answer the following questions.

i How long was it between the start of Viking raids and the Norman invasion of England?

ii What came first, the First Crusade or the Magna Carta?

iii How long before today was the Peasant Revolt?

b **I can determine the origin of a source**

Where is Source 1 from and when was it made?

c **I can recognise continuity and change**

Rank these from most changed to least changed, from the Medieval Period until now:

- technology
- family life
- warfare
- school
- food.

d **I can recognise a cause and an effect**

Match the causes and effects:

Europe became more peaceful after about 1000 CE.	Nobles forced him to sign the Magna Carta, which included equality before the law.
Joan of Arc helped the French win battles in the Hundred Years War.	Stone castles became more common.
Motte-and-bailey castles were easy to attack with fire.	The Church became wealthy.
People had to tithe one-tenth of their income to the Church.	The English burned her at the stake.
The King of England was above the law.	Towns grew.

e **I can recognise historical significance**

Rank these from most to least historically significant:

- Christians visiting the Holy Land
- Joan of Arc being killed
- the Black Death
- the Norman Conquest of England
- the Peasant Revolt.

Step 2

a **I can place events on a timeline**

Place these events in order from earliest to latest:

The Great Famine, lasting two years, ended in 1317

111 years after Charlemagne was made emperor, the Viking Rollo was made the first Duke of Normandy

Gothic architecture starts being built in 1135

In 800 CE, Charlemagne was crowned Emperor

The Hundred Years War between France and England began in 1337 (although it was actually 116 years long)

b I can list specific features of a source

Looking at Source 1, describe each of the three main figures in detail.

c I can describe continuity and change

What is one thing that has changed a lot from the Middle Ages until the present? Describe the change, giving details about the Middle Ages and the present day.

d I can determine causes and effects

i What caused the Black Death to spread from Asia to Europe?

ii What effect did the large number of deaths from the plague have?

e I can explain historical significance

Why is the increase in the size and number of towns important historically?

Step 3

a I can create a timeline using historical conventions

Create a timeline using the events from Step 2a using correct historical conventions. Make sure you have equal spacing between years.

b I can find themes in a source

What do you think the artistic theme of Source 1 might be? Read the History How-to section for some ideas about the kinds of things that might be themes for a source. Give evidence from the artefact to back up your answer.

c I can explain why something did or did not change

Why do you think workers earn better money in Australia today than they did in medieval Europe? Give evidence from the past and the present in your response.

d I can explain why something is a cause or an effect

The end of the Roman Empire caused Europe to be a much more dangerous place. Why?

e I can apply a theory of significance

Answer the following questions about the Black Death:

i How important was it at the time?

ii How deeply were people affected?

iii How many people did it affect?

iv For how long were they affected?

v Is it still relevant to today?

Step 4

a I can distinguish causes, effects, continuity and change from looking at timelines

Looking at the timeline on page pages 58–59, what two events can you see that are linked by cause and effect? Explain how one caused the other.

b I can use my outside knowledge to help explain a source

What do you know about medieval Europe that would help you to explain the roles of the three figures in Source 1? Include evidence from the artefact in your explanation.

c I can analyse patterns of continuity and change

Looking at the timeline on pages 58–59, what is something that stayed relatively the same during the Middle Ages? Why do you think this is? Give evidence that includes statements from the timeline.

d I can analyse cause and effect

List four effects of peasants having to live in small houses made from mud and straw and with no running water. Rank them from most significant effect to least significant effect. Write a few sentences explaining why you ranked your most significant effect higher than the least.

e I can analyse historical significance

Feudalism was important in medieval Europe. Give three reasons why. Which is the most important reason? Which of your three reasons supports or is linked to one of the other reasons? Are any of your reasons similar to each other or are all three completely separate?

Masterclass

Step 5

a **I can describe patterns of change**

Looking at the timeline on pages 58–59, what overall themes can you see? Is there a particular type of event that occurs regularly? What is it and why do you think events like this were common in the Middle Ages?

b **I can use the origin of a source to explain its creator's purpose**

Source 1: Why do you think the artefact was designed artistically? Support your answer with knowledge about medieval Europe and evidence from the artefact. Check page 199 of the History How-to section for help with possible purposes a creator might have.

c **I can evaluate patterns of continuity and change**

Children in the Medieval Period were made to work from a very young age. Comment on how a young child's life then differed from that of a young child in Australia today.

d **I can evaluate cause and effect**

Do you think the feudal system caused Europe to become more peaceful, or did Europe becoming more peaceful allow the feudal system to take hold? Give historical evidence to support your answer.

e **I can evaluate historical significance**

The Norman Conquest of England in 1066 is considered an important historical event. Do you agree? Support your response with historical evidence.

Historical writing

1 Structure

Imagine you are given the essay topic: 'Explain how William the Conqueror won the Battle of Hastings'. Write an essay plan for this topic. Include at least three main paragraphs.

2 Draft

Using the drafting and vocabulary suggestions on page 214, draft a 400–600 word essay responding to the topic.

3 Edit and proofread

Use the editing and proofreading tips on page 215 to help you edit and proofread your draft.

Historical research

4 Organise and present information

Imagine you are doing a large research project: 'Warfare in the Middle Ages'. Write a contents page for this project. There should be an introduction, conclusion, at least four main sections and many subsections. Number your chapters.

Capstone

How can I understand medieval Europe?

In this chapter, you have learnt a lot about medieval Europe. Now you can put your new knowledge and understanding together for the capstone project to show what you know and what you think.

In the world of building, a capstone is an element that finishes off an arch, or tops off a building or wall. That is what the capstone project will offer you, too: a chance to top off and bring together your learning in interesting, critical and creative ways. You can complete this project yourself, or your teacher can make it a class task or a homework task.

mea.digital/GHV8_H2

Scan this QR code to find the capstone project online.

H3

Mongol expansion

HOW DID GENGHIS KHAN BUILD THE WORLD'S LARGEST EMPIRE?

page 104

cause and effect

page 108

WHY WAS THE MONGOL ARMY SO POWERFUL?

economics + business

page 116

WHAT MAKES A BUSINESS SUCCESSFUL?

key individual

page 118

HOW DID KUBLAI KHAN BECOME THE FIRST FOREIGNER TO RULE CHINA?

How can I understand the Mongol expansion?

The Mongolian century, as the 1200s are known, was a time of great change. The great Genghis Khan unified the Mongols and created the largest land empire in history. His grandson Kublai Khan integrated with the Chinese and formed the Yuan Dynasty. While the Mongol army devastated Central Asia's population, the Mongols also fostered cultural development in the arts, ensured peace among their peoples, respected different cultures and religions, and improved trade.

Source 1

Handbag for carrying flint, as used by Mongol horsemen. [Made with metal, silver, coral and gems. Mongolia, undated.]

learning ladder

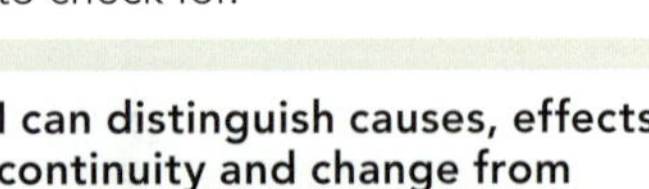

Step	Chronology	Source analysis	Continuity and change
step 5	**I can describe patterns of change** I read timelines and see the 'big picture'. I group timeline events about the Mongol Empire and see if they show patterns of change. I know typical historical patterns to check for.	**I can use the origin of a source to explain its creator's purpose** I combine knowledge of when and where a source was created to answer the question, 'Why was it created?'	**I can evaluate patterns of continuity and change** I answer the question, 'So what?' about patterns of continuity and change. I weigh up ideas and debate the importance of a continuity or a change.
step 4	**I can distinguish causes, effects, continuity and change from looking at timelines** I read Mongol Empire timelines and find events linked by cause and effect. I also find things that are the same or different from then until later times.	**I can use my outside knowledge to help explain a source** I have enough outside knowledge about the Mongol Empire to help me explain a source.	**I can analyse patterns of continuity and change** I see beyond individual instances of continuity and change in the Mongol Empire and identify broader historical patterns, and I explain why they exist.
step 3	**I can create a timeline using historical conventions** When given a set of events, I construct a historical timeline using the dates, and using correct terminology, spacing and layout.	**I can find themes in a source** I look a bit closer into a source and find more than just features. I find themes or patterns in the source.	**I can explain why something did or did not change** I answer the question, 'Why?' something changed or stayed the same between historical periods.
step 2	**I can place events on a timeline** When given a list of events and developments about the Mongol Empire, I put them in order from earliest to latest, the simplest kind of timeline.	**I can list specific features of a source** I look at Mongol Empire sources and list detailed things I can see in them.	**I can describe continuity and change** I have enough content knowledge about two different historical periods to recognise what is similar or different about them, and can describe it.
step 1	**I can read a timeline** I read timelines with Mongol Empire events on them and answer questions about them.	**I can determine the origin of a source** I can work out when and where a Mongol Empire source was made by looking for clues.	**I can recognise continuity and change** I recognise things that have stayed the same and things that have changed from the Mongol Empire until now.

Source 2

In 1210, Jin envoys from Zhongdu (later Beijing) arrived in Mongolia to demand that Genghis Khan recognise the new leader of the Jin. Genghis Khan spat on the ground and fired a volley of insults at the new leader. By 1215, the Mongol army had crushed the Jin and burned Zhongdu to the ground. [Miniature created c. 1430, from *Jami' al-tawarikh* (*The Universal History*), by Rashid-al-Din Hamadani, c. 1307. From the collection of the Bibliothèque Nationale de France.]

Warm up

Chronology

1 Genghis Khan and Kublai Khan were both great rulers of the Mongol Empire. Which of these influential leaders ruled first? (See the timeline on pages 100–01.)

Source analysis

2 Look at Source 2. When and where was the painting made?

Continuity and change

3 Which of these things have stayed the same since the Mongol Empire?
  a The communication system known as the *Yam*, which used horse relay stations
  b The nomadic lifestyle of some Mongolians
  c Weaponry

Cause and effect

4 Source 2: What do you think caused the fear that the Mongol army inspired?

Historical significance

5 The Mongols controlled a vast empire, one in which merchants could travel safely. Why do you think this was significant for trade?

Cause and effect

I can evaluate cause and effect
I answer the question, 'So what?' about cause and effect. I weigh up ideas and debate the importance of a cause or an effect.

I can analyse cause and effect
I don't just see a cause or an effect as one thing. I determine the factors that make up causes and effects.

I can explain why something is a cause or an effect
I can answer 'How?' or 'Why?' a cause led to an effect in the Mongol Empire.

I can determine causes and effects
Applying what I have learnt about the Mongol Empire, I can decide what the cause or effect of something was.

I can recognise a cause and an effect
From a supplied list, I recognise things that were causes or effects of each other in the Mongol Empire.

Historical significance

I can evaluate historical significance
I answer the question, 'So what?' about things that are supposedly historically important. I weigh up events and cast doubt on how important things are.

I can analyse historical significance
I separate out the various factors that make something historically important.

I can apply a theory of significance
I know a theory of significance. I use it to rank the importance of Mongol Empire events.

I can explain historical significance
I answer the question, 'Why?' about things that were important in the Mongol Empire.

I can recognise historical significance
From the history of the Mongol Empire, I can work out which are important.

Why learn about Mongol expansion?

With an **empire** spanning from 1206–1368, the reputation of the Mongols has changed significantly in recent times. Famed for their bloodthirsty conquests, fearsome warriors and deadly tactics, historians now view Mongols as having contributed much to the political unity of Russia and China, and to developing important trade routes that spread goods, people and ideas. However, in places like the Middle East, the legacy of destruction the Mongols brought is still considered catastrophic.

Source 1

Genghis Khan leads his Mongol army to chase down the enemy. [Image created c. 1430 for Rashid-al-Din Hamadani, *Jami' al-tawarikh* (*The Universal History*) c. 1307, Bibliothèque Nationale de France, France.]

Source 2

Film still of Kublai Khan (Genghis Khan's grandson) and Italian explorer Marco Polo, from the 2014 television series *Marco Polo*

key ideas timeline.

1206 CE – Genghis Khan is made ruler of the Mongols

1215 CE – Mongols destroy Jin capital, Zhongdu

1242 CE – Mongols withdraw from Europe

Mongols invade Europe 1223–1241 CE

Life of Kublai Khan 1215–1294 CE

Life of Genghis Khan 1162–1227 CE

Source 3

Painting depicting the Vietnamese defeating the Mongols at the Battle of Bach Dang in 1288

Source 4

This helmet is from around 1350–1450; armour from this period is very rare. The style of decoration with flat strips of gold shows Mongolian influence on its design.

Learning ladder H3.1

Show what you know

1 Which century did the Mongol Empire span?

2 What was the relationship between Genghis Khan and Kublai Khan?

Chronology

Step 1: I can read a timeline

3 How many years separated the beginnings of the rules of Genghis Khan and Kublai Khan?

Step 2: I can place events on a timeline

4 Using information from this page, put these events in order from earliest to latest.

Kublai Khan becomes Mongol ruler

The Ming Dynasty reclaims China and the Mongol Empire ends

Mongol army conquers Zhongdu, capital of medieval China

Genghis Khan dies, 1227

Step 3: I can create a timeline using historical conventions

5 Create your own timeline showing the events from question 4.

Step 4: I can distinguish causes, effects, continuity and change from looking at timelines

6 What is the common theme in the entries shown on the timeline and what do you think was the cause of this?

HOW TO

Chronology, page 194

1258 CE — Mongol Empire invades Poland

1258 CE — Mongols capture Baghdad

1264 CE — Kublai Khan becomes ruler of Mongol Empire

1274 CE — First failed Mongol invasion of Japan

1281 CE — Second failed Mongol invasion of Japan

1351–68 CE — Peasant rebellions against the Yuan Dynasty

Marco Polo travels in China 1271–1288 CE

Yuan Dynasty 1271–1368 CE

Mongol Empire 1206–1368 CE

What was life like in Mongolia?

Mongolia's harsh landscape and cold, windy, dry climate made life difficult. People adopted a nomadic lifestyle, herding their animals from one place to another.

Harsh life in the Steppe

The amazing expansion of the Mongol Empire began with **nomadic** herders on the high, cold grasslands known as **steppes** in Mongolia. In this dry region with few trees, cold winds swept across the plains forcing temperatures as low as −40°C in winter. In summer, temperatures can reach as high as 38°C in the Gobi Desert region.

The grasslands are covered in snow and ice during the long winters and can turn to straw during baking summers. However, the short spring season provides good pasture for horses, goats, yaks and sheep.

Nomadic way of life

The harsh steppe environment forced the **Mongols** to become nomads, as growing crops was almost impossible. They moved their herds of animals around in search of the best pasture and weather. In the spring and summer, the herds stayed on the open plains to graze, and in winter they were herded to protected valleys. Mongol family clans came together as a tribe, which was led by a chief or **khan**. There could be fierce conflict between tribes.

Families lived in portable circular tents called **yurts**. They were covered in felt and made from wool with a wooden door. A hole was left open at the top to allow air to circulate, light to come in and the chimney pipe to poke through. Heavy woollen carpets line the walls next to where people sleep on beds, which serve as couches during the day. In the middle of the yurt is a small charcoal stove and a dining table and chairs.

Source 1

Bayan-Ölgii is the westernmost and highest province of Mongolia. Even today, the Mongols herd goats in the traditional manner, moving their portable tents called yurts and using horses as transport. The region is famous for the traditional practice of hunting on horseback with trained eagles.

Mongolia's environments

Source 2

Map of Mongolia's environments

Mongol horses

Nomadic Mongols tamed and trained wild horses. They shot wild game with their bows and arrows while riding on horseback. These skills made them excellent warriors.

Horses were also used to transport tents and belongings. Female horses produced milk that the Mongols used to make yoghurt and cheese. This provided the main source of nutrition for the Mongols in the bleak steppe conditions. Mongols also trained eagles to find and hunt prey such as foxes and hares.

Bayan-Ölgii climate graph

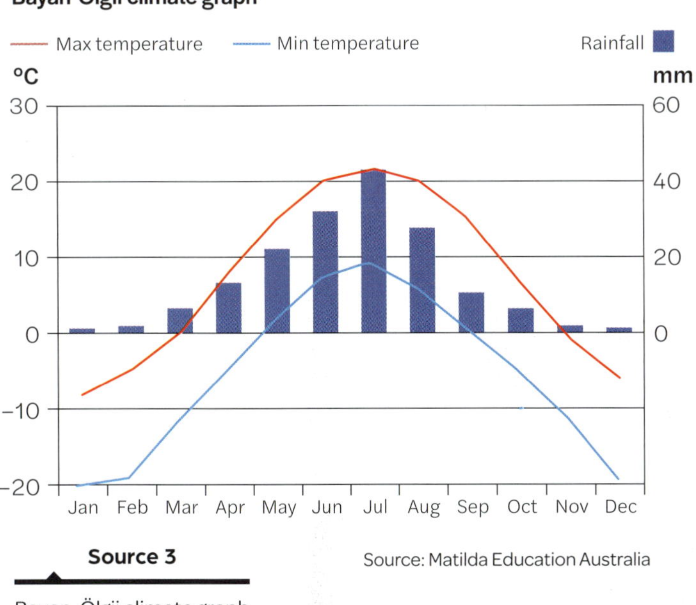

Source 3

Source: Matilda Education Australia

Bayan-Ölgii climate graph

Learning ladder H3.2

Show what you know

1 What is the difference between a nomad and a traditional farmer?

2 Source 3: In which months of the year is the maximum temperature 0°C or colder?

3 How are a clan, a tribe and a khan different?

4 Look at Source 1. Describe the physical conditions of this place in detail.

Cause and effect

Step 1: I can recognise a cause and an effect

5 Match the following causes and effects:

Growing crops was difficult in Mongolia	Grass can turn to straw
Mongols could shoot arrows on horseback	Mongols became nomads
Temperatures can reach 38°C in summer	They were excellent warriors

Step 2: I can determine causes and effects

6 What caused Mongols to have to herd their animals into valleys in winter?

Step 3: I can explain why something is a cause or an effect

7 Why were Mongols nomadic?

Step 4: I can analyse cause and effect

8 What have you learnt about the Mongols that might suggest they would make fierce conquerors?

HOW TO

Cause and effect, page 203

How did Genghis Khan build the world's largest empire?

Genghis Khan, or Temujin as he was first named, was born in 1162 on the freezing plains of Mongolia. The son of a tribal chief, he would grow to unite the warring Mongolian tribes and establish the second-largest land empire in history.

When his father was killed by a rival tribe called the Tatar tribe, the nine-year-old Temujin tried to claim his rightful position as chief of his clan. The clan rejected him and cast his family out into the bleak Mongolian plains to live on berries, rats and birds. Temujin killed his older half-brother when he discovered he was hiding food. He became the head of the family and saved them from starvation.

At the age of 16 Temujin decided the best way forward was to create alliances. He later wrote, 'A man who seeks power needs friends with power'. He married a girl from a powerful tribe to gain influence. When she was kidnapped by the Merkit tribe, Temujin convinced another tribe to fight with him to free her. They overran the powerful Merkit tribe and freed his wife.

Through conflict and negotiations, Temujin gradually gained the respect, fear and partnership of other Mongolian tribes. He avenged his father's death by crushing the Tatar tribe, boiling the leaders alive and ordering the death of every male taller than a wagon's axle. By 1206, at the age of 44, Temujin finally united the Mongolian tribes under his leadership. Temujin was now the lord of the Mongolian tribes and given the title Genghis Khan – 'Genghis' meaning universal and 'Khan' being the Mongolian title for a ruler.

Source 1

Genghis Khan developed a fearsome reputation as a bloodthirsty leader who would not let anything get in his way. His loyal army slaughtered enemies in horrifying ways, which sent a terrifying message to others to let the Mongol army in without resistance.

Source 2

Bronze statue of the great emperor – Genghis Khan, at the Palace of Mongolian Government, in Ulaanbaatar, the capital city of Mongolia.

Conquering Asia

From his humble beginnings, Genghis Khan grew to become a great military leader. The Mongolian lands were his, but Genghis Khan had bigger plans. His determination to succeed combined with a charismatic personality, intelligence, courage and ruthlessness saw Genghis Khan lead his ferocious Mongol army on the largest military expansion in history. In only 20 years, they conquered most of northern China and Central Asia.

Over 100 years, Genghis Khan and his successors built the largest **empire** the world had seen, stretching from China in the east to the European country of Hungary in the west. The Mongol Empire at its greatest had grown to more than 30 million square kilometres – about the size of the continent of Africa and twice the size of the Roman Empire.

Building a war machine

Unlike other empire builders before him, such as Julius Caesar of the Roman Empire, Genghis Khan didn't have a well-trained army to call on. He pulled together nomadic tribes into a disciplined military machine that swept through Asia.

Genghis Khan's army was built by promoting those with talent, rather than those with a senior position in society. Known as a **meritocracy**, ordinary men could climb the ranks to command armies if they had the skills to do so. Genghis Khan established a large group of capable leaders to whom he could delegate responsibility. These trusted and skilled commanders were also completely loyal to their leader.

The other key to Genghis Khan's leadership of his army was strict discipline within each unit of his army. Disobedience led to punishment. If one person in the *arban* (a ten-man unit) fled from a battle, he and all other members of the unit were executed. If the entire *arban* deserted a battle, then all members of the zuut (a hundred-man unit) were put to death.

One of the weaknesses of Mongol tribes was that when they won a battle, they would stop to **loot** the camp of the enemy, allowing enemy soldiers to escape. Genghis Khan made it clear to his troops that there would be no looting until the enemy troops were eliminated. In 1202 as Genghis Khan prepared to take revenge against the Tartar tribe for the murder of his father, he issued strict instructions to his troops:

> 'If we are victorious, let no-one take booty right then. It will be shared out later. If we must retreat, we will return to the starting point and, once we have formed again, we will attack with vigour. Anyone who fails to return to the formation will be decapitated.'

Source 3

Borja Pelegero, in *National Geographic History*, Aug/Sept 2015

Source 4

This modern artwork shows the aftermath of the Battle of Kalka River in 1223. The Mongol general Jebe on horseback orders the execution of the Russian prince Mstislav of Kiev, who is soon to join the pile of other captured nobles in the background. [Pavel Ryzhenko, *The Battle of Kalka* (1996).]

Creating fear

Like the **Vikings** before him, Genghis Khan used fear as a weapon. His troops killed many and used the most brutal tortures imaginable. This helped to make future conquests easier, as other groups became too afraid to fight the Mongols. Genghis Khan offered his enemies a simple choice – surrender or die.

Genghis Khan's growing army spread terror in a calculated way. The more resistance they encountered from an opponent, the more horrible the treatment the opponent received from the Mongol army. Terrified witnesses to this brutality were allowed to escape to spread the word not to oppose the Mongols.

Genghis Khan's cruelty was unprecedented. At the Battle of Kalka River in 1223, the Mongol army defeated the Russian forces, wiping out 90 per cent of the soldiers. Mstislav of Kiev and his army retreated and surrendered in return for their safe treatment. Once they surrendered, however, the Mongol army slaughtered the Russian force. They executed Mstislav of Kiev and buried the remaining noble prisoners alive under a victory platform, enjoying a victory feast on top of them as the nobles suffocated beneath them.

The achievements of Genghis Khan

Despite his cruel methods, Genghis Khan's military campaigns were incredible. He conquered more than double the area of any other person in history. His vast empire joined Asia and Europe and he built profitable trading routes (see pages 122–23) to help finance his rule.

The Mongol Empire did not crumble after Genghis Khan's death in 1227. His family continued to expand the Mongol Empire. Within decades, his grandson Kublai Khan (see pages 118–21) ruled all of China, and had expanded the Mongol Empire to be the largest that had ever existed.

Source 5

Genghis Khan leads his Mongol army to chase down the enemy. [Image created c. 1430 for Rashid-al-Din Hamadani, *Jami' al-tawarikh* (*The Universal History*) c. 1307, Bibliothèque Nationale de France, France.]

Learning ladder H3.3

Show what you know

1 How big was the Mongol Empire at its peak?
2 How was being promoted in the Mongol army different to other armies of the time?
3 Describe the conditions for a soldier in the Mongol army.
4 List the different individuals and groups defeated by Genghis Khan that are mentioned in this section.
5 Explain what is meant by 'A man who seeks power needs friends with power'. How might this wisdom be used in modern life?
6 Discuss with a partner whether, in your opinion, Genghis Khan's great achievements outweigh his brutal methods. Justify your answer with historical evidence.

Source analysis

Step 1: I can determine the origin of a source

7 When was Source 4 made?

Step 2: I can list specific features of a source

8 Describe in detail all the people in Source 4. Which group do you think each belonged to?

Step 3: I can find themes in a source

9 Look at Source 4. List five adjectives to describe the Mongols. What do you think the artist is trying to show about the Mongols as warriors?

Step 4: I can use my outside knowledge to help explain a source

10 Look at Source 5. What offensive and defensive equipment can you see the Mongols using? What kind of materials do you think were used to create this equipment?

HOW TO

Source analysis, page 197

Why was the Mongol army so powerful?

The Mongol army consistently beat much larger armies with their discipline, special breed of horses, innovative new equipment and superior riding skills.

Mongol cavalry

The Mongol army was made up of **cavalry** – soldiers who fight on horseback. Their Mongol horses, riding skills and weapons gave them the advantage against slower, less-agile enemies they encountered.

Mongols began riding and hunting at an early age and became very skilled. The tough Mongol horses made excellent warhorses. They were small, easy to mount and dismount, and would forage for their own food. They were resilient and could travel long distances in both extremely hot and cold climates. Each warrior would care for a small herd of three to five horses, to ensure they always rode a fresh horse into battle.

Source 1

A mounted Mongol archer. Members of the Mongol cavalry were armed with powerful bows that could fire an arrow more than half a kilometre. A lifetime of training and special wide stirrups allowed the rider to stand and turn to shoot as the horse galloped forward. [Creator unknown, *A Mongol horseman with a composite bow* (c. 13th century).]

Source 2

Soldiers in the Mongol cavalry were perfectly equipped for battle.

Horses
Mongol horses were small, fast and well balanced, allowing Mongol archers to stand and turn to fire on enemies. They were fit, could be ridden for up to 40 kilometres each day and could withstand the extremes of the Mongolian climate.

Order and discipline

Each leader reported to the level above, allowing the Mongol army to attack as a large *tumen* or divide into *zuuts* to encircle enemies. Commanders were chosen for their talent, not their position in society:

- *Khagan* – the equivalent of the emperor, who had supreme power and final decision rights
- *Kheshig* – the imperial guard, who ensured the safety of the *Khagan*
- ***Noyan*** – a general in charge of a ***tumen***, a unit of 10 000 soldiers
- *Mingghan* – a group of 1000 soldiers, led by a commander
- ***Zuut*** – a unit of 100 soldiers with an elected leader
- ***Arban*** – a unit of 10 soldiers with an elected leader

Stirrups
The 1100-year-old metal stirrup shown was a game-changing Mongol invention. The strong and wide metal stirrups allowed the rider to stand and fire arrows as the horse twisted and turned.

Sword
The metre-long curved scimitar was used for slashing attacks on horseback.

Quiver
Soldiers carried two quivers containing more than 60 arrows.

Bow
Bows were made in layers of birch wood, deer sinew and sheep horn to give the flex needed to fire straight and long. Mongol arrows could travel over 500 metres.

Arrow
Different arrows were used in different situations. Whistle arrows produced a whistling sound designed to terrify enemies. Flaming arrows were used to burn down villages.

Sneaky tactics

The Mongol army developed many clever tactics for different situations. A favoured tactic was the fake retreat, where they would charge the enemy and then retreat, to lead the enemy into an ambush. They also used encirclement, where a light cavalry would quickly encircle the enemy from a distance and then shower them with arrows.

Learning ladder H3.4

Show what you know

1 Why was fighting from horseback an advantage?

2 What made the Mongol horses superior?

3 What was the benefit of the Mongol army having the structure it did?

4 Look at Source 1. Is this a primary or a secondary source? Does this mean it is unbiased?

Cause and effect

Step 1: I can recognise a cause and an effect

5 Match the following causes and effects:

Enemies were terrified	Special breed of horse
Mongols could stand to fire arrows while riding	Use of whistling arrows
Mongols could fight in any conditions	Mongols invented new stirrups

Step 2: I can determine causes and effects

6 What skills and abilities did Mongol warriors have that would still be useful for a modern soldier?

Step 3: I can explain why something is a cause or an effect

7 Mongols began riding and hunting at a very young age. Explain how this is a causal factor linked to their military skill.

Step 4: I can analyse cause and effect

8 Imagine you are narrating a history documentary about the Mongol empire. Write a voice-over script for the scene shown in Source 2. Explain how the horses, weapons and skill of the warriors all combined to contribute to the Mongols' military dominance.

HOW TO
Cause and effect, page 203

How did Genghis Khan's empire expand?

Genghis Khan unified the Mongolian tribes by 1206, following years of bitter fighting. The ruthless soldiers terrorised enemies throughout Asia and into Europe and built the largest empire that had ever been seen.

Attacks on China

Genghis Khan first looked south to China to expand his empire and obtain resources not found in Mongolia. The Mongols were dependent on China for many goods including grain, metals and manufactured items such as fabrics. The Jin Dynasty, which controlled northern China, and the Xia Dynasty from northwest China tried to limit the amount of trade the Mongols could conduct in the early 13th century. Unable to obtain the goods they needed, the Mongols raided and eventually invaded these two kingdoms.

1 Battle of Badger Mouth, 1211

The Mongols sent 90 000 troops, nearly their entire army, in a military campaign against the Jin Dynasty. The Jin sent an official to negotiate a peace settlement. Genghis Khan persuaded the official to surrender and provide information about the Jin army. Genghis Khan sent his general to launch a surprise cavalry charge through the Badger Mouth passage. Because of the mountainous terrain, the Mongols had to fight on foot. The Mongol army slaughtered the entire Jin army, leaving bodies strewn over hundreds of kilometres.

The expanding Mongolian Empire under Genghis Khan, 1206–27

Source 1

The expanding Mongolian Empire under Genghis Khan, 1206–27

Source 2

The Mongol army rides into battle in the 2007 movie, *Mongol*.

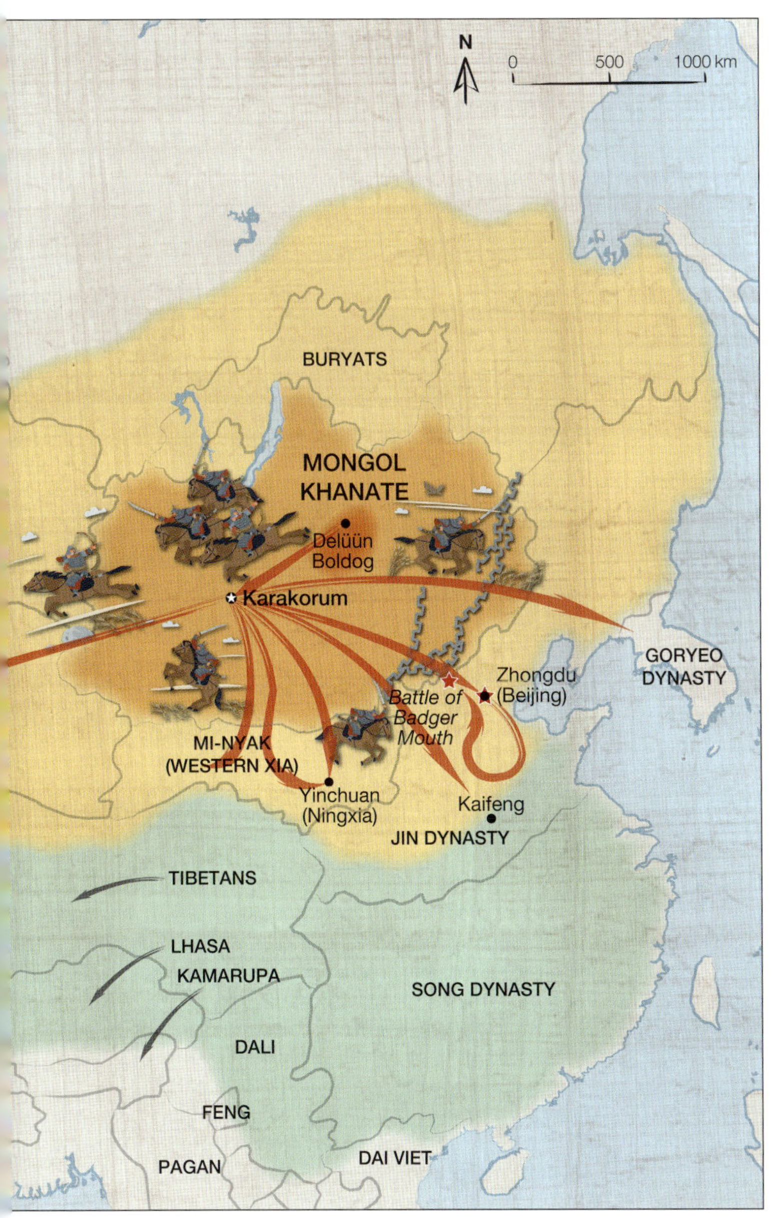

Source: Matilda Education Australia

2 Battle of Zhongdu, 1215

By 1213, Genghis Khan's troops controlled the Jin territory up to the Great Wall of China, the ancient wall built to keep out Mongol raiders. He then decided to break his invading army into three smaller armies. Each broke through the wall and rejoined to attack Zhongdu (modern-day Beijing), the Jin capital. Here the Mongol army used **siege** tactics by surrounding the city and cutting off its supply routes. With the population starving, the Jin eventually surrendered in 1215. The people were slaughtered and the city was looted and burned.

Attacks on Central Asia

Genghis Khan did not originally intend to expand into Central Asia. He sent the ruler of the Khwarezmid Empire a message to seek trade with them:

> 'I am master of the lands of the rising sun while you rule those of the setting sun. Let us conclude a firm treaty of friendship and peace.'
>
> Source 3
>
> Genghis Khan

The Khwarezm ***shah*** (king) reluctantly agreed to a peace treaty and trade with the Mongols. A trade caravan was sent, but the Mongol traders were killed and their goods were stolen. Genghis Khan sent three officials to protest to the *shah*, but he had their heads cut off and returned to Genghis. In 1219, the Mongol army of 200 000 men rode west and raided several cities in the Khwarezmid Empire. Using captured prisoners as a human shield, Genghis Khan's army stormed the capital city of Samarkand in 1220.

3 Battle of the Indus, 1221

With the death of the *shah* at Samarkand, his son took control and pulled together what was left of the Khwarezm army in Afghanistan. Genghis sent forces to hunt them down, but suffered a rare defeat. The furious Mongol leader headed to the battlefront himself and crushed the Khwarezm army in battle at the Indus River.

Attacks on Europe

Following the defeat of the Khwarezmid Empire, Genghis' trusted *noyan* (general) Jebe asked him for a chance to continue his conquests. Jebe Noyan and Subutai Noyan each led a *tumen* of 10 000 men (see page 109) to Russia.

4 Battle of Kalka River 1223

Outnumbered four to one, the Mongol army utterly destroyed a combined force of Russian principalities (see page 106). The battle marked the end of the longest ever cavalry raid where the Mongol army rode nearly 9000 kilometres over a period of three years.

Source 4

Genghis Khan's army attacking the Jin army at the Great Wall of China in 1215. The Mongol army broke through the defences and then moved to the capital of the Jin dynasty at Zhongdu, where they used siege tactics to capture the city. [Alberto Salinas, *Genghis Khan marched an army of 200,000 into China* (c. 20th century), lithograph, private collection.]

Source 5

The Great Wall of China still stands today. In its 2700-year history, only Genghis Khan's Mongol army was able to break through. Intelligence from Jin informants led the Mongols to secret passages through the wall and clever tactics such as the fake retreat (see page 109) were used to defeat the Jin army on the wall. When the wall was breached, the Mongol army rode into the Jin capital of Zhongdu and captured it in 1215. Genghis Khan's army was joined by new Chinese recruits with knowledge of explosives and new weapons.

Learning ladder H3.5

Show what you know

1 Why did the Mongols want to expand south?

2 Is Source 3 a primary or secondary source? What makes you say that? What about the image in Source 2? Is the image a primary or a secondary source? Explain why.

3 Was the Khwarezm *shah* wise in his actions? Explain your reasons for your answer. If you think he was unwise, suggest what else he could have done instead.

4 How similar do you think the personalities of Genghis Khan and the Khwarezm *shah* were, based on their actions? Give evidence to support your answer.

Continuity and change

Step 1: I can recognise continuity and change

5 Which of these things have stayed the same until the present day?

- a The Great Wall of China is an amazing feat of engineering.
- b Mongolian expansion into Europe
- c A land border between Mongolia and China

Step 2: I can describe continuity and change

6 Look at Source 1. Using geographic terms such as north, south, east, west, inland, coastal and landforms, describe the expansion of the Mongolian Empire in 1206–27.

Step 3: I can explain why something did or did not change

7 What do the battles of Badger Mouth and Zhongdu have in common? How are they different?

Step 4: I can analyse patterns of continuity and change

8 Analyse the different military tactics Genghis Khan used to expand his empire. Select one of his tactics and explain it in detail.

HOW TO

Continuity and change, page 200

How do you manage an enormous empire?

Genghis Khan managed the massive Mongol Empire with skilled administrators, strict rules, severe punishments and a relay communication system which enabled messages to travel quickly.

Genghis Khan's empire

Genghis Khan was a brilliant, merciless leader who built a vast empire, slaughtering millions in his quest to rule the world. If Genghis' great Mongol Empire was to survive and expand, he needed to work with the people he had conquered and provide structures for the **empire** that, in his time, stretched for 13 500 000 square kilometres.

Genghis Khan spared the lives of skilled workers and hand-picked talented military leaders to work for his empire. He allowed his conquered people to follow their own religions. With the support of religious leaders, Genghis could foster good relations with newly conquered people. However, he also insisted on loyalty to himself and no other. Any disobedience could mean death.

Governing the empire

The Mongols had to adapt to manage the huge empire they now controlled. The Great **Khan** sat on top of the government structure as the Universal Ruler with absolute power. He was advised by a prime minister, known as the ***beqlare-beq***, and his ministers, known as ***viziers***.

Genghis Khan was smart enough to understand that he couldn't rule such a vast empire alone. To help him make decisions, he established councils of advisers drawn from the different nations and tribes he had conquered.

Genghis Khan replaced the leaders of conquered lands with loyal military governors called ***basqaqs***. They administered the territory, collected taxes and ensured people did not rebel.

Meritocracy

The Mongol Empire was a **meritocracy**, where all positions were chosen based on merit. Positions of responsibility were given according to great loyalty or bravery in conflict.

However, this meritocracy did not extend to Genghis Khan and his family. He decreed that only a member 'the Golden Family' could rule over the empire.

A code to live by

Genghis Khan established laws and a code to live by to let all people in his empire know what was expected of them. Known as the ***Yassa***, the code set out rules for people in the Mongol Empire.

Some of the laws and guidelines set out in the *Yassa* were:

- all people must present their daughters to the Khan at the start of each year so that he may choose some for himself or his family
- it is forbidden to use honorary titles, and only names are to be used
- all religions are to be respected and no preference is to be given to any one of them
- anyone who lies, practices sorcery, or spies on others will be put to death
- people who take goods on credit and don't pay for them on three occasions will be put to death
- anyone supplying food or clothing to a prisoner without permission will be put to death
- a person who finds a runaway slave or prisoner and does not return him will be put to death
- assault or abuse of women was forbidden.

Rules were strictly enforced and the fear of punishment made it safer to travel throughout the empire. It was commonly said that 'a maiden bearing a nugget of gold on her head could wander safely throughout the realm'.

Communication

For Genghis Khan, quick and reliable communication across such a large empire was critical. He set up the ***Yam***, a system of relay stations every 30–50 kilometres across the empire. Each station had a building or tent with beds and food, along with holding pens and food for horses. Relay riders would hand a message to a fresh rider, or have a meal and take a fresh horse to continue on. This enabled messages to travel 200–300 kilometres per day.

Every rider carried a metal pendant called a ***paiza*** to show they were messengers of the Khan. It was engraved with a message: 'By the power of eternal heaven, [this is] an order of the Emperor. Whoever does not show respect [to the bearer] will be guilty of an offence.'

As well as using the *Yam* to pass messages, the system also allowed military commanders and officials to move quickly throughout the empire.

Source 1

Paiza were medallions carried by *Yam* riders and diplomats to show they were official representatives of the Khan.

Source 2

The Genghis Khan Equestrian Statue is a 40-metre-tall stainless-steel statue near the Mongolian capital of Ulaanbaatar.

Learning ladder H3.6

Show what you know

1 What does the statue in Source 2 suggest about Genghis Khan? What was the artist trying to show about him?

2 How did the Mongols ensure quick communication across the empire?

3 What freedoms did the Mongols allow?

4 In what ways did Genghis Khan favour his own family?

Continuity and change

Step 1: I can recognise continuity and change

5 What were crimes under the Mongols that are still crimes today?

Step 2: I can describe continuity and change

6 Look at the guidelines of the *Yassa*. Which of these are still guidelines in today's society? Why?

Step 3: I can explain why something did or did not change

7 People are still generally promoted based on talent rather than birth. Why do you think this has not changed?

Step 4: I can analyse patterns of continuity and change

8 Why were the *Yam* and *Yassa* so critical for the orderly continuance of the Mongol Empire? Give examples to support your claims.

HOW TO

Continuity and change, page 200

economics+business

What makes a business successful?

Genghis Khan pioneered many of the key strategies that make businesses successful today. Like small modern start-up businesses, Genghis began with few resources and had be innovative to compete.

Lessons from the past

Businesses can learn a great deal from successful leaders of the past such as Genghis **Khan**. From a poor outcast who worked hard just to survive in the harsh Mongolian landscape, he united fighting tribes and went on to build the largest **empire** that had ever been.

Many of Genghis Khan's most successful strategies and actions can be applied to modern business today:

- He united warring groups and moulded them into a disciplined team.
- He saw talent in others and promoted those with skills ahead of those with titles.
- Through careful planning and practice, he tried to beat his enemies psychologically before fighting began and was therefore able to defeat much larger armies.

Source 1

With much smaller armies than most enemies, Genghis Khan developed tactics such as the fake retreat to break up the larger armies and use his advantage of lighter and faster cavalry.

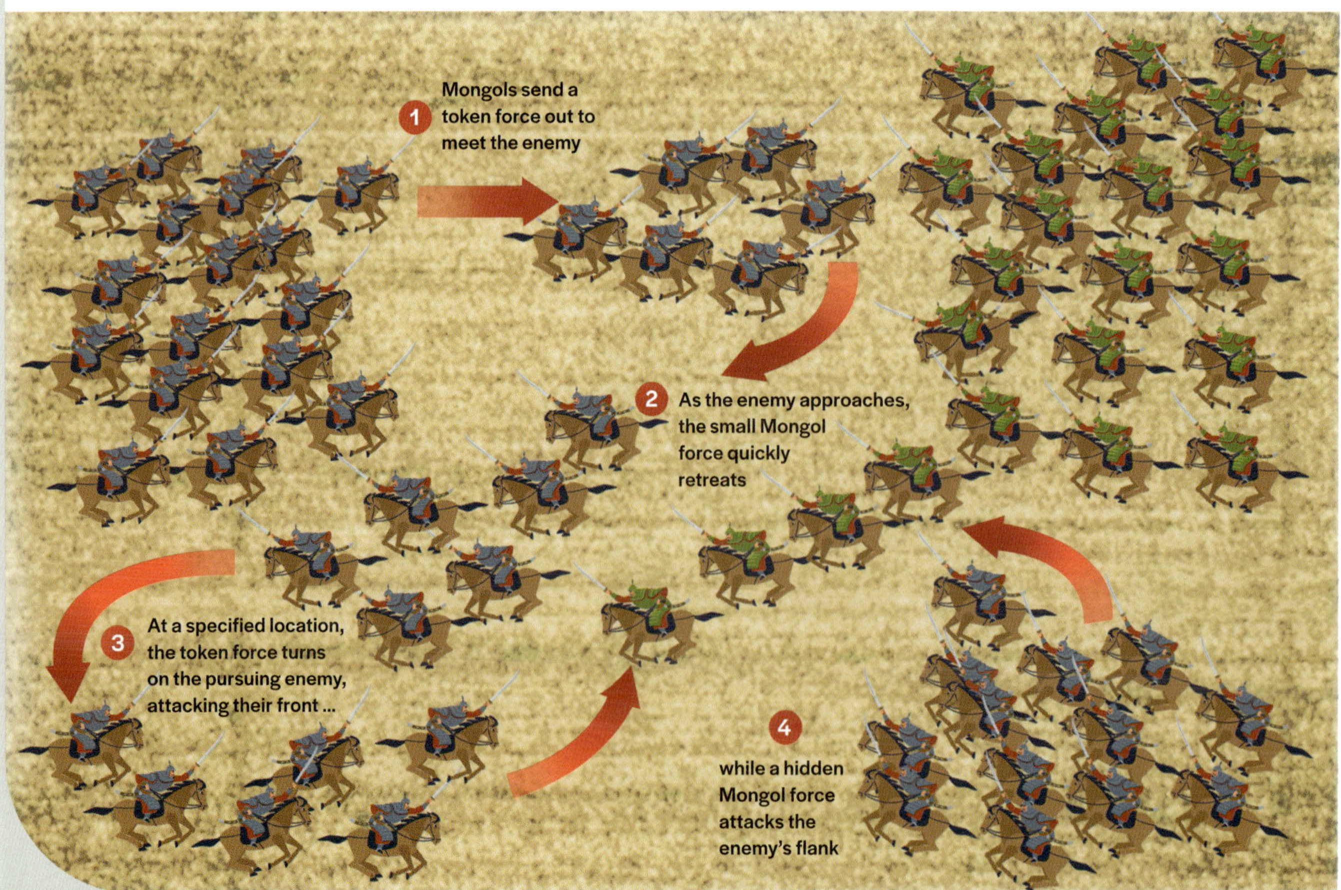

Source 2

At the age of 16, South Australian teenager Nathan Woodrow started his own clothing label, Ryde Clothing, from his bedroom. Nathan drew original skating, surfing and riding clothing designs. He started paying $25 per shirt to have them printed and ended up losing $5 on every shirt he sold. Fast-forward 12 months and he was sending his drawings to a designer in Queensland via Instagram, and doing his own screen printing from a shipping container on his parents' rural property and making money.

- He established the ***Yam*** relay communication system (see page 115) as he knew that good communication was essential to manage his vast empire.

The following table shows five characteristics of a successful modern start-up and gives examples from both Genghis Khan and an Australian start-up business, Ryde clothing, founded by Nathan Woodrow.

Successful business strategies for a start-up	Genghis Khan, ruler of the Mongol Empire	Nathan Woodrow, business owner of Ryde
1 Deliver something that people want	He put an end to constant fighting between Mongol tribes and brought peace and wealth.	He came up with cool original drawings to appeal to his target market of teenagers.
2 Use the skills of others to give you the edge	He spared the lives of skilled workers and talented military leaders.	He worked with a designer via Instagram to complete his rough clothing designs.
3 Communicate constantly with your customers	With armies spread over thousands of kilometres, he established horse relays to move messages quickly.	He stays in touch with his customers through his website and social media.
4 Embrace technology to find better and faster ways to operate	He captured and harnessed new technologies; e.g. Chinese explosives.	He purchased his own screen-printing machine when using professional printers became too expensive.
5 Build an engaged customer base through word of mouth	He spread fear to encourage other enemies to surrender.	Many active teenagers wear and share Ryde designs through social media channels.

Learning ladder H3.7

Economics and business

Step 1: I can recognise economic information

1 What lessons can businesses learn from Genghis Khan?

Step 2: I can describe economic issues

2 What do start-up businesses have in common with Genghis Khan?

Step 3: I can explain issues in economics

3 What was similar and what was different about the communication methods of Genghis Khan and Nathan Woodrow?

Step 4: I can integrate different economic topics

4 How did both Genghis Khan and Nathan Woodrow use innovation to overcome a shortage in resources? How did they use technology to give them an advantage?

Step 5: I can evaluate alternatives

5 Imagine you are starting your own business. Think of a product idea and show how you could incorporate the five successful business strategies discussed.

H3.8

key individual

How did Kublai Khan become the first foreigner to rule China?

Becoming the Great Khan

Kublai was born in 1215 as the grandson of the first Mongol emperor Genghis Khan (see pages 104–07). Following the death of Genghis in 1227, his son Ogodei finished conquering the Jin Dynasty and expanded into Europe, conquering as far west as Hungary, and raiding even further. When Ogodei died in 1241, there was a power struggle for many years before Kublai's brother Mongke was elected Great Khan in 1251.

Mongke appointed Kublai, now in his thirties, as governor of the Chinese territories. At that time, only northern China was under Mongol rule with the south ruled by the powerful Song Dynasty. For many centuries, the Chinese viewed themselves as the most important civilisation in the world. The Chinese regarded the Mongols as **barbarians**, so ruling the Chinese was going to be a difficult task for Kublai.

Source 1

Kublai Khan brought the empire to its greatest height by defeating the Song Dynasty in southern China to create the Yuan Dynasty. He reigned for 30 years, crushing rebellions against him to ensure that all knew he was the Great Khan, ruler of the Mongol Empire. This is a film still from the 2014 television series *Marco Polo*.

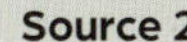

Source 2

A trebuchet attack on a Chinese city by the Mongol army from an illuminated manuscript by Rashid al-Din in the early 14th century. [Rashid-al-Din Hamadani, 'Mahmud ibn Sebuktegin attacks the fortress of Zarang', *The Universal History*, c. 1307 CE, private collection.]

Kublai had always been interested in Chinese culture and studied the Ancient Chinese philosophies of **Confucianism** and **Buddhism**. His respect for the customs and traditions of the Chinese people he governed helped Kublai gain their acceptance. Kublai also appointed Chinese advisers to help him rule.

Mongke asked Kublai to join him in an attack on the Song Dynasty to bring all of China under Mongol control. In 1259, Mongke died during the campaign and Kublai was forced to return to the Mongol capital of Karakorum. When he arrived, he discovered that his brother Ariq had claimed the title of **Great Khan**. Kublai challenged Ariq, and a four-year **civil war** broke out between the brothers. In 1264 Kublai's army finally won and he became the Mongol Empire ruler – Kublai Khan.

Conquering China

After becoming the Great Khan, Kublai set about conquering the Song Dynasty in southern China. Kublai Khan was greatly outnumbered. The Song Dynasty had a huge population of 70 million people and an army many times the size of the Mongol force. Following some good progress, the Mongol forces were stalled in 1268 when they reached the city of Xiangyang on the Han River. The high walls that surrounded the city were 7 metres thick and 5 kilometres long.

The Mongols had brought **trebuchets** (see page 78) that could sling missiles for a distance of 100 metres. These weapons had been used successfully against the Jin Dynasty (see pages 110–11). The Song expected this attack and widened the moat around the city to 150 metres to ensure missiles wouldn't hit their walls. The Mongols then used **siege** tactics (see page 78) against the Song. A Mongol fleet of 5000 ships blocked supplies from the Han river and set up camps along roads to stop supplies coming overland. The Chinese had also planned for this and had huge stores of food.

The siege lasted for five years. Eventually the Mongols called on two Persian engineers from the other side of their empire to build a mangonel (see page 78) that could sling a 300-kilogram missile up to 500 metres. The mangonels threw huge rocks and thundercrash bombs filled with gunpowder into Xiangyang and other cities of the Song. Soon the Song Dynasty was defeated.

The Yuan Dynasty

In 1271 Kublai Khan declared the beginning of the Yuan Dynasty, and crowned himself as the first Yuan emperor of China. He moved the Mongol capital of Karakorum to Dadu, the modern-day city of Beijing. Kublai Khan built a huge walled palace in Dadu and a southern palace in Xanadu. The great adventurer Marco Polo (see page 123) described the Imperial Palace of Kublai Khan:

> 'The hall of the palace is so large that it could easily dine 6000 people; and it is quite a marvel to see how many rooms there are besides. The building is altogether so vast, so rich and so beautiful, that no man on Earth could design anything superior to it.
>
> **Source 3**
>
> *The Travels of Marco Polo*, Book 2, Chapter 10

When the Song were completely defeated in 1276, Kublai Khan became the first foreigner to rule all of China. His next task was to integrate two completely different cultures under his rule as the Great Khan. The Great Wall of China separated two very different ways of life – the nomadic Mongolian herders north of the wall and the sophisticated Chinese cities south of the wall.

Kublai combined the best parts of Mongol and Chinese administration. He brought Confucian scholars into his court to help blend Mongol and Chinese traditions. He established a 14-member General Secretariat of trusted officials to recommend people for administrative positions, to draft new laws and enforce his laws.

Kublai Khan also set about to improve communication and trade throughout China. He expanded the *Yam* postal service, which was already the fastest in the world (see page 115). Two and a half million workers restored and extended China's Grand Canal. The canal and paved highway next to it became a 1600-kilometre link between north and south.

The Yuan emperor also provided help to peasant farmers, built schools, gave merchants low-interest loans and set a fair tax rate for the people. Kublai supported Asian arts and allowed people to follow their own religions. He encouraged the development of China and its people and provided the foundations of a grand empire.

The four khanates of the Mongol Empire, c. 1294

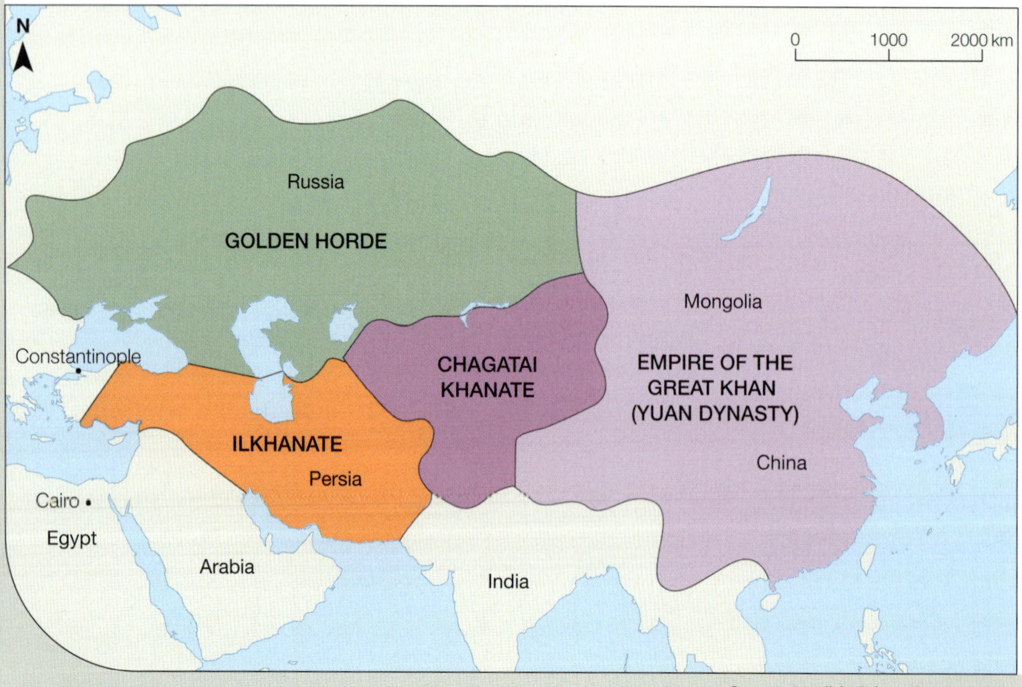

Source: Matilda Education Australia

Social hierarchy

To ensure the Mongols remained in control, Kublai established a social hierarchy based on race, similar to the social hierarchy used in the Jin Dynasty. The Yuan Dynasty contained four levels:

1. **Mongols** – the elite in society who were given the most important government jobs. Mongols did not pay tax and were given large estates of land to be worked by Chinese peasants.
2. ***Semuren*** – non-Chinese allies from the west of the empire were appointed as government officials. They did not pay tax.
3. ***Hanren*** – the inhabitants of northern China.
4. ***Nanren*** – the inhabitants of southern China, the former Song Dynasty.

Laws were different for each class. Mongols enjoyed lenient laws while laws for the *Hanren* and *Nanren* were harsh.

The death of Kublai Khan

Kublai's dream to rule the world would never be realised. He tried unsuccessfully on two occasions to invade Japan and also failed to take Java (modern-day Indonesia).

Kublai Khan sank into depression following the deaths of his favourite wife and only son. He became an alcoholic and obese. Kublai fell ill and died in 1294 at the age of 79.

The civil unrest that emerged after Mongke Khan's death had seen the empire begin to break into four separate **khanates**. This was reinforced by Kublai Khan's death, as none of his heirs could match his talent as a leader. Uprisings against Mongol rule in China soon began and, by 1368, the Yuan Dynasty had been overthrown.

Source 4

From 1260, the Mongol Empire had begun to fracture into four khanates that pursued their own interests.

Source 5

The Rebellion of Nayan

In 1287 Kublai Khan's uncle, a Mongol prince called Nayan, tried to overthrow him. Nayan and many other Mongols had become concerned with Kublai's passion for Chinese culture and disregard for traditional Mongol ways. Kublai Khan, aged 72, led his troops into battle against Nayan on the grasslands of Mongolia. He was too overweight to ride a horse, so he travelled in a palanquin atop four elephants. Nayan had pulled together an army of 300 000 to challenge his nephew. Kublai summoned a force of 460 000, surprising and quickly defeating Nayan's forces. According to Mongol law, as Nayan was royal he had to be executed in a bloodless way. Nayan was wrapped in carpet and trampled to death by horses and carts.

Learning ladder H3.8

Show what you know

1 What tactics did the Mongols use to defeat the Chinese at Xiangyang?

2 How did Kublai Khan integrate Chinese and Mongol ways in his empire?

3 From Source 4, write a physical description of the various sections of the Mongol Empire. Use geographic terms such as north, south, east, west, inland, coastal and landforms.

4 Look at Source 2. What offensive (attacking) and defensive strategies can you see?

Historical significance

Step 1: I can recognise historical significance

5 Rank these in order of historical importance:

Kublai Khan's achievements

Kublai Khan's death

Kublai Khan's name

Step 2: I can explain historical significance

6 Why was it important for Kublai Khan to conquer all of China?

Step 3: I can apply a theory of significance

7 Source 2: How important to the people of China was Kublai Khan's victory against the Song in 1276? Give examples of how people were affected by the new Mongol leader.

Step 4: I can analyse historical significance

8 Discuss what lessons a modern ruler can learn from Kublai Khan's achievements and the failures he suffered during his reign.

Historical significance, page 207

How did the Mongols expand trade?

Trade across Asia and with Europe greatly increased as the Mongol Empire provided safe travel and favourable conditions for merchants to travel long distances. Kublai Khan changed China's poor opinion of merchants.

Growth of trade under Genghis Khan

As Genghis Khan began to build the Mongol Empire, he quickly realised that his soldiers would need swords, spears and bows and arrows along with leather armour, saddles and other gear for horses. He needed to take trade to the next level.

To help trade grow in his expanding **empire**, Genghis provided protection for merchants, who could now travel safely from Europe in the west to Asia in the east. Merchants were exempt from taxes and were given loans to help establish their businesses.

The East–West trade routes along the ancient **Silk Road** linked Europe to Asia for the first time. Not only did goods move along this route, but also new ideas and inventions.

Continued growth under Kublai Khan

As a Mongol, Kublai Khan appreciated the importance of trade and held merchants in high regard. The Chinese, on the other hand, placed merchants at the lowest level of society; their attitude was that merchants did not make anything and only worked for their own gain.

Kublai Khan set out to change this view and improve merchants' social status. He invited merchants from India, Central Asia, Persia and Europe to come to his new seat of power.

Kublai Khan gave these foreign merchants a medallion called a *paiza* (see page 115) that allowed them to travel safely through the empire, be exempt from paying taxes, and to use the *Yam* relay stations to rest and eat.

Source 1

Trade routes in the Mongol Empire

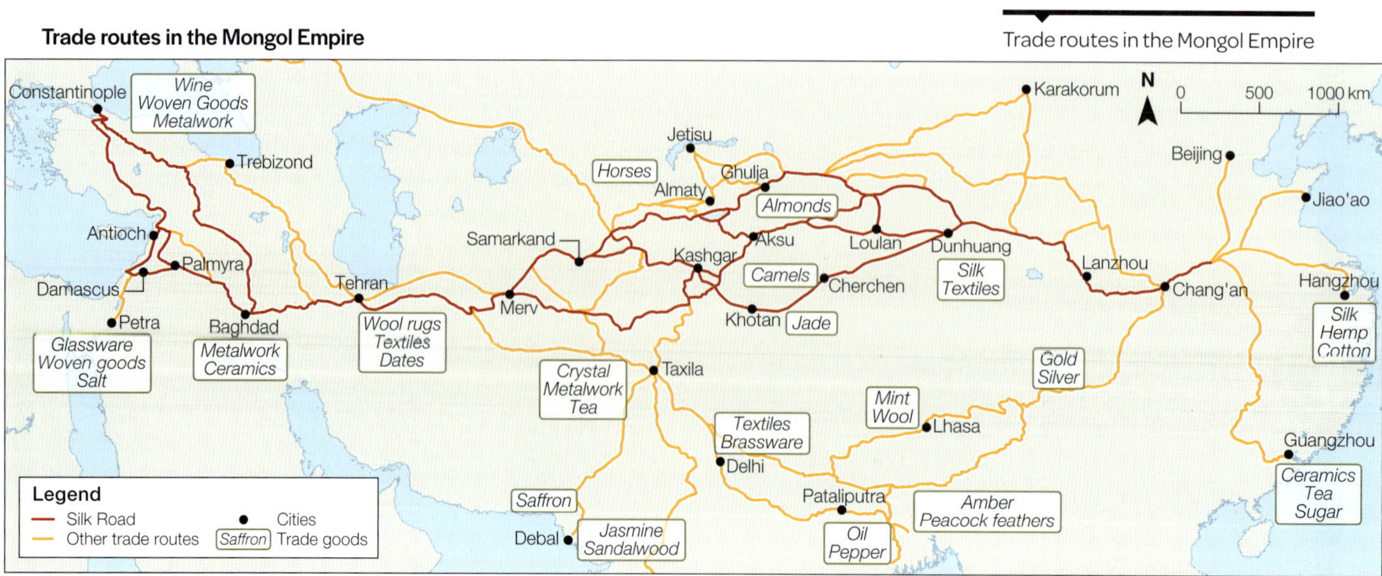

Source: Matilda Education Australia

Source 2

Kublai Khan and Marco Polo from the 2014 television series *Marco Polo*. In 1271, at only 17 years old, Marco Polo travelled along the Silk Road to China with his father and uncle. Marco Polo stayed in China for 17 years. He learned to speak Mongolian and Chinese and soon became a special representative of Kublai Khan. He was sent on special missions to China, India, Vietnam, Burma and Sumatra.

Marco Polo

Safe passage for merchants made the long travel across Asia easier for European traders like Marco Polo. Much of what we know about the Mongol Empire in the time of Kublai Khan comes from Marco Polo. He described the palaces, customs, inventions and trade in great detail.

Marco Polo was amazed to find that paper currency had replaced metal coins. He wrote:

> 'With these pieces of paper, they can buy anything and pay for anything. And I can tell you that the papers that reckon as 10 bezants [gold coins] do not weigh one ... all these pieces of paper are issued with as much solemnity and authority as if they were of pure gold or silver ... the chief officer deputed by the Khan smears the seal entrusted to him with vermilion, and impresses it on the paper, so that the form of the seal remains imprinted upon it in red; the money is then authentic. Anyone forging it would be punished with death.'

Source 3

The Travels of Marco Polo, Book 2, Chapter 24

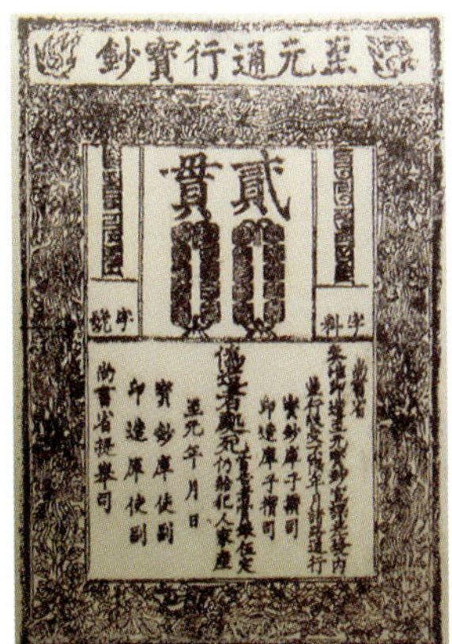

Source 4

A Yuan dynasty banknote, 1287 CE, from China/Mongolia.

Learning ladder H3.9

Show what you know

1. Why did Mongols become such great traders?
2. What benefits did the Silk Road bring?
3. How did the Chinese and Mongols differ in their views about merchants?
4. How did Genghis and Kublai Khan promote trade?
5. Looking at the map in Source 1, what were the main trading items in the westernmost parts of the Silk Road?
6. Source 3: What is Marco Polo impressed by in the excerpt? Use quotes from his own words in your answer.

Historical significance

Step 1: I can recognise historical significance

7. Put these facts in order from most to least historically important:

   Genghis Khan protects merchants.

   The Silk Road links Asia to Europe.

   Kublai Khan holds merchants in high regard.

   People start to use paper money.

Step 2: I can explain historical significance

8. Look at Sources 3 and 4. How significant was the move to paper money in the Mongol Empire? Is it still important today?

Step 3: I can apply a theory of significance

9. How deeply do you think everyday people would have been affected by the increase in trade between Asia, the Middle East and Europe after the Mongol Era? Are these trade networks still important today?

Step 4: I can analyse historical significance

10. How significant were Marco Polo's journeys from Europe to China? Give at least one reason that they were significant and another that they were not. Which do you think is more persuasive?

Historical significance, page 207

Why did the Mongol Empire fall?

When Kublai Khan died, the huge Mongol Empire could no longer be controlled by one Great Khan, so it split into four. The once invincible Mongol army proved to be no match for the Japanese and Vietnamese armies, as well as the uprisings from peasants in China. The Yuan Dynasty fell in 1368, ending the Mongol Empire.

Problems ruling the Mongol Empire

After his fierce Mongolian warriors stormed their way through Asia and Europe to establish the greatest **empire** the world had ever seen, Genghis Khan put in place laws and administration to undertake the difficult task of ruling a vast empire with many different cultures (see pages 114–15).

While Genghis Khan was in power, the Mongol Empire remained stable, but after his death in 1227 there were many changes of leadership and much fighting within his family over the leadership of the empire. It wasn't until his grandson Kublai took charge in 1264 that the Mongol Empire would have strong, consistent leadership.

Source 1

In 1288, the Vietnamese defeated the Mongols at the Battle of Bach Dang. They destroyed or captured 400 Mongol ships and sent the massive army of more than 300 000 Mongols into retreat. This was the third and final attempt by Kublai Khan to take Vietnam.

Failed Mongol invasions

Kublai Khan's feared cavalry, which had earlier taken all of China, now failed to take much ground in the rest of south-east Asia. His cavalry struggled to make much progress in Vietnam's rugged, tropical jungles. In three campaigns, the Mongols only managed to take a small amount of ground at great cost financially and with the loss of tens of thousands of soldiers.

The Mongol army twice tried to take control of Japan, but on both occasions were defeated by the **samurai** (see pages 142–45). In the second attack in 1281, a typhoon wiped out most of the 4400 ships in the Mongol fleet.

The serious defeats of the Mongols by the Vietnamese and Japanese made other Asian states more confident about facing the Mongols in battle. The Mongols were no longer as feared as they were under the rule of Genghis Khan.

The fall of the empire

From 1260, the Mongol Empire had begun to split into four **khanates**. After Kublai died in 1294, his heirs fought for power and the khanates remained divided. Each khanate operated independently and blended local customs and religions with those of the Mongols. The khanates were:

- the Golden Horde in the north-west
- Chagatai in Central Asia
- Il khanate in the south-west
- the Yuan Dynasty in the east.

In 1338, massive flooding hit farms along the Yellow River. Peasants who had been devastated by the floods were called on to repair dam walls. More and more Chinese began to rebel against the Mongol government.

The Yuan Dynasty ended in 1368, when the last Mongol emperor was driven out of China. Following the loss of soldiers from failed invasions into south-east Asia and the division of the Mongol Empire, the Mongol army could no longer hold off the growing number of uprisings.

The Ming Dynasty reclaimed China, thereby ending the Mongol Empire. The other khanates continued, with Chagatai the last to fall in 1658.

Learning ladder H3.10

Show what you know

1 Why weren't the Mongols able to secure victory in Vietnam or Japan?

2 Why did the Mongol Empire eventually fall?

3 Look at Source 1. What can you learn from the Battle of Bach Dang *just* from looking at this picture? Include discussion of military equipment, tactics and the outcome.

4 Why do you think that the 1338 floods made Chinese people want to rise up against the Mongols? Use your knowledge from this whole chapter to help you answer this question.

Chronology

Step 1: I can read a timeline

5 Look at the timeline on pages 100–01. How long did the Yuan Dynasty rule in China?

Step 2: I can place events on a timeline

6 Place these events in chronological order:

1294 Kublai Khan dies

1259–1263 Civil war in the Mongol Empire

1271 Yuan Dynasty begins in China

1264 Kublai Khan becomes Mongol ruler

Step 3: I can create a timeline using historical conventions

7 Take at least five of the dates from this section and put them in a timeline, using correct historical conventions. Refer to the History How-to section on page 195 to see what those conventions are.

Step 4: I can distinguish causes, effects, continuity and change from looking at timelines

8 Looking at the timeline on pages 100–01, create and complete a table like the following:

Cause		Effect	
Date	Event	Date	Event

Chronology, page 194

Masterclass

Learning Ladder

Work at the level that is right for you or level-up for a learning challenge!

Source 1

Creator unknown, manuscript from Firdawsi's *Shahnama* (c. 1300–25), ink and watercolour on paper, The British Museum, England.

Step 1

a **I can read a timeline**

Look at the timeline on page 100–01.

i How long did the Mongol Empire last?

ii How long ago did Genghis Khan become ruler of the Mongols?

iii How many years were there between the beginning of the Mongol Empire and the end of the Yuan Dynasty?

b **I can determine the origin of a source**

From when and where is Source 1?

c **I can recognise continuity and change**

Which of these things has changed the most, and which has changed the least from the Mongol Age until now?

i The empire ruling Asia

ii Horses used in Mongolia

iii Nomadic farming in Asia

iv Bows and arrows for weapons

v Female horses produce milk

d **I can recognise a cause and an effect**

Put these causes and effects in the correct place in the diagram.

i Mongol warriors could shoot with bows and arrows while on horseback.

ii The Mongol army was over 100 000 warriors strong.

iii The Mongols conquered many other armies.

iv Many people feared the Mongol army.

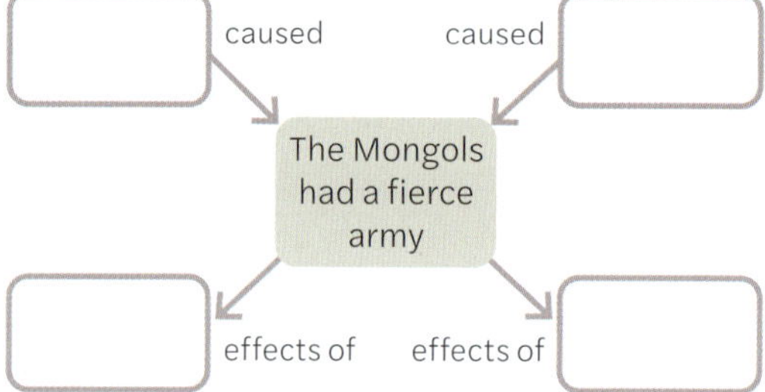

e **I can recognise historical significance**

Rank these aspects of History from most to least important:

i clothing ii trade iii people's names iv war.

Step 2

a **I can place events on a timeline**

Put these events in order from earliest to most recent (note that there are six dates to include):

- The Khwarezmid Empire was destroyed in 1221.
- 15 years after the destruction of the Khwarezmid Empire, the Mongols invade Korea.
- 1264, Kublai becomes Great Khan. He died 30 years later.
- The Ming Dynasty overthrows the Mongol Yuan Dynasty in 1368. It had stood for 97 years.

b I can list specific features of a source

Describe in detail what is occurring in the bottom panel of Source 1. Who do you think the figures represent and what are they doing?

c I can describe continuity and change

Look at Source 1 on page 102. Conduct research online to determine if some Mongols are still nomadic today, and if so, note any similarities or differences between ancient and modern nomads. Has the nomadic population increased or decreased over time?

d I can determine causes and effects

i What was the effect of the Mongols having the *Yam* communication network?

ii Why did the Mongol army fail to conquer Japan?

e I can explain historical significance

Why were Mongolian horses so important?

Step 3

a I can create a timeline using historical conventions

Using the dates from Step 2a, place the events and developments on a timeline that obeys historical conventions (listed on page 195).

b I can find themes in a source

Look at the image in Source 1. What do you think the text is about? Give evidence from the picture in your answer.

c I can explain why something did or did not change

Why is Asia a series of nation states now, rather than one big empire like it was during the 13th century?

d I can explain why something is a cause or an effect

Why did the Mongols give promotions to people in the army based on talent instead of noble birth?

e I can apply a theory of significance

Think about the death of Genghis Khan.

i How important was it to people at the time?

ii How many were affected by his death?

iii How deeply were those people affected?

iv For how long were people affected by his death?

v How important is his death today?

Step 4

a I can distinguish causes, effects, continuity and change from looking at timelines

Look at the timeline on pages 100–01. What is the link between the Mongol Empire and the Yuan Dynasty? Reference the timeline and the dates on it in your answer.

b I can use my outside knowledge to help explain a source

Source 1: Consider the time and location this painting was created. You may need to research where the modern country of Iran is. Who would have controlled this area of land at the time it was created? What kind of agriculture do you think is taking place, based on your knowledge of who would have lived in this area at the time?

c I can analyse patterns of continuity and change

The Mongol Empire was at its largest under Kublai Khan, before it slowly disintegrated to nothing. Why?

d I can analyse cause and effect

List the various effects of the increase in trade under the Mongols.

i Effect on everyday people

ii Effect on future people

iii Effect on the economy at the time

iv Effect on the environment

v Effect on the power of the empire

e I can analyse historical significance

Kublai Khan merged Mongolian and Chinese ways in the Yuan Dynasty. Copy and complete the following table.

Importance to History	Importance to Mongolians at the time	Importance to Chinese people today	Importance to sharing of ideas	Importance to food

Masterclass

Step 5

a **I can describe patterns of change**

Many historians discuss the pattern of 'rise and fall' of empires. Describe the rise and fall of the Mongol Empire with reference to the timeline on pages 100–01 and other relevant dates from this chapter.

b **I can use the origin of a source to explain its creator's purpose**

Think about when and where Source 1 was produced. Think about the activity you can see in the bottom panel, and the fact that there is writing above it. With this in mind, why do you think Source 1 was created?

c **I can evaluate patterns of continuity and change**

The Mongol Age caused major upheaval in the 13th century but was over soon after that. Does this mean we shouldn't bother investigating this period? Give reasons to back up your answer.

d **I can evaluate cause and effect**

Some historians believe the Mongols were able to conquer so much land because they were so brutal and everyone was afraid of them. Others think it was because of their superior military technology, such as cavalry archers. Which do you think is more likely? Back up your answer with historical evidence.

e **I can evaluate historical significance**

The Mongol era happened about 800 years ago. Even though they achieved major feats, their reign was too long ago to mean anything today. Do you agree with this statement? Give historical examples in your answer.

Historical writing

1 Structure

Imagine you are given the essay topic, 'Why do historians believe the Mongols were more than just bloodthirsty conquerors?'. Write an essay plan for this topic. Include at least three main paragraphs.

2 Draft

Using the drafting and vocabulary suggestions on page 214, draft a 400–600 word essay responding to the topic.

3 Edit and proofread

Use the editing and proofreading tips on page 215 to help edit and proofread your draft.

Historical research

4 Organise and present information

Imagine you are doing a large research project: 'Mongol technology and inventions'. Write a contents page for this project. There should be an introduction, conclusion, at least four main sections and many subsections. Number your chapters.

Capstone

How can I understand Mongol expansion?

In this chapter, you have learnt a lot about Mongol expansion. Now you can put your new knowledge and understanding together for the capstone project to show what you know and what you think.

In the world of building, a capstone is an element that finishes off an arch, or tops off a building or wall. That is what the capstone project will offer you, too: a chance to top off and bring together your learning in interesting, critical and creative ways. You can complete this project yourself, or your teacher can make it a class task or a homework task.

mea.digital/GHV8_H3

Scan this QR code to find the capstone project online.

H4

Japan under the *shoguns*

How can I understand Japan under the *shoguns*?

Japan is a fascinating and ancient nation. Situated on our doorstep, with close trading ties to Australia, Japan has a long and rich history of visual, literary and performing arts. The country endured many political highs and lows. It was a divided land before three great *shoguns* united the nation, despite many civil wars. Eventually the country was united in peace under the Tokugawa *Shogunate*.

learning ladder

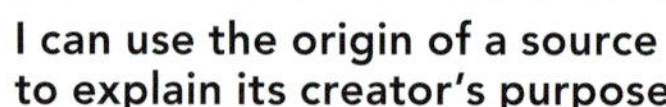

Step	Chronology	Source analysis	Continuity and change
step 5	**I can describe patterns of change** I read timelines and see the 'big picture'. I group timeline events about Japan under the *shoguns* and see if they show patterns of change. I know typical historical patterns to check for.	**I can use the origin of a source to explain its creator's purpose** I combine knowledge of when and where a source was created to answer the question, 'Why was it created?'	**I can evaluate patterns of continuity and change** I answer the question, 'So what?' about patterns of continuity and change. I weigh up ideas and debate the importance of a continuity or a change.
step 4	**I can distinguish causes, effects, continuity and change from looking at timelines** I read timelines of Japan and find events linked by cause and effect. I also find things that are the same or different from then until later times.	**I can use my outside knowledge to help explain a source** I have enough outside knowledge about Japan under the *shoguns* to help me explain a source.	**I can analyse patterns of continuity and change** I see beyond individual instances of continuity and change in Japan under the *shoguns* and identify broader historical patterns, and I explain why they exist.
step 3	**I can create a timeline using historical conventions** When given a set of events, I construct a historical timeline using the dates, and using correct terminology, spacing and layout.	**I can find themes in a source** I look a bit closer into a source and find more than just features. I find themes or patterns in the source.	**I can explain why something did or did not change** I answer the question, 'Why?' something changed or stayed the same between historical periods in Japan.
step 2	**I can place events on a timeline** When given a list of events and developments about Japan under the *shoguns*, I put them in order from earliest to latest, the simplest kind of timeline.	**I can list specific features of a source** I look at a medieval Japanese source and list detailed things I can see in it.	**I can describe continuity and change** I have enough content knowledge about two different historical periods to recognise what is similar or different about them, and can describe it.
step 1	**I can read a timeline** I read timelines with medieval Japanese events on them and answer questions about them.	**I can determine the origin of a source** I can work out when and where a medieval Japanese source was made by looking for clues.	**I can recognise continuity and change** I recognise things that have stayed the same and things that have changed in Japan.

Warm up

Source 1

Hanging scroll painting [Gan Rei, *Mt Fuji from Miho no Matsubara* (c. 1840–1883), painted silk, The British Museum.]

Cause and effect	Historical significance
I can evaluate cause and effect I answer the question, 'So what?' about cause and effect. I weigh up ideas and debate the importance of a cause or an effect.	**I can evaluate historical significance** I answer the question, 'So what?' about things that are supposedly historically important. I weigh up events and cast doubt on how important things are.
I can analyse cause and effect I don't just see a cause or an effect as one thing. I determine the factors that make up causes and effects.	**I can analyse historical significance** I separate out the various factors that make something historically important.
I can explain why something is a cause or an effect I can answer 'How?' or 'Why?' a cause led to an effect in medieval Japan.	**I can apply a theory of significance** I know a theory of significance. I use it to rank the importance of medieval Japanese events.
I can determine causes and effects Applying what I have learnt about Japan under the *shoguns*, I can decide what the cause or effect of something was.	**I can explain historical significance** I answer the question, 'Why?' about things that were important in Japan under the *shoguns*.
I can recognise a cause and an effect From a supplied list, I recognise things that were causes or effects of each other in Japan under the *shoguns*.	**I can recognise historical significance** When shown a list of things from Japanese history, I can work out which are important.

Chronology

1 1603 CE marked the beginning of the Edo Period, in which Tokugawa *shogun* controlled a unified Japan. Japan isolated itself from the rest of the world until 1853 CE when Admiral Perry visited Japan and forced it to open up to foreign trade. Which of these events occurred first?

Source analysis

2 Examine Source 1 and its caption. The source was painted by Japanese artist Gan Rei, and depicts Mount Fuji. Looking at the coastline and the boats in the source, what can you say about when it was painted?

Continuity and change

3 Still looking at Source 1, how have the landscape and coastline around Mount Fuji changed over time? What clues in the image led you to your answer?

Cause and effect

4 Which of these things caused the end of the **samurai**?
 a The Mongol invasions of Japan
 b Emperor Meiji banned them
 c The end of the Edo Period

Historical significance

5 Looking at the timeline on pages 132–33, why was the Edo Period significant?

Why learn about the history of Japan?

Japan's legendary Heian Period was the birthplace of its astonishing artistic legacy: *noh* theatre, dance, music and poetry. Lady Murasaki wrote the world's first novel more than 1000 years ago. During the Edo Period, Japan isolated itself from outside influences, helping its unique and fascinating culture to flourish. Learning about this captivating culture also helps us to understand its huge influence on world affairs in the first half of the 20th century.

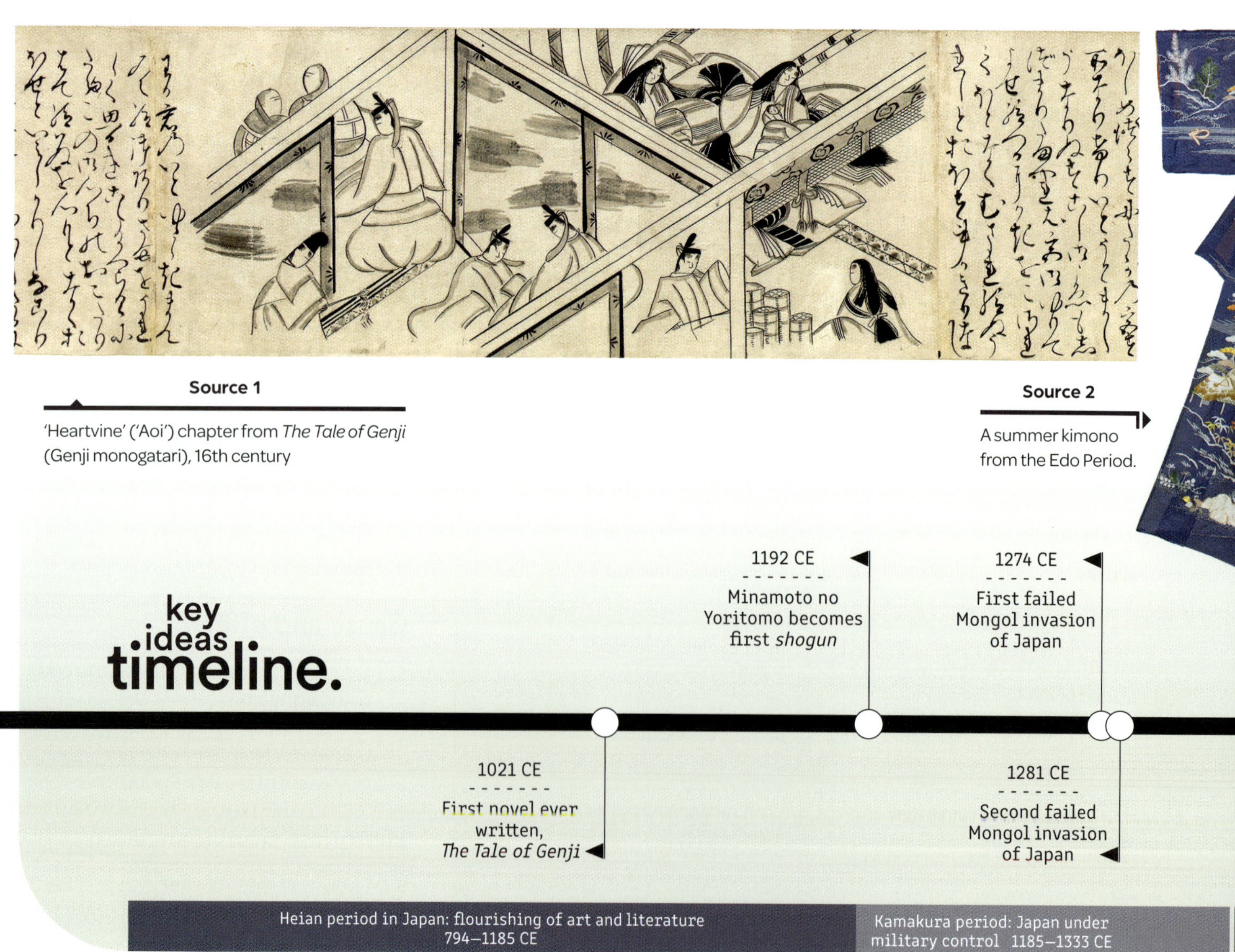

Source 1

'Heartvine' ('Aoi') chapter from *The Tale of Genji* (Genji monogatari), 16th century

Source 2

A summer kimono from the Edo Period.

Source 3

A modern Japanese garden

Source 4

Colour lithograph of Toyotomi Hideyoshi (1536-98)

Learning ladder H4.1

Chronology

Step 1: I can read a timeline

1 Look at the timeline.
   a What is the longest period on the timeline?
   b How much longer after Christianity was introduced to Japan did Admiral Perry arrive?
   c How long ago did Admiral Perry first arrive in Japan?

Step 2: I can place events on a timeline

2 Put these dates in reverse order from most recent to oldest:

First novel ever written

Korea was invaded by Japan

Christianity was introduced to Japan

Edo period begins

Step 3: I can create a timeline using historical conventions

3 Using correct conventions, create a timeline with the events from question 2 on it.

Step 4: I can distinguish causes, effects, continuity and change from looking at timelines

4 How are the events of 1573 and 1590 related? Was there a relationship between Nobunaga and Hideyoshi? (See pages 148–49.)

HOW TO

Chronology, page 194

1549 CE
Christianity introduced to Japan

1573 CE
Oda Nobunaga overthrows Ashikiga *shogunate* and begins unification of Japan

1590 CE
Toyotomi Hideyoshi completes unification of Japan

1592 CE
First failed Japanese invasion of Korea

1853 CE
Admiral Perry visits Japan and forces the Japanese to open up to foreign trade

1905 CE
Japan wins first battle with Western power, defeating Russia in a naval encounter

Sengoka period: warring states in Japan 1467–c.1600 CE

Muromachi period: Ashikiga *shoguns* control Japan 1336–1573 CE

Edo period: *shogun* controls an isolated Japan 1603–1867 CE

The Meiji Restoration 1868–1914 CE

What was life like before the *shoguns*?

Clans fought over the small areas of fertile farmland in the mountainous country of Japan. Eventually the Yamato clan controlled Japan and their leader became the Emperor. However, he still relied on the support of lords and their warriors.

Geography of Japan

Japan is an **archipelago** – a chain of more than 3000 islands that are the peaks of underwater mountains. Travel over mountainous land was difficult, so small communities, isolated and separate from one another, developed in valleys and along the coast.

Japanese clans

Initially, instead of one Japanese culture, small, isolated societies made up of different ***uji*** developed across Japan. The *uji* were family groups related by blood or by marriage. Each *uji* was led by a chief who received a tax of food in return for protection. Wars between rival *uji* broke out regularly over access to the scarce farmland.

Around 300–400 CE the powerful Yamato tribe controlled the Osaka plain, one of the few large agricultural areas in Japan. As the Yamato's military strength grew, other clan leaders promised loyalty to the Yamato chief and recognised him as their supreme leader.

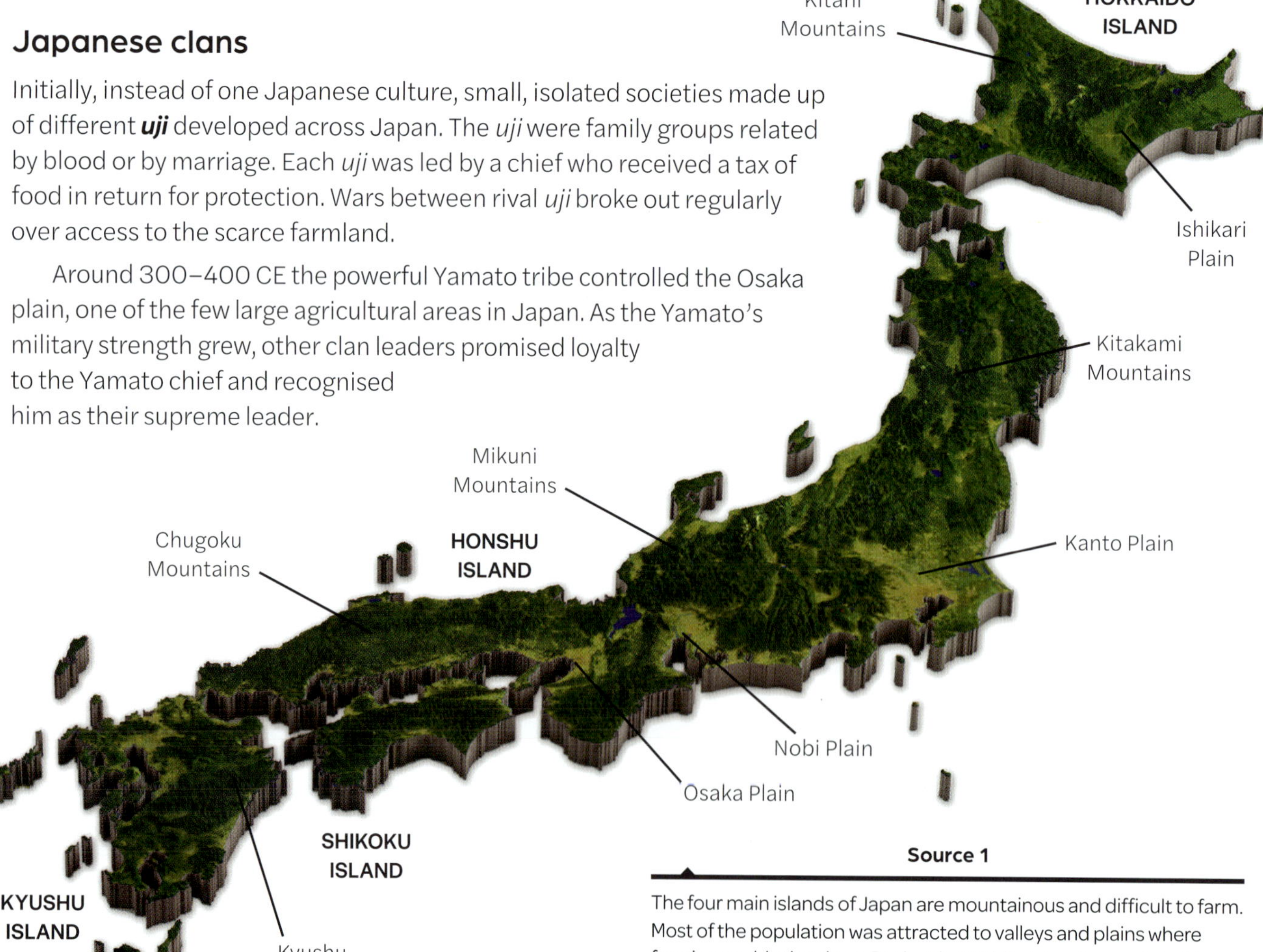

Source 1

The four main islands of Japan are mountainous and difficult to farm. Most of the population was attracted to valleys and plains where farming could take place. During the Middle Ages, the capital of Japan was first at Nara and then at Kyoto, both located on the Osaka Plain.

Source 2

The Kyoto Imperial Palace was the ruling palace of the Emperor of Japan between 794 and 1868. The Imperial Palace was reconstructed in 1855 after fire destroyed the previous palace.

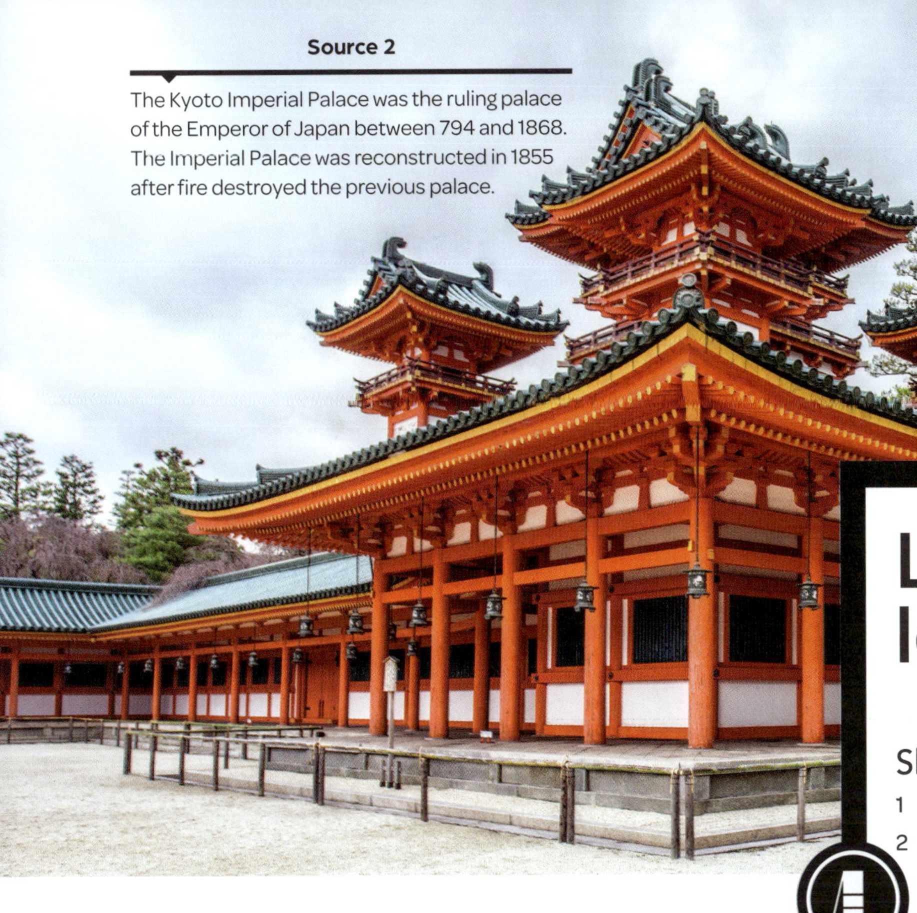

Between the 4th and 9th centuries, Japan's many clans gradually came to be unified under a central government, with the head of government an **emperor** or empress.

All land belonged to the emperor and **peasants** paid taxes to him in goods such as rice or in labour. Rule over Japan's mountainous landscape and different clans was difficult and required strong military support and administration.

The Heian Period

In 794, Emperor Kammu moved the official capital from Nara to Heian-kyo (modern-day Kyoto). This marked the start of the Heian Period that would last for nearly 400 years. The emperor became increasingly isolated from his subjects. He lived in a luxurious palace known as the Imperial Court and appointed advisers to undertake the day-to-day running of Japan.

The emperor granted land to nobles known as ***daimyo*** as a reward for helping him maintain control. The *daimyo* employed warriors to protect their land from rivals. These warriors developed their own social class – the **samurai** (see pages 142–45). Emperor Kammu appointed his leading samurai to maintain control throughout the country. This office became known as ***shogun***, and would soon come to rule Japan for hundreds of years.

Learning ladder H4.2

Show what you know

1 Why did people fight over land?

2 How did the Yamato chief become emperor?

Cause and effect

Step 1: I can recognise a cause and an effect

3 Match the following causes and effects:

Clan chiefs would protect peasants	Other leaders pledged loyalty to their chief
Daimyo helped the ruler maintain control	Families would pay a tax of food or labour
Only 15 per cent of Japan can be farmed	The emperor granted *daimyo* land
Rebel *daimyo* clans challenged the emperor	The emperor used *samurai* for protection
The Yamato clan became powerful	There was competition for land

Step 2: I can determine causes and effects

4 Source 1: What effect did the lack of good farmland have?

Step 3: I can explain why something is a cause or an effect

5 What was the effect of the emperor's need for protection and control?

Step 4: I can analyse cause and effect

6 What was the cause of Japan developing as separate *uji* rather than a single culture? How did the Yamato *uji* unite a number of clans?

Cause and effect, page 203

Why did the *shoguns* take control?

Powerful noble families known as the *daimyo* had always influenced the imperial family. In 1192, the role of *shogun* ensured that the *daimyo* and their samurai armies would formally control Japan.

Power behind the throne

During the Heian Period from 794 to 1185, Japan was ruled by an **emperor** from the **Imperial Court** in Heian-kyo (modern-day Kyoto). However, the real power was held by the powerful Fujiwara ***daimyo*** clan. They were a noble family that controlled most of the key roles in the Imperial Court and were governors of many of Japan's provinces. Many Fujiwara women married emperors.

When an emperor died and his son was too young to rule, a member of the Fujiwara clan would rule as **regent** until the child became old enough. In 858 Fujiwara Yoshifusa became Japan's first regent for his grandson, the nine-year-old emperor. Instead of handing over power when the young emperor became old enough to rule, Yoshifusa stayed in power. This began a period of many centuries where military leaders ruled Japan.

Wrestling for power

Towards the end of the Heian Period, the struggle for power between the largest *daimyo* clans intensified. The *daimyo* class had become very powerful through:

- receiving land from the emperor in return for their loyalty, which made the *daimyo* very wealthy
- developing armies of **samurai** (see pages 142–45) to protect them and expand their territories
- influencing the imperial family through marriage and acting as regents for young emperors.

In 1068, the Emperor Go-Sanjo reduced the power of the Fujiwara clan by appointing rival Minamoto clan members to important roles in government. Go-Sanjo's mother and wives were from the Minamoto clan, so he and his heirs were loyal to them.

In 1180, the Taira clan took control of government when Taira no Kiyomori, the grandfather of the two-year-old emperor, ruled as regent in his name. The Minamoto clan rebelled against the Taira clan, leading to a **civil war** that lasted three years. In 1183 the young emperor and the Taira clan were forced to flee the Imperial Palace in Kyoto.

The first *shogun*

In 1183 the Minamoto clan placed Go-Toba, a three-year-old member of the imperial family, as emperor. The Japanese emperor was simply a puppet, while the strong Minamoto clan ruled behind the throne. In 1192, Emperor Go-Toba appointed Minamoto no Yoritomo, the head of the Minamoto clan, as the supreme leader of the Japanese armed forces. The title of this role was the ***shogun***.

Yoritomo set up his centre of administration, known as a ***shogunate***, in Kamakura while the emperor stayed in Kyoto. The real governing power quickly shifted to the *shogunate* and, for the next 700 years, Japan would be ruled by *shoguns*, with emperors becoming ceremonial and religious leaders.

Source 1

Minamoto no Yoritomo becomes *shogun* in 1192 and sets up his *shogunate* in Kamakura while the emperor stays in Kyoto. *Shoguns* would rule Japan for the next 700 years. [Painting by H. M. Burton from Hutchinson's *History of the Nations*, published 1915.]

The Kamakura *shogunate*

The *shogunate* in Kamakura ruled Japan for 150 years, with different clans taking the role of *shogun*. Minamoto no Yoritomo introduced **feudalism** to Japan (see pages 138–39). The feudal system set out rights and responsibilities for each class in society. For example, the *daimyo* lords were granted land in return for supporting the *shogun*. The *daimyo* gave land to samurai warriors in return for their protection.

During the Kamakura *shogunate*, the Mongol army twice tried to take control of Japan under the leadership of Kublai Khan (see pages 118–21). The *daimyo* were called on to supply their samurai armies to the *shogun* to protect Japan. On both occasions the Mongols were defeated.

Learning ladder H4.3

Show what you know

1 How did the first *shogun* take control of Japan?

2 Source 1: Who was the first *shogun*?

3 What does 'ruled behind the throne' mean?

4 Describe the difference between the *shogun* and a samurai, from looking at Source 1.

Historical significance

Step 1: I can recognise historical significance

5 Which of these figures was most historically significant?
- Emperor Go-Toba
- Minamoto no Yoritomo

Step 2: I can explain historical significance

6 Explain why Fujiwara Yoshifusa's decision to retain power instead of handing it over to his grandson was historically significant.

Step 3: I can apply a theory of significance

7 Use Partington's model of significance on page 207 to assess Emperor Go-Toba. Was he significant?

Step 4: I can analyse historical significance

8 How did the *daimyo* class become so powerful?

HOW TO

Historical significance, page 207

How was Japanese society organised?

In the Middle Ages, Japanese society was organised under the feudal system, where all people had rights and obligations. Society was organised in a strict hierarchy – from the *shogun* and rich *daimyo* down to the lowly merchants. The wealthy classes offered land and protection to peasants in return for their labour.

Feudalism

Under the rule of the *shoguns*, **feudalism** in Japan was similar to the model in Europe (see pages 66–67). Feudalism required a strict hierarchy of social classes organised from most to least important. People knew what was expected of them and, when you were born into a class, that is where you stayed.

Feudalism worked on a system of rights and responsibilities for each class in society. The social system became an important part of Japanese society after the *shogun* Tokugawa Ieyasu (1542–1616) divided the country into provinces. Each of these provinces was run by a lord called a *daimyo*.

The peasant *shogun*

Toyotomi Hideyoshi (see pages 148–49) broke the strict class rules of Japanese society. He was born into a peasant family in 1536 and quickly rose to become a samurai, *daimyo* and eventually the ruler of all Japan in 1585.

Although Hideyoshi ruled as Japan's military dictator, he did not take the title of *shogun*. This title was reserved for descendants of Minamoto no Yoritomo, the first *shogun*.

Source 1

Samurai who were loyal to their *daimyo* were rewarded with land to build grand castles on (see pages 146–47). Himeji Castle was built by the samurai Akamatsu Norimura in 1333.

Emperor
The emperor was at the highest level of Japanese society. He lived in the Imperial Palace in Kyoto. The emperor was the spiritual head of the country, and after 1185 his role was mainly ceremonial.

Shogun
The *shogun* was the de facto leader of Japan, and owned a quarter of the land in the country. *Shoguns* ruled Japan from 1192 until 1868. The *shogun* commanded the military forces and ran the government, known as the *shogunate*. The *shogun* was formally appointed by the emperor.

Daimyo
Land not owned by the *shogun* was divided into regions ruled by powerful lords known as *daimyo*. Each *daimyo* hired an army of samurai warriors to protect his family and property. These armies were made available to the *shogun* when he needed them. The *daimyo* received taxes and labour from peasants living in his province.

Samurai
The samurai were Japanese warriors who swore an oath of loyalty to serve and protect their *daimyo*. They followed a code of conduct called the *Bushido* that highlighted honour, courage, skill in martial arts and loyalty to the *daimyo*. ***Ronin*** were samurai with no allegiance to a *daimyo*.

Peasants
Peasants were farmers or fishers, and made up 80 per cent of the population. Japanese peasants were considered a higher class than they were in medieval Europe (see pages 66–67) because they produced food, which was needed by people of all classes.

Merchants and craftspeople
Merchants traded goods and lent money. They had a low social status because they were seen to profit from the hard work of others. Despite receiving little respect, merchants were often wealthy. Craftspeople made goods and tools, but received less status because their goods were not regarded as essential for life.

Outcasts
There were two groups of outcasts at the bottom of society. The *eta* performed tasks considered distasteful such as tanning hides, butchering animals, preparing bodies for burial and executing prisoners. The *hinin* were seen as undesirable people, such as street cleaners, beggars, former criminals and street performers.

Source 2

The groups in feudal Japan's society

Hideyoshi had himself adopted by the Fujiwara clan and used their title of *kampaku*. He installed a teenager as emperor and took control behind the scenes. Hideyoshi was challenged by *daimyo* who believed his lack of nobility did not give him the authority to lead, but he crushed their challenges with his military might.

Learning ladder H4.4

Show what you know

1 What were samurai required to do?
2 What did peasants get in return for working and paying rent?
3 Which types of people were ranked lower than peasants?
4 Why did Toyotomi Hideyoshi never take the title of *shogun* despite becoming the ruler of Japan in 1585?
5 From looking at Source 1, what features of the castle would have helped people defend against invaders?

Continuity and change

Step 1: I can recognise continuity and change

6 Rank these in order from most changed to least changed from medieval Japan to the present. You may need to do some online research about modern Japan to help you.

The leadership of Japan
The role of the samurai
The use of castles
Eating rice
Clothing

Step 2: I can describe continuity and change

7 How did life change from before the *shoguns* (pages 134–35) to the time of feudalism?

Step 3: I can explain why something did or did not change

8 How did Hideyoshi break Japanese class rules?

Step 4: I can analyse patterns of continuity and change

9 Why did feudalism promote stability and continuity in medieval Japan?

Continuity and change, page 200

What was the role of women in medieval Japan?

Women were expected to serve men and run and protect the household. Some single women trained as entertainers known as geishas.

A lifetime of service

Women in medieval Japan were expected to serve men throughout their lifetime. A girl obeyed her father; a wife obeyed her husband; and if a woman became a widow, she obeyed her son. A married woman was expected to bear her husband a son to carry on the family name.

Women from higher social classes had the least freedom. Their husband was chosen as a business deal between families to create an alliance and improve their social status and wealth. While women in lower social classes faced more hardships, they had more freedom to choose their partners than noblewomen. In addition, women in medieval Japan were not allowed to divorce or remarry if widowed.

Samurai women

Samurai women were trained to fight just as the male samurai were (see pages 142–45), but used small daggers or poles with blades at the end called *naginata*. They were prepared to do whatever it took to protect their family and honour the family name. Samurai women sometimes fought alongside men, like the warrior Tomoe Gozen, described in Source 1 in the *Tale of Heike*, which was compiled in 1240 from the oral accounts of monks.

Samurai women rarely fought in battles. With their samurai husbands away for long periods in combat, it was up to the women to protect their homes and children and even to take revenge where required. Other roles of samurai women were to oversee the crops and food supplies and to manage the servants.

Geisha women

With white make-up, red lips and silk kimonos, **geisha** are still associated with Japan today. Since around 600 CE, geishas have been entertainers, highly trained in Japanese dance, poetry and playing musical instruments.

Geishas were highly respected in Japanese society and were even used by the powerful ***daimyos*** to spy on or even assassinate other lords. Before becoming a geisha, a young girl was a ***maiko***. She would begin her training to become a geisha around the age of six. The *maiko* would learn how to sing, play instruments, talk with men, conduct the traditional tea ceremony and perform traditional dances.

Geisha women were expected to be single. They remained unmarried and were viewed as independent workers who earned their own money from entertaining clients. Any geisha who did marry had to leave the geisha house.

'She handled unbroken horses with superb skill; she rode unscathed down perilous descents. Whenever a battle was imminent, Yoshinaka sent her out as his first captain, equipped with strong armour, an oversized sword, and a mighty bow; and she performed more deeds of valour than any of his other warriors.'

Source 1

Tale of Heike, written before 1330 CE. Translation: Helen Craig McCullough, *Heike Monogatari*, Stanford University Press, 1988, p. 291

Source 2

A 19th-century Japanese artist has painted a scene from the Battle of Awazu in 1184: Tomoe Gozen kills her enemy while on horseback. She fought in the conflicts that led to the first *shogunate* in Japan. [Yōshū Chikanobu, *Tomoe Gozen with Uchida Ieyoshi and Hatakeyama no Shigetada* (1899), woodblock print.]

Source 3

Geisha are Japanese women who earn a living through entertaining mainly male audiences. They were highly regarded in medieval Japanese society. Geisha wear silk kimonos, and distinctive white and red make-up:

1 Apply white face-powder paste to your face, chest and neck
2 Draw in eyebrows with pencil and apply red eye shadow to corner of eyes
3 Apply red lipstick as shown in these images.

Learning ladder H4.5

Show what you know

1 What was the purpose of marriage in the higher social classes?
2 How were male and female samurai different?
3 What skills did geisha have?
4 Looking at Source 1, what can you learn about the skills and equipment female samurai had?
5 What was the key role played by samurai women in medieval Japan and what skills were needed to perform this?
6 Contrast the lives of geisha and samurai women in medieval Japan.

Continuity and change

Step 1: I can recognise continuity and change

7 What evidence is there that women had lower status than men in medieval Japan?

Step 2: I can describe continuity and change

8 Who are the geisha and how long have they been a part of Japan's culture?

Step 3: I can explain why something did or did not change

9 Why do you think women had to stop being geisha once they got married?

Step 4: I can analyse patterns of continuity and change

10 Why has the role of women in most societies changed so much since medieval times?

HOW TO

Continuity and change, page 200

How did the samurai help keep control in feudal Japan?

During the time of the *shoguns* in Japan, the warrior class known as the samurai were a powerful force. The word samurai means 'one who serves'. From around 900 to 1600 CE, these highly trained and loyal warriors enforced law and order for their *daimyo* and fought for the *shogun*.

The mighty samurai

As part of the **feudal system**, the **samurai** gave their services as warriors to the ***daimyo*** in return for grants of land. Samurai not only kept the peace in their lord's lands, but they were also called on to protect the lands from invading forces or attack a rival *daimyo*.

The samurai were very good at fighting and highly skilled in **martial arts**. For the samurai, a battle was a matter of personal and family honour. When two opposing samurai fought, they charged at each other on horseback, slashing with their long swords. When they dismounted, they began hand-to-hand combat with short knives. The loser was beheaded. The heads of enemy officers were presented to their commanders and displayed in public to remind people of the samurai's power.

To be a samurai, you had to be born into a samurai family. Rigorous training began around age five, where boys learned fighting and military strategy along with reading, writing, calligraphy, poetry and **Zen Buddhism** (see page 150).

One of the key martial arts learned was ***kendo*** – with this method students learned fast movements and sword techniques. Archery was practised using live animals such as dogs as targets and unarmed combat often led to broken bones. Samurai girls were also trained in martial arts to protect their homes (see pages 140–41).

Bushido

Samurai were expected to live by a strict ethical code called ***Bushido***, the Way of the Warrior. The code stressed loyalty to one's master, respect for superiors, ethical behaviour and courage at all times. The most respected samurai warrior was the one that severed the most heads in battle.

The *Bushido* taught the samurai to choose his own death rather than to suffer any dishonour in defeat. Samurai warriors often chose the ritual suicide known as ***seppuku*** rather than facing humiliation at the hands of the enemy. The *Bushido* required complete self-discipline, even when facing death (see Source 2).

Source 1

Rigorous training to become a samurai began around age five, where boys began learning martial arts and an understanding of Zen Buddhism, a Chinese philosophy promoting humility and balance.

> 'Bushido is nothing but charging forward, without hesitation, unto death. A bushi in this state of mind is difficult to kill even if he is attacked by 20 or 30 people ... In a normal state of mind, you cannot accomplish a great task. You must become like a person crazed and throw yourself into it as if there were no turning back. Moreover, in the Way of the martial arts, as soon as discriminating thoughts arise, you will already have fallen behind.'

Source 2

Yamamoto Tsunetomo, *Hagakure: the Book of the Samurai*, c.1709–16

Source 3

Samurai weapons and armour

Ninja warriors

Ninja were also highly trained warriors. They were specially trained spies and assassins who appeared around 1460. Ninja were hired by *daimyo* to carry out secretive missions such as spying on or murdering a rival lord. Their main activity was to sneak into enemy castles and report back on activity or sabotage the castle by setting a fire or causing an explosion.

Ninja did not have allegiance to a particular *daimyo*, and devoted their lives to perfecting the art of ***ninjutsu***. The skills and traditions of the ninja were passed on from ***sensei*** (teacher) to student. Like samurai, ninja began training at an early age. They learned the '18 disciplines', including how to fight with weapons such as throwing stars.

Ninja were extremely fit and could run for long periods, jump long distances, scale high walls and fight without weapons.

Source 4

The 18 disciplines of the ninja

1 Spiritual refinement

2 Hand-to-hand combat

3 Sword fighting

4 Fighting with spears

5 Fighting with a *naginata*

6 Fighting with a staff

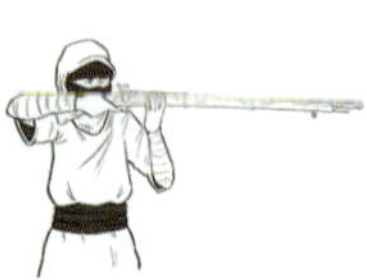

7 Using guns and explosives

8 Fighting with a sickle and chain

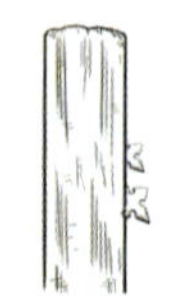

9 Using throwing stars

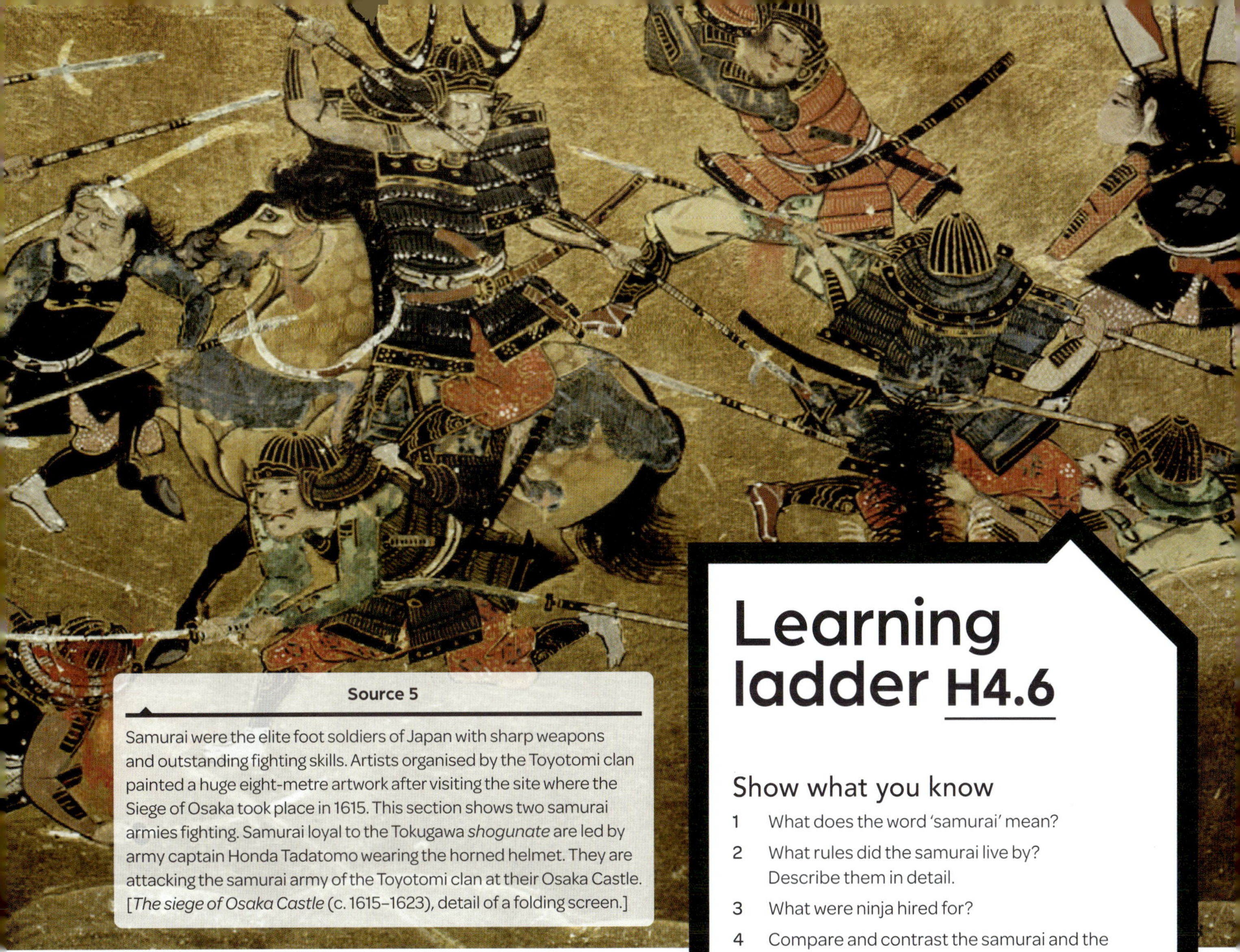

Source 5

Samurai were the elite foot soldiers of Japan with sharp weapons and outstanding fighting skills. Artists organised by the Toyotomi clan painted a huge eight-metre artwork after visiting the site where the Siege of Osaka took place in 1615. This section shows two samurai armies fighting. Samurai loyal to the Tokugawa *shogunate* are led by army captain Honda Tadatomo wearing the horned helmet. They are attacking the samurai army of the Toyotomi clan at their Osaka Castle. [*The siege of Osaka Castle* (c. 1615–1623), detail of a folding screen.]

10 Archery

11 Water crossing techniques

12 Using disguises

13 Disappearing techniques

14 Using concealed weapons

15 Climbing walls

16 Medicines and first-aid

17 Spying techniques

18 Astronomy

Learning ladder H4.6

Show what you know

1 What does the word 'samurai' mean?

2 What rules did the samurai live by? Describe them in detail.

3 What were ninja hired for?

4 Compare and contrast the samurai and the ninja. What do they have in common and what is different?

Source analysis

Step 1: I can determine the origin of a source

5 When and where is Source 5 from?

Step 2: I can list specific features of a source

6 Look at Source 5. Describe in detail everything you see in this image. Only describe what you see, not what you believe is taking place. Include details of clothing, equipment and number and position of individuals.

Step 3: I can find themes in a source

7 Rank the 18 disciplines of the ninja in in Source 4 in order of how important you think they would have been for a ninja to be able to complete their work.

Step 4: I can use my outside knowledge to help explain a source

8 Looking at Source 3, what materials would the Japanese have needed to produce the weapons and armour in this image?

HOW TO

Source analysis, page 197

How were castle towns designed?

Shoguns and powerful *daimyos* lived in highly protected castles designed to keep out the samurai armies of rival *daimyo*.

Castle fortresses

Daimyo built **castles** all over Japan to show their power and protect them from their enemies. The *daimyo* and their **samurai** protectors built castle fortresses surrounded by moats, stone walls and mazes. Most attacks on castles came from other *daimyo*. Japan was embroiled in **civil war** for hundreds of years, where *daimyo* warlords fought for control.

The first castle towns were surrounded by moats and wooden fences. From the 1300s onwards, the *daimyo* began to construct stone walls around timber castles. Medieval castles in Japan looked different to those in Europe (see pages 74–75), but shared many features:

- located on top of a hill
- rows of tall walls for protection
- moats surrounding them to stop tunnelling
- steep stairways to make attack difficult
- portholes to shoot arrows at the enemy
- a solid main gate to slow or trap the enemy.

Castle towns

The entire town was built with defence in mind. Japanese castle towns, such as Hemeji castle in Source 2, were generally surrounded by a moat with only a few entrances into the town. Each entrance had a strong gate and was guarded by high-ranked samurai who lived in their own large houses. Lower-ranked samurai lived along the main roads inside the castle town.

Once inside the main gate of the castle town, the enemy had to contend with winding streets and a series of gates and walls designed to slow and trap them. These mazes made attack difficult. The *daimyo* and his samurai could retreat to the ***donjon*** – the main tower of the castle which contained weapons and a food supply – when under attack.

Siege of Osaka Castle

Osaka Castle was built for Toyotomi Hideyoshi (see pages 148–49) in 1583. Following his death in 1598, power was seized by Tokugawa Ieyasu from Hideyoshi's son, who was still a child and based at Osaka Castle. He stood in Ieyasu's way to rule all of Japan.

Source 1

Samurai storm the gate of Hirado Castle, the seat of the Matsura clan. [Storm on gate of Hirado, *History of Japan*, Acrylic Illustration.]

Source 2

The Himeji Castle was surrounded by a large moat and featured a complex maze of walls and gates designed to slow down the enemy and protect the seven-storey *donjon*.

The Siege of Osaka was a series of battles undertaken by the Tokugawa ***shogunate*** against the Toyotomi clan in 1614 and 1615. Tokugawa troops filled in a moat, dug under the castle walls and tore down the outer wall of Osaka Castle. In a final attack, cannons and fire burnt the Osaka Castle to the ground and the Tokugawa clan ruled for the next 250 years.

Learning ladder H4.7

Show what you know

1 Why were castles built?

2 Why were castles often built on high ground and with moats?

3 Why was the Siege of Osaka castle important historically?

4 Looking at Source 2, describe all the features of Himeji castle that would make it hard to conquer.

5 What could an invading army do to overcome the defences of fortresses like Osaka Castle?

Chronology

Step 1: I can read a timeline

6 When did the Siege of Osaka take place?

Step 2: I can place events on a timeline

7 Put these events in chronological order:

The end of the feudal age in Japan in 1868

The period of 'warring states' that started in 1467

The Siege of Osaka in 1614–15

The need for castles arose from about 1300 CE

Step 3: I can create a timeline using historical conventions

8 Using correct conventions, create a timeline with the events from question 7 on it.

Step 4: I can distinguish causes, effects, continuity and change from looking at timelines

9 Look at the timeline on pages 132–33. Suggest at least three events that are continuous. Give a brief explanation as to why these events represent continuity, not change.

HOW TO

Chronology, page 194

key individual

How did Hideyoshi rise to lead Japan?

Toyotomi Hideyoshi, the son of a peasant farmer, rose from a low social class to become a samurai and then a powerful *daimyo*, who would unify Japan and put an end to 150 years of civil war.

Rise of the peasant leader

When Toyotomi Hideyoshi was born in 1536, Japan was made up of a series of provinces controlled by local ***daimyo***. They no longer feared the ***shogun***, Japan's military leader. Life was chaotic, with lords battling one another for power, peasant rebellions, and bandits and pirates running amok. Amid the chaos, Hideyoshi was born in a small village in Owari Province.

At 22 he joined the army of Oda Nobunaga, a ruthless daimyo who was set on conquering all of Japan. Hideyoshi began as Nobunaga's sandal-bearer, the person who looked after the warlord's shoes. He quickly proved himself as capable of more and served in Nobunaga's army. His intelligence and courage led him to be made a general at the age of 30. The peasant boy had quickly risen to be a samurai noble.

Hideyoshi became one of Oda Nobunaga's leading generals. He captured many castles using innovative tactics such as digging tunnels under castle walls and diverting a river to flood another. In 1582, Nobunaga died and a fight erupted for leadership of the Oda clan. Hideyoshi won and installed Nobunaga's grandson as the new Oda daimyo. Some lords refused to acknowledge the authority of a man of peasant birth, but Hideyoshi quickly put down challenges such as the one from Oda general Shibata Katsuie at the Battle of Shizugatake in 1583.

Hideyoshi unifies Japan

Hideyoshi had himself adopted by the Fujiwara clan. He took the title of ***kampaku***, not *shogun*, and installed the teenage Go-Yozei as his **puppet emperor**.

In 1588, Hideyoshi called a meeting of his *daimyo* at his new palace in Kyoto. In the presence of the young emperor, Hideyoshi had the *daimyo* pledge their allegiance to him as *kampaku*. By 1590, Hideyoshi had defeated the Shimazu and Hojo clans to unify Japan for the first time.

Hideyoshi was now the supreme commander and quickly introduced a number of measures to bring peace and stability to Japan. Hideyoshi would not allow wars between *daimyo* or uprisings from peasants. To achieve control, Hideyoshi froze the social order so that the class you were born in was the class you died in. In the new social order:

- towns were banned from allowing farmers to become merchants or craftspeople
- to prevent peasant uprisings, no Japanese citizens (except for samurai) could own a weapon
- no-one could hire a *ronin* (a samurai without a master) to fight on their behalf
- to keep an eye on *daimyo* lords, each lord had to spend every second year in the capital and they were ordered to leave their wives and children there permanently
- a **census** of population and land was held to accurately set taxation rates.

Expansion of Japan

Hideyoshi put together an army of 140 000 and 2000 ships to invade Korea. With help from China, Korea repelled Japan's samurai army on two occasions. During the second invasion in 1598, with his troops fighting a losing battle, Hideyoshi died. His heir was just five years old, so Hideyoshi had put in place a trusted Council of Five Elders to rule until the child came of age.

Source 1

The Battle of Shizugatake Pass in 1583 occurred in the Sengoku period in Japan between supporters of Toyotomi Hideyoshi and Shibata Katsuie. [*Battle of Shizugatake Pass 1583*, coloured lithograph print, 1904.]

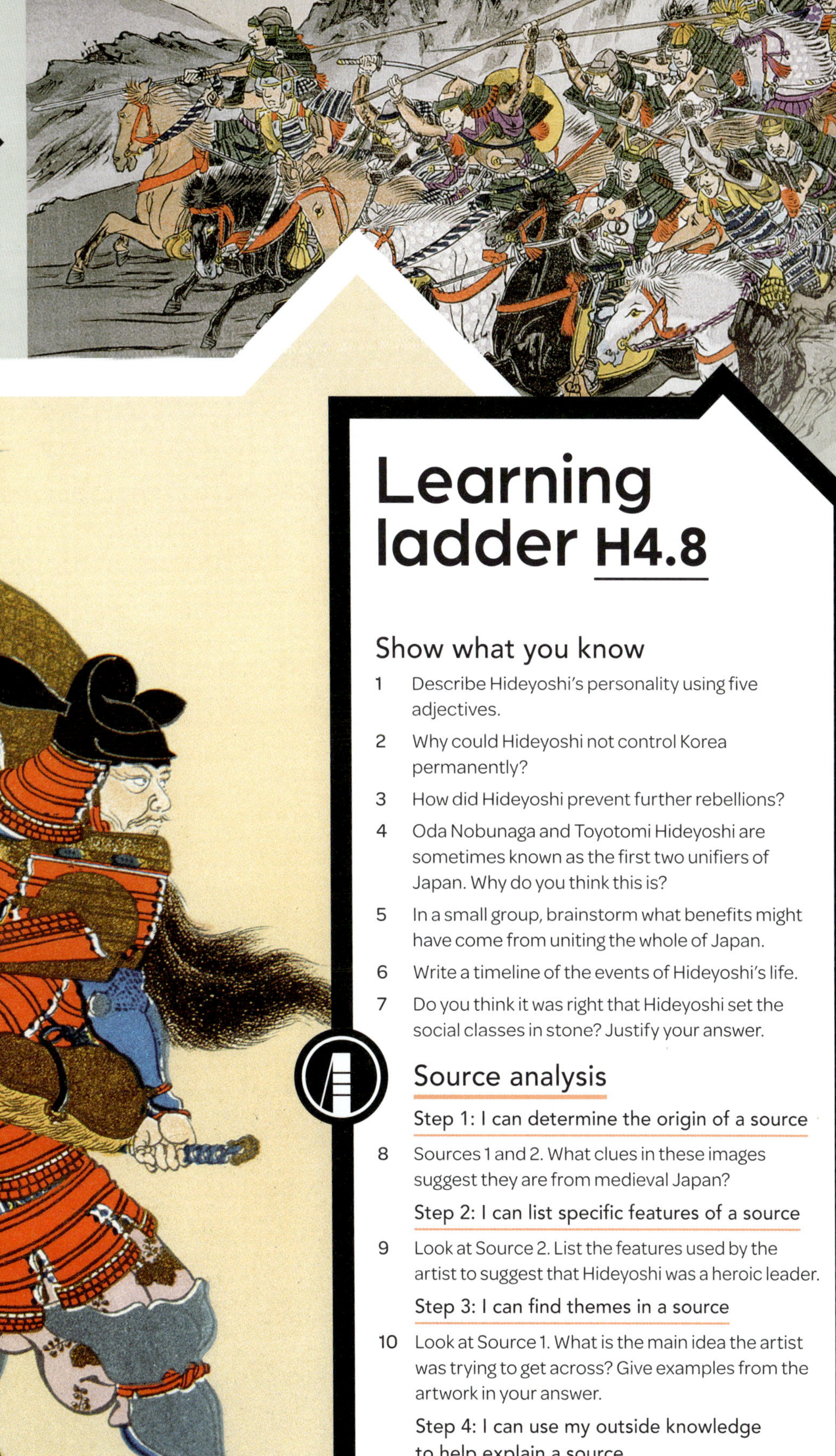

Source 2

A colour lithograph of Toyotomi Hideyoshi (1536–98)

Learning ladder H4.8

Show what you know

1 Describe Hideyoshi's personality using five adjectives.

2 Why could Hideyoshi not control Korea permanently?

3 How did Hideyoshi prevent further rebellions?

4 Oda Nobunaga and Toyotomi Hideyoshi are sometimes known as the first two unifiers of Japan. Why do you think this is?

5 In a small group, brainstorm what benefits might have come from uniting the whole of Japan.

6 Write a timeline of the events of Hideyoshi's life.

7 Do you think it was right that Hideyoshi set the social classes in stone? Justify your answer.

Source analysis

Step 1: I can determine the origin of a source

8 Sources 1 and 2. What clues in these images suggest they are from medieval Japan?

Step 2: I can list specific features of a source

9 Look at Source 2. List the features used by the artist to suggest that Hideyoshi was a heroic leader.

Step 3: I can find themes in a source

10 Look at Source 1. What is the main idea the artist was trying to get across? Give examples from the artwork in your answer.

Step 4: I can use my outside knowledge to help explain a source

11 With what you have learnt about Hideyoshi's character, how do you explain the expression on Hideyoshi's face in Source 2? Why do you think the artist painted him in this way?

HOW TO Source analysis, page 197

How does Japanese art and culture reflect beliefs?

Art and culture, gardens and martial arts are greatly influenced by Japan's ancient Shinto religion and religions introduced from Asia such as Buddhism.

Religions of Japan

Shinto and **Buddhism** are the two major religions of Japan. Shinto is Japan's ancient religion, while Buddhism was imported from mainland Asia in the 6th century CE. The two religions have lived in harmony and are both ingrained in Japanese art and culture.

Followers of Shinto believe in sacred ancestor spirits known as ***kami***, which took on many natural forms such as the Sun, hills, rocks, lakes, rivers and trees. All people, animals, places and things are believed to possess *kami*. Buddhists see life as a cycle where you are born, you die and are reborn. If you live a good life, you will come back as a better person or thing in your next life.

Gardens of Japan

Garden design is closely connected to the religion and philosophy of Japan. Shinto- and Buddhism-inspired spiritual garden styles aspire to create spaces where people can reflect and meditate.

Zen Buddhism

Zen Buddism's mixture of Buddhism and **Taoism** (a Chinese philosophy promoting humility and balance) set believers on a path to understand the meaning of life. It was introduced into Japan during the 12th century. Zen gardens involve carefully placed rocks and moss surrounded by sand or gravel that is raked to represent ripples in water. They were to be used as a tool to help meditation about the true meaning of life.

Source 1

An example of Japanese garden design

Source 2

Sumo wrestlers, known as *rikishi*, need to develop great weight for strength, fast reflexes and great balance.

Martial arts

Both Zen Buddhism and Shinto have been closely related to the development of Japanese martial arts, such as judo, karate, *kendo*, *ninjutsu* and **sumo**. **Samurai** warriors embraced Zen Buddhism in martial arts and its principles of self-discipline and rising above a fear of death. Shinto also influenced martial arts and often provided the reason to fight for ancestor spirits.

Sumo wrestling

Sumo wrestling is steeped in Shinto traditions and symbols, with referees dressing in the style of Shinto priests. Before sumo wrestling became a professional sport during the time of the Tokugawa *shogunate*, it was performed in the grounds of a Shinto shrine or temple. The ring or ***dohyo*** is still considered sacred, symbolising the Earth. The day before a *sumo* tournament, sumo judges, called ***gyoji***, perform a ring-blessing ceremony. Dressed in the white robes of a Shinto priest, the *gyoji* purify the *dohyo* with salt. *Sake*, a traditional Japanese alcoholic drink, is poured around the ring as an offering to the gods.

Learning ladder H4.9

Show what you know

1 Summarise the two main belief systems of Japan in your own words.

2 How are martial arts linked to Shintoism?

3 Would you prefer to follow Buddhism or Shinto? Explain your reasons with reference to each.

4 Imagine you are writing a sacred text, including both Shinto and Buddhist beliefs. Write five rules that people should follow.

5 Write a short interview with a sumo wrestler. Ask him at least four questions.

Cause and effect

Step 1: I can recognise a cause and an effect

6 How did traditional Zen gardens help Buddhists in their faith?

Step 2: I can determine causes and effects

7 Looking at Source 1, what features of the garden make it helpful for meditation and calm? How is it different from a Western garden?

Step 3: I can explain why something is a cause or an effect

8 Why might Buddhism make people want to lead a better life?

Step 4: I can analyse cause and effect

9 How does Japanese art and culture (such as Zen gardens and sumo wrestling) reflect Japanese belief systems?

Cause and effect, page 203

economics+business

Why did Japan isolate itself from the world?

Today, businesses regularly engage in overseas trade. They operate under government rules set to ensure the trade is legal and that any fees are paid. Trade was quite different during the Tokugawa *shogunate*. Japan was virtually shut off and the shogun controlled all foreign trade.

The Tokugawa *shogunate*

Following the death of Hideyoshi (see pages 148–49) in 1598, Tokugawa Ieyasu established the Tokugawa *shogunate.* Beginning in 1603 CE, Ieyasu set up his *shogunate* in Edo (modern-day Tokyo) and within 100 years the small fishing town was transformed into the biggest city in the world. The Edo Period would last for 264 years.

Why did the *shogun* isolate Japan?

The Tokugawa *shoguns* were suspicious that foreigners could change Japan's traditions. They were particularly concerned about the spread of **Christianity**, first introduced by Catholic Portuguese and Spanish traders in the mid-16th century.

There was a belief that Christianity was undermining Japan's traditional **Shinto** and **Buddhist** religions (see pages 150–51). Christianity was banned, and to limit exposure to foreign ideas, all Japanese citizens were forbidden to travel overseas.

The Tokugawa *shoguns* isolated Japan from the rest of the world for over 220 years through the government policy of ***sakoku***. The policy was enacted by *shogun* Tokugawa Iemitsu, Tokugawa Ieyasu's grandson. Only Dutch and Chinese traders were allowed to trade with Japan.

Trade in and out of Japan could only take place through the port of Nagasaki, a city controlled by the Tokugawa clan. Merchants had to pay fees and taxes for the right to take part in foreign trade through Nagasaki. This enabled the Tokugawa clan to become wealthier and reduced the power of other ***daimyo*** clans.

Source 1

A Dutch ship at the Dejima trading post in the bay of Nagasaki. [*Nagasaki: Dejima Island* (1804), painting.]

Source 2

William Adams was the first Englishman to reach Japan. In 1600 he arrived as part of a Dutch expedition and was not allowed to leave Japan. As an adviser to *shogun* Tokugawa Ieyasu, Adams warned Ieyasu about the influence of the Catholic Church through the Portuguese and Spanish traders. [Wood engraving (c. 1900).]

Sakoku edict of 1635

Source 3 contains excerpts from a letter containing 17 official orders written by *shogun* Iemitsu to the two commissioners of the port city of Nagasaki.

1 Japanese ships are strictly forbidden to leave for foreign countries.

2 No Japanese is permitted to go abroad. If anyone attempts to do so secretly, he must be executed. The ship so involved must be impounded and its owner arrested, and the matter must be reported to the higher authority.

3 If any Japanese returns from overseas after residing there, he must be put to death.

4 If there is any place where the teachings of the [Catholic] priests is practised ... you must order a thorough investigation.

5 Any informer revealing the whereabouts of the followers of the priests must be rewarded accordingly.

Source 3

Some of the orders from the *Sakoku* edict of 1635. [Shogun Iemitsu, *The Edict of 1635 Ordering the Closing of Japan: Addressed to the Joint Bugyō of Nagasaki* (1635).]

The US demands trade

In 1853, a fleet of United States ships commanded by Admiral Matthew Perry demanded that the Japanese open up to trade. Under threat from ships armed with powerful cannons, Japan reluctantly agreed to trade with Western countries by signing the Treaty of Kanugawa in 1854, ending the long period of isolation.

Learning ladder H4.10

Economics and business

Step 1: I can recognise economic information

1 a Look at Sources 1 and 2: Why was Tokugawa wary of foreigners?

b What action did he take to limit contact with other countries?

Step 2: I can describe economic issues

2 What was the policy of *sakoku*? Give three examples of how the policy isolated Japan from the rest of the world.

3 Add a further three rules to the *sakoku* edict that would further keep foreign influences out of Japan.

Step 3: I can explain issues in economics

4 What was the Tokugawa *shogunate* trying to achieve with their policy to isolate Japan from the rest of the world?

5 Why do you think they still needed to trade with the Dutch and Chinese?

6 How did the Tokugawa clan benefit from controlling trade with the outside world?

Step 4: I can integrate different economic topics

7 Research what tariffs are and why they are used. Give one example of a tariff in operation today.

8 Compare tariffs to the isolation of Japan by the Tokugawa *shogunate*. What do the two have in common and how are they different?

Step 5: I can evaluate alternatives

9 Was it a good thing that Admiral Perry opened Japan up to outside trade? Support your answer with reasons.

How did the emperor take back control from the *shoguns*?

The Meiji period was a time of rapid modernisation in Japan. Following more than 200 years in isolation during the Tokugawa *shogunate*, rule in Japan returned to the emperor and a small group of powerful advisers

The fall of the *shoguns*

Following the Treaty of Kanagawa signed between Admiral Matthew Perry and the Tokugawa *shogunate* in 1854 (see page 153), Japanese ports opened up to trade with America. The Japanese did not want to trade with the United States, but they were scared of Perry's 'Black Ships'. After 220 years of isolation, the Japanese had never seen huge steamships before, and thought the American ships were 'giant dragons puffing smoke'.

The ***daimyo*** began to lose faith in the ***shogun***, whose primary responsibility as supreme military leader was to protect Japan. In 1858, Japan was pressured into setting low **tariffs** on goods imported from America. Soon after, the Russians, French and British also forced the Japanese to sign similar treaties.

With reduced power and failing public confidence in the *shogunate*, the Choshu, Satsuma and Tosa clans forced the resignation of *shogun* Tokugawa Yoshinobu. He handed back power to Emperor Meiji in 1867. The emperor moved from his Imperial Palace in Kyoto to the *shogun's* former palace of Edo Castle. The city of Edo was renamed Tokyo, and a new age of Japan began.

The Meiji Restoration

The transfer of power from the *shogun* to the emperor is called the Meiji Restoration. Emperor Meiji was only 15 years old when he came to power, so the government was run by advisers from the clans that had brought the emperor back to power.

In 1871, all 304 provinces under the control of *daimyo* were returned to the emperor. They were combined into 75 larger administrative **prefectures**, each controlled by a governor.

Source 1

A steam engine at Tokyo in 1870. Railway lines increased from just 29 kilometres of track in 1872 to more than 11 000 kilometres of track by 1914. [Utagawa Kuniteru, *The Takanawa Steam Railway, Tokyo*, Edo-Tokyo Museum, 1870.]

Source 2

A portrait of Emperor Meiji and the imperial family in 1900. In an effort to modernise Japan, Emperor Meiji adopted many Western ideas, including their fashions. The adults wear Western clothes, while the children are dressed in traditional Japanese attire. [Torajirō Kasai, *The Japanese Imperial Family* (1900), chromolithograph, 45.5 x 61.9 cm, Library of Congress.]

The new administration took away most of the privileges of the samurai class (see pages 142–45). Many samurai were still loyal to their local *daimyo* instead of the country, so the Meiji government took away their right to carry swords in public.

The government introduced military service (**conscription**) for all men, to develop a national army open to all classes. The new laws ended the old feudal system where the samurai provided protection for the *shogun* and *daimyo*.

Modernising Japan

Emperor Meiji and his government moved rapidly to modernise Japan with visits to the United States, Europe and Asia to gather industrial, technological and military information.

Japan needed to quickly increase its production capacity. It built iron smelters, coal mines, shipyards and spinning mills. Huge numbers of Japanese migrated from the countryside to the new industrial regions. A railway system was built to connect the new industrial centres with Japan's ports.

The adoption of many Western weapons and technologies strengthened Japan's military and the Imperial Japanese Navy was established in 1869. Japan began its overseas expansion with successful wars with China in 1894–95 and Russia in 1904–05.

In just a few decades, the Meiji government transformed Japan into a powerful modern nation.

Learning ladder H4.11

Show what you know

1 What brought the emperor back to power?

2 What steps did the new government take to modernise?

3 Looking at Source 2, what evidence suggests the Japanese were becoming more Western?

4 Looking at Source 1, what new and old technologies can you see in the image? Why do you think the artist includes both?

Historical significance

Step 1: I can recognise historical significance

5 Why is the period from 1868 called the Meiji Restoration?

Step 2: I can explain historical significance

6 Japan modernised and westernised very quickly. With a partner, discuss whether you think this benefited the Japanese people. Use evidence to support your answer.

Step 3: I can apply a theory of significance

7 Answer these questions about the modernisation of Japan under Emperor Meiji:
   a Why was the period of 1870–1905 important?
   b How deeply did it affect people at the time?
   c How many people were affected?
   d For how long did the modernisation influence Japan?
   e Is it still relevant today?

Step 4: I can analyse historical significance

8 Emperor Meiji's restoration was the end of rule by the *shoguns*. How significant was this? Give at least three reasons, discussing them in order from most to least important.

Historical significance, page 207

Masterclass

Learning Ladder

Work at the level that is right for you or level-up for a learning challenge!

Source 1

Katsushika Hokusai *Chiho Sokuryo no zu* (1848), woodblock print on paper, The British Museum.

Step 1

a **I can read a timeline**

Look at the timeline on page pages 132–33 and answer these questions:

i How long ago was the first novel written?
ii Which was the shortest of the periods listed?
iii How long after the end of the Heian Period ended did the Edo Period begin?

b **I can determine the origin of a source**

Where, when and by whom was Source 1 created?

c **I can recognise continuity and change**

Rank these in order from most to least changed in Japan. You might need to research modern Japan to help you answer.

- Religion
- Geography
- Technology
- Form of government
- Clothing

d **I can recognise a cause and an effect**

Match the causes and effects in the following:

Admiral Perry demanded Japan open up to trade.	The Emperor became rich.
Daimyo offered protection.	Japan began to modernise after seeing how advanced the rest of the world was.
Peasants paid taxes in rice or labour.	Peasants moved onto their land.
Samurai followed the *Bushido* code.	The people lost faith in the *shogun* and brought the emperor back into power.
The *shogun* was unable to protect Japan against the military technology of the Americans.	They were loyal, ethical and courageous.

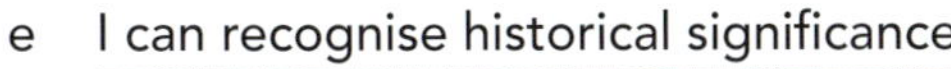

e I can recognise historical significance

Rank these in order from least to most historically important:

- return of the emperor to power in the 1600s
- Japanese gardens
- peasants being banned from owning swords
- Admiral Perry opening up Japan to foreign trade
- only 15 per cent of Japan could be farmed.

Step 2

a I can place events on a timeline

Put these events in order from earliest to most recent:

- The Onin War began in 1467 and lasted for 10 years.
- Buddhism came to Japan in 538.
- Christianity was introduced in 1549 but was banned 65 years later.
- Mount Fuji had a major eruption in 1707.
- The Mongol invasion of 1281 failed because of a typhoon.

b I can list specific features of a source

List at least 10 details in Source 1.

c I can describe continuity and change

What is one thing that changed a lot from ancient Japan to feudal Japan? Explain your answer using historical evidence.

d I can determine causes and effects

i What caused Japan to modernise in the late 1800s?

ii What was the effect of castle fortresses being built?

e I can explain historical significance

Why are the belief systems of Shinto and Buddhism so important to Japan?

Step 3

a I can create a timeline using historical conventions

Using the events in Step 2a, create a timeline using correct historical conventions. Remember to use equal spacing for equal years.

b I can find themes in a source

Taking in the scene in Source 1, what do you think is the main subject of the print? Give examples from the print to back up your answer.

c I can explain why something did or did not change

The requirement to show respect for one's elders and the dead is something that has remained fairly unchanged in Japanese society since ancient times. Why do you think this is?

d I can explain why something is a cause or an effect

How have Japanese religious beliefs affected their garden design?

e I can apply a theory of significance

How important would the unification of Japan under Toyotomi Hideyoshi have been at the time? How many people did it affect? Is that unification still important today?

Step 4

a I can distinguish causes, effects, continuity and change from looking at timelines

Looking at the timeline on pages 132–33, which two events are linked by cause and effect? Explain why.

b I can use my outside knowledge to help explain a source

Source 1: what role in Japanese society do you think the people in the picture played? What do you think they are doing in the image? Why might this have been important in the late Edo Period?

c I can analyse patterns of continuity and change

Technology in Japan improved greatly after 1853. Why was this?

d I can analyse cause and effect

Why did Japanese clans in medieval times fight for control of land?

e I can analyse historical significance

List at least three reasons why it was historically important that Japan was forced to open up to foreign trade. Write a short paragraph discussing the three reasons, including a discussion of which was most important.

Masterclass

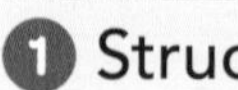

Step 5

a **I can describe patterns of change**

Have a look at the list of common historical patterns of change on page 196. Which of these do you think applies to Japanese history? Justify your response.

b **I can use the origin of a source to explain its creator's purpose**

Research what period in Japanese history the print in Source 1 was made. Why do you think the print was produced?

c **I can evaluate patterns of continuity and change**

Over centuries, power in Japan became more and more concentrated in the hands of a few leaders. Is this a good or bad thing? Give reasons for your answer.

d **I can evaluate cause and effect**

Women had less power than men in feudal Japan. Why was this? Was this a good or a bad thing? Explain your answer.

e **I can evaluate historical significance**

Was the unification of Japan under the *shoguns* really that important? Explain your answer.

Historical writing

1 Structure

Imagine you are given the essay topic, 'Explain how feudal Japanese society was structured'. Write an essay plan for this topic. Include at least three main paragraphs.

2 Draft

Using the drafting and vocabulary suggestions on page 214, draft a 400–600 word essay responding to the topic.

3 Edit and proofread

Use the editing and proofreading tips on page 215 to help you edit and proofread your draft.

Historical research

4 Organise and present information

Imagine you are doing a large research project: 'Modernising Japan from 1870'. Write a contents page for this project. There should be an introduction, conclusion, at least four main sections and many subsections. Number your chapters.

Capstone

How can I understand Japan under the *shoguns*?

In this chapter, you have learnt a lot about Japan under the *shoguns*. Now you can put your new knowledge and understanding together for the capstone project to show what you know and what you think.

In the world of building, a capstone is an element that finishes off an arch, or tops off a building or wall. That is what the capstone project will offer you, too: a chance to top off and bring together your learning in interesting, critical and creative ways. You can complete this project yourself, or your teacher can make it a class task or a homework task.

mea.digital/GHV8_H4

Scan this QR code to find the capstone project online.

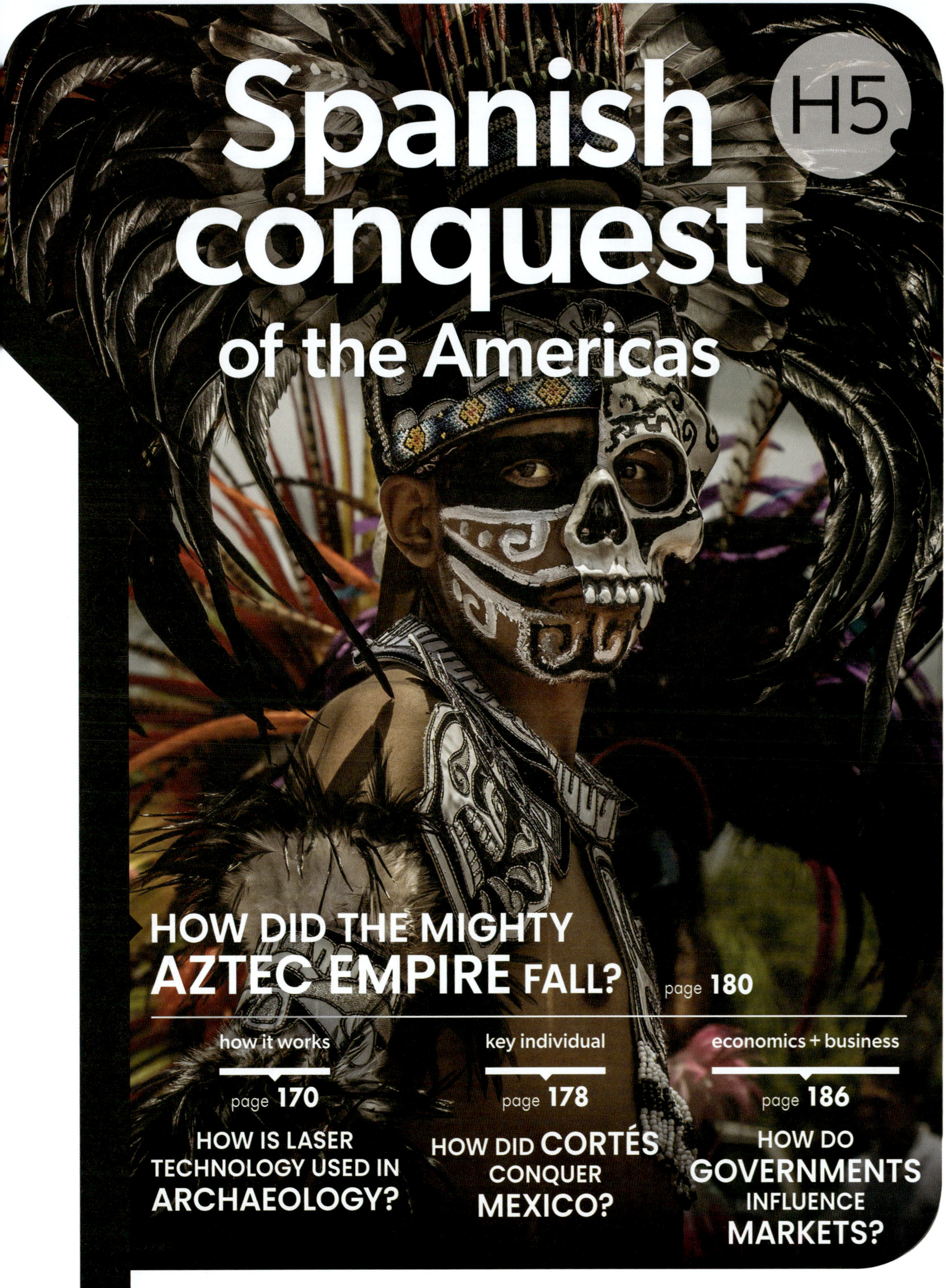

Spanish conquest of the Americas

H5

HOW DID THE MIGHTY **AZTEC EMPIRE** FALL?

page 180

how it works

page 170

HOW IS LASER TECHNOLOGY USED IN **ARCHAEOLOGY?**

key individual

page 178

HOW DID **CORTÉS** CONQUER **MEXICO?**

economics + business

page 186

HOW DO **GOVERNMENTS** INFLUENCE **MARKETS?**

How can I understand the Spanish conquest?

Christopher Columbus' voyage from Europe to the Americas in 1492 marked the beginning of possibly the most devastating meeting of two cultures the world has ever seen. In less than a century, it is estimated that over 90 per cent of the inhabitants of the New World were dead, from violence, starvation or disease. What the Spanish conquerors found in the New World were unique and fascinating cultures, the Aztec and the Inca.

learning ladder

Step	Chronology	Source analysis	Continuity and change
step 5	 **I can describe patterns of change** I read timelines and see the 'big picture'. I group timeline events about the ancient and medieval Americas and see if they show patterns of change. I know typical historical patterns to check for.	**I can use the origin of a source to explain its creator's purpose** I combine knowledge of when and where a source was created to answer the question, 'Why was it created?'	**I can evaluate patterns of continuity and change** I answer the question, 'So what?' about patterns of continuity and change. I weigh up ideas and debate the importance of a continuity or a change.
step 4	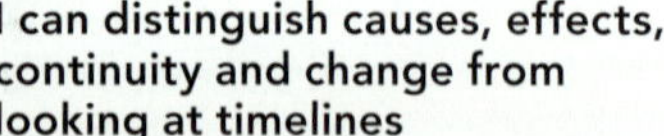 **I can distinguish causes, effects, continuity and change from looking at timelines** I read timelines of the Americas and find events linked by cause and effect. I also find things that are the same or different from then until later times.	**I can use my outside knowledge to help explain a source** I have enough outside knowledge about the ancient and medieval Americas to help me explain a source.	**I can analyse patterns of continuity and change** I see beyond individual instances of continuity and change in the ancient and medieval Americas and identify broader historical patterns, and I explain why they exist.
step 3	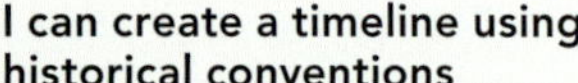 **I can create a timeline using historical conventions** When given a set of events, I construct a historical timeline using the dates, and using correct terminology, spacing and layout.	**I can find themes in a source** I look a bit closer into a source and find more than just features. I find themes or patterns in the source.	**I can explain why something did or did not change** I answer the question, 'Why?' something changed or stayed the same between historical periods in ancient and medieval America.
step 2	**I can place events on a timeline** When given a list of events and developments about the ancient and medieval Americas, I put them in order from earliest to latest, the simplest kind of timeline.	**I can list specific features of a source** I look at an ancient or medieval American source and list detailed things I can see in it.	**I can describe continuity and change** I have enough content knowledge about two different historical periods to recognise what is similar or different about them, and can describe it.
step 1	**I can read a timeline** I read timelines with ancient and medieval American events on them and answer questions about them.	**I can determine the origin of a source** I can work out when and where an ancient or medieval American source was made by looking for clues.	**I can recognise continuity and change** I recognise things that have stayed the same and things that have changed from the ancient or medieval Americas until now.

Source 1

Mosaic covered wooden disc that was probably used as a ceremonial shield for the Aztecs. Made from pine with a mosaic of turquoise and three types of shell. The design is thought to depict the Aztec universe. [c. 1400–1521. Discovered in Mexico, 1866.]

Warm up

Chronology

1 In what century did Christopher Columbus sail to the Americas?

Source analysis

2 Look at Source 1. When and where was the shield made? When and where was it discovered?

Continuity and change

3 The Spanish invaders destroyed many original objects from the Aztec civilisation. The object in Source 1 survived. How are objects important for the continuity of history?

Cause and effect

4 List the things that caused the deaths of over 90 per cent of Indigenous Americans after 1492.

Historical significance

5 Looking at your answer to question 4, which of these would you hypothesise was the most historically significant?

Cause and effect	Historical significance
I can evaluate cause and effect I answer the question, 'So what?' about cause and effect. I weigh up ideas and debate the importance of a cause or an effect.	**I can evaluate historical significance** I answer the question, 'So what?' about things that were supposedly historically important in the Americas. I weigh up events and cast doubt on how important things are.
I can analyse cause and effect I don't just see a cause or an effect as one thing. I determine the factors that make up causes and effects.	**I can analyse historical significance** I separate out the various factors that make something historically important.
I can explain why something is a cause or an effect I can answer 'How?' or 'Why?' a cause led to an effect in the ancient and medieval Americas.	**I can apply a theory of significance** I know a theory of significance. I use it to rank the importance of ancient and medieval American events.
I can determine causes and effects Applying what I have learnt about the ancient and medieval Americas, I can decide what the cause or effect of something was.	**I can explain historical significance** I answer the question, 'Why?' about things that were important in the ancient and medieval Americas.
I can recognise a cause and an effect From a supplied list, I recognise things that were causes or effects of each other in the ancient or medieval Americas.	**I can recognise historical significance** When shown a list of things from ancient or medieval American history, I can work out which are important.

Why learn about the Spanish conquest of the Americas?

Humans first crossed into North America from Asia via a land bridge during the **Ice Ages**, between 35 000 and 15 000 years ago. Early peoples gradually spread southwards, all the way to modern-day Mexico in North America and then down to Central America. Historians refer to this area as **Mesoamerica**.

The **Olmec**, **Maya**, and **Aztec** are three of the great Mesoamerican civilisations. Along with the South American civilisation of the **Inca**, these four distinct civilisations rose and fell at different times and for different reasons. The Olmecs, Mayans and then the Aztecs settled in different regions of the area we now call Mexico. The Inca built their cities in modern-day Peru. How the Spanish devastated the Inca and Aztec civilisations is one of history's most vital lessons, and certainly one of its most intriguing.

Source 2

Tenochtitlan

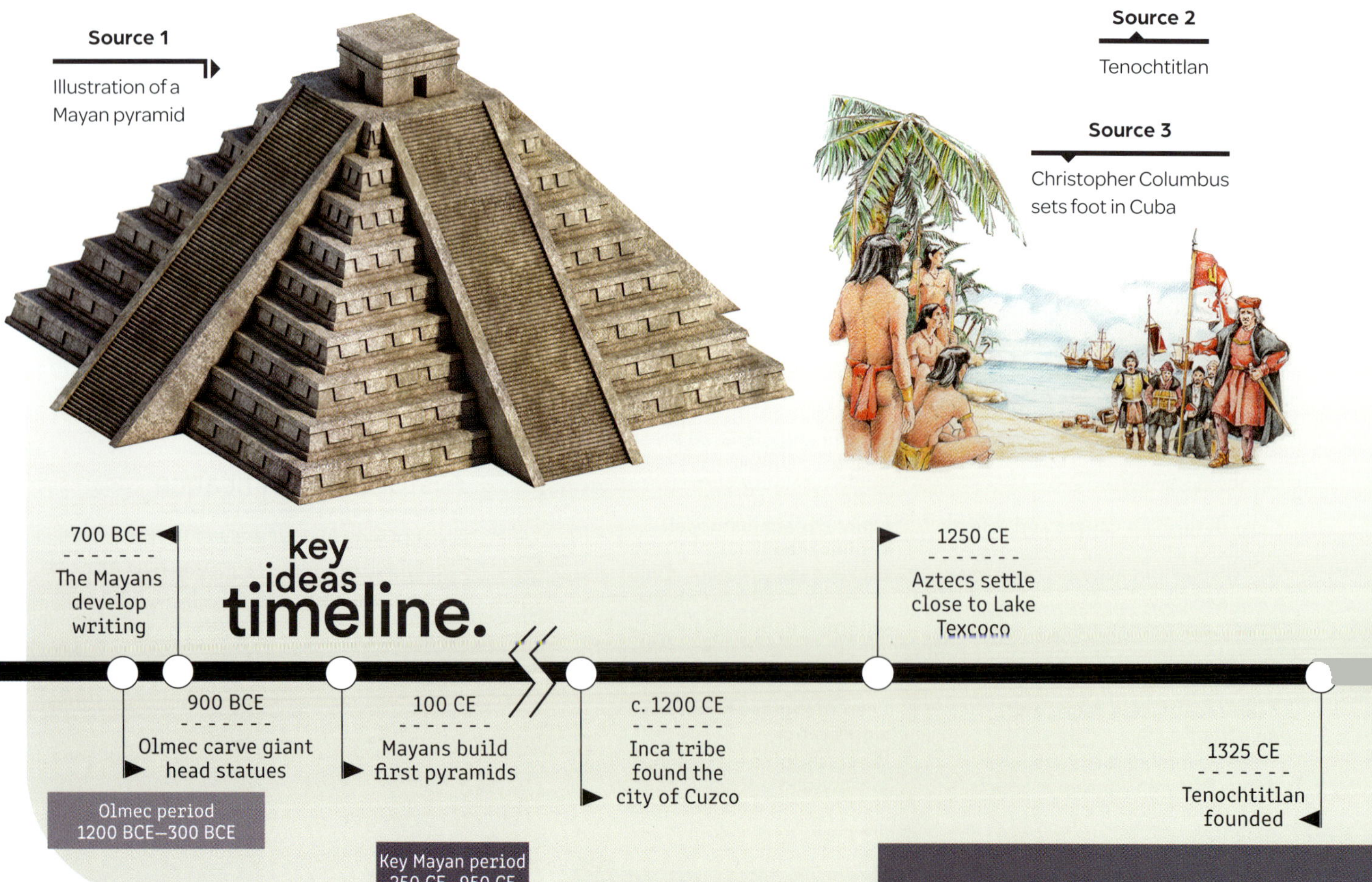

Source 1

Illustration of a Mayan pyramid

Source 3

Christopher Columbus sets foot in Cuba

Learning ladder H5.1

Chronology

Step 1: I can read a timeline

1 How long had the Aztecs been in Mexico before they founded Tenochtitlan?

2 In what order did the four Mesoamerican civilisations mentioned on this page settle in the Americas?

Step 2: I can place events on a timeline

3 Put these events in order from earliest to most recent.

Early Mayan settlements 1800 BCE

The Maya develop writing in 700 BCE

The Maya build their first pyramids 100 CE

The Maya begin to farm in 600 BCE

Step 3: I can create a timeline using historical conventions

4 Using the conventions discussed on page 195, create a timeline with the events from question 3 on it.

Step 4: I can distinguish causes, effects, continuity and change from looking at timelines

5 Using examples from the timeline, what can you suggest was the cause of the fall of the mighty Aztec Empire. Give examples of how long the empire lasted and how quickly it fell.

HOW TO

Chronology, page 194

Source 4

A painting from 1829 of the attack on an Aztec temple by Hernán Cortés

1350 CE – Causeways and canals built at Tenochtitlan

1492 CE – Christopher Columbus sails to the Americas

1502 CE – Aztec Empire at its most powerful under Moctezuma II

1521 CE – Cortés conquers the Aztecs

1572 CE – Pizarro conquers the Incans

Inca Empire 1438–1533 CE

Spanish conquest of Inca Empire 1532–1572 CE

Aztec Empire 1250–1521 CE

What was life like in the ancient Americas?

The Olmecs established the first great civilisation in Mexico and in Central America. Their culture greatly influenced the later cultures of the Mayans and the Aztecs.

Source 1

Olmec necklace made of jade, rock crystal, pyrite and shell. [From Gulf Coast, Mexico, c. 1200–300 BCE.]

Olmecs

The **Olmecs** were the first major civilisation from the Americas, in existence from around 1200 BCE until 300 BCE. They built villages and cities on the swampy land near the Gulf of Mexico.

People came together to trade in cities and take part in religious ceremonies at the stepped-pyramid temples built for their gods, which were later adopted by the Maya (see pages 166–69) and the **Aztecs** (see pages 172–75).

Not a lot is known about the Olmec people, but we do know they created beautiful jewellery and realistic sculptures of people and animals. These craftworks were made from basalt, clay, jade, obsidian and greenstone. Many of these resources were found at some distance from Olmec centres, suggesting that they had a large trading network.

Source 2

Civilisations of Mesoamerica

Source: Matilda Education Australia

Source 3

A replica of an Olmec head from the Chankanaab National Park, in Mexico. The statues are thought to represent Olmec rulers, and vary in size from 6–50 tonnes.

Learning ladder H5.2

Show what you know

1 For roughly how long did the Olmec civilisation last?

2 Roughly how long ago did humans come to the Americas?

3 Why did people come together at pyramid temples?

4 What do you think might be one way the Olmecs were able to move the stone for their head statues long distances?

5 Looking at Source 2, write a detailed description of the location of the Aztec and Mayan civilisations.

6 Examine Source 3. Give a physical description of this head, using lots of adjectives.

7 With a partner, discuss why you think not a lot is known about Olmec people.

Historical significance

Step 1: I can recognise historical significance

8 Which of the following is the most historically significant?
   - a People came to the Americas via a land bridge 35000 to 15000 years ago.
   - b The Olmecs built pyramids.
   - c The Olmecs carved giant heads out of rock.
   - d The Olmecs lived on swampy land.
   - e The Olmecs made beautiful jewellery.

Step 2: I can explain historical significance

9 Why was it important that the Olmec people came together at pyramids?

Step 3: I can apply a theory of significance

10 What evidence is there that Olmec cities were important to the people at the time?

Step 4: I can analyse historical significance

11 How relevant is the Olmec civilisation today? How important was it to other Mesoamerican cultures?

HOW TO

Historical significance, page 207

The mystery of the giant Olmec heads

The Olmecs are best known for the giant heads they carved from stone. So far, 17 heads have been found with each of them wearing a helmet, perhaps like those worn in the ball game ***ulama*** (see page 175). Each giant head looks similar, but they have different facial features, leading historians to suggest that the sculptures are of Olmec rulers.

The giant heads were carved from basalt around 900 BCE. One of the giant heads at San Lorenzo has traces of red paint, suggesting that the statues may once have been painted. The statues weigh between 6 and 50 tonnes and measure up to 3.4 metres high.

Archaeologists are puzzled by how the Olmecs moved the huge stones needed and how they created the statues. Basalt quarries were located in the Tuxtla Mountains (Sierra de los Tuxtlas), around 70 kilometres away from Olmec settlements. In the city of San Lorenzo, a giant head was lifted to the top of a 45-metre-high plateau.

The Olmec did not have cattle to pull the load and did not have wheeled vehicles. Some historians suggest that they floated the heavy stones on rafts or dragged them over wooden rollers.

How did the Maya build a civilisation in the jungle?

The Mayans, or Maya, developed sophisticated cities in the dense jungles of Central America. They lived in separate city-states with grand palaces and pyramid temples. Around 950 CE, the cities were abandoned and reclaimed by the rainforest.

Life in the rainforest

Between 250 and 900 CE, the **Mayans** built elaborate cities with pyramid temples and palaces in the thick jungles of Central America. The Maya took full advantage of the rainforest's resources. Forest animals such as deer, puma, jaguar and crocodiles were hunted for their meat and hides.

Limestone was mined for buildings, obsidian was used for weapons and tools, and jade was used for sculpture and jewellery. The long green tail feathers of the male quetzal bird were so valuable that it was forbidden to kill the bird. Quetzal and other forest bird feathers were used to make elaborate costumes and headdresses for nobles.

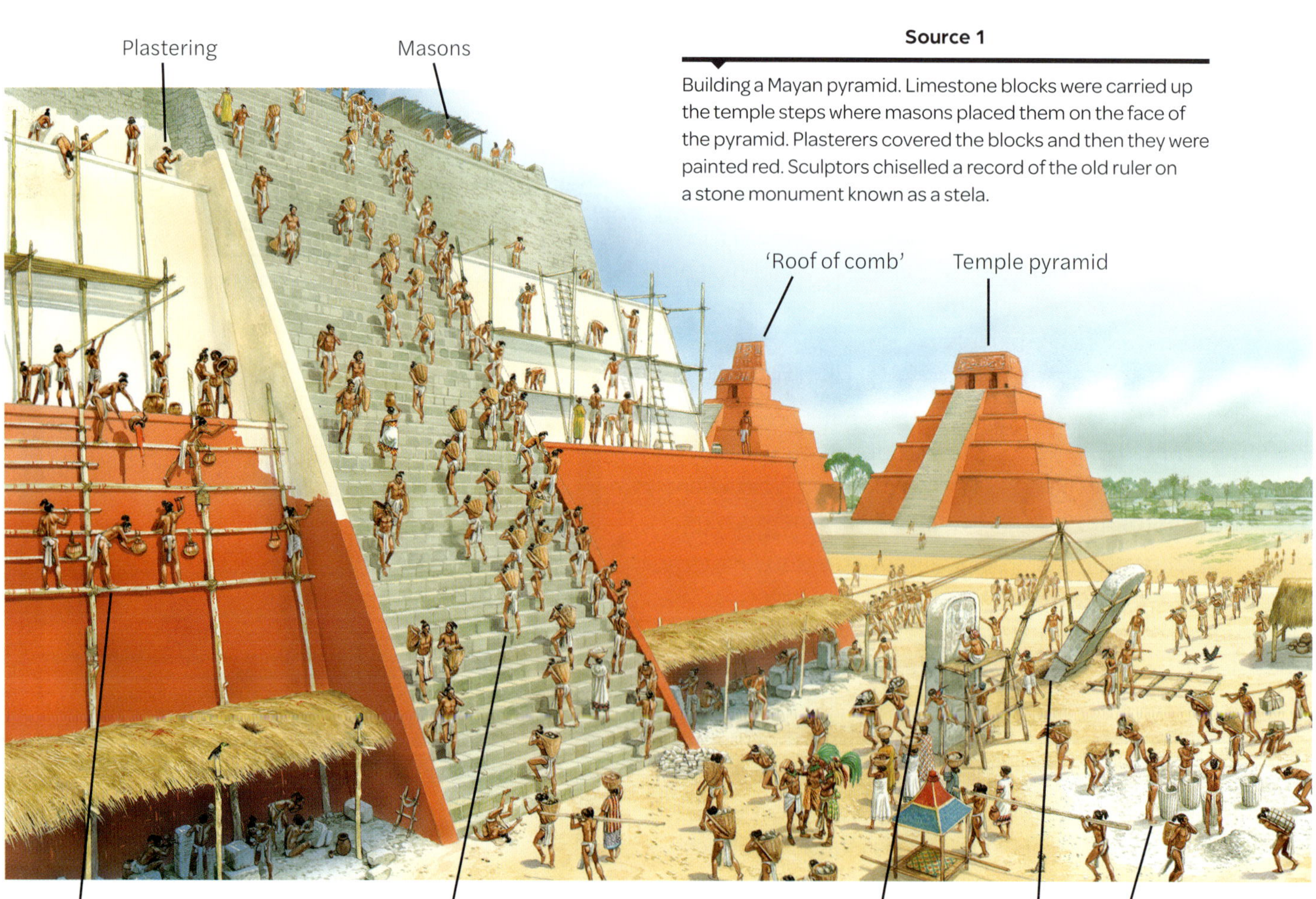

Source 1

Building a Mayan pyramid. Limestone blocks were carried up the temple steps where masons placed them on the face of the pyramid. Plasterers covered the blocks and then they were painted red. Sculptors chiselled a record of the old ruler on a stone monument known as a stela.

Source 2

Between 1839 and 1942, the archaeologists John Lloyd Stephens and Frederick Catherwood surveyed Mayan sites in Mexico and Central America. In 1843, the two released a book that captured the magnificent cities of the Mayan Empire. This illustration shows a pyramid in Chichen Itza that was being slowly reclaimed by the rainforest around it. [19th century lithograph, 'El Castillo, Chichen Itza', by Frederick Catherwood.]

Maya city-states

Areas of rainforests were cleared and swamps were drained to build Mayan cities. Early cities were constructed about 300 BC. **Archaeologists** have uncovered large Mayan cities with pyramid temples, palaces, open plazas and courts for the ball game ***ulama*** (see page 175).

Unlike the civilisations that would follow – the **Aztecs** (see pages 172–75) and the **Inca** (see pages 184–85) – the Maya were never one unified empire ruled from a capital city. The Maya Empire was a series of independent **city-states** linked by language, trade and culture.

Each city-state controlled the area around their city and sometimes other smaller cities nearby. City-states would compete for power and resources with other city-states in the **empire**.

Tikal (see page 171) was the largest city of the Maya civilisation with a population of 100 000 people. Chichen Itza was another large Mayan city with massive pyramids, large temples and the largest *ulama* court found. Chichen Itza is now protected as a World Heritage Site and 1.4 million people visit the site each year.

Pyramids for the gods

Great stepped-pyramid temples were built for the gods in Mayan cities. The temples were used for religious ceremonies, including human sacrifices to the gods (see page 174).

Each city-state was controlled by a ruler who was thought to be descended from the gods. When a ruler died, his body, along with food and precious goods, was buried within a pyramid temple. The new ruler had hundreds of workmen build another pyramid temple over the top of the old one. There were no wheeled vehicles or cattle, so large stone blocks had to be moved by hand.

Farming in the forest

Away from the cities, most Mayans were farmers. They cleared fields in the rainforest called ***milpas*** where they planted corn, beans and squash. Corn was used to make a flat bread called **tortilla**. Because of the poor-quality rainforest soils, the Maya rested each *milpa* after the harvest to let the soil regenerate. The same practice is used in Central America today.

The Mayans discovered how to use cocoa pods to make an unsweetened hot-chocolate drink enjoyed by all social classes. Cocoa was often consumed in religious ceremonies and other celebrations.

Markets and trading

Stalls were set up in the city plaza for traders to sell their goods. The Mayans used cocoa beans as currency – the same beans they used to make their drinking chocolate.

Key market traders

- Farmers from nearby villages sold fruit, vegetables and tobacco.
- Hunters sold deer, puma, jaguar and other skins along with caged birds and monkeys.
- Fishers sold live turtles and shellfish along with dried fish.
- **Artisans** sold goods such as woven fabrics, toys, tools, candles and musical instruments.

Trading routes connected the city–states of the Maya Empire with other peoples in Central America. People cut tracks through the rainforest to transport goods overland by foot. Canoes connected cities around the Yucatan Peninsula.

Maya writing system

The Maya were the only early people of **Mesoamerica** to have a complete writing system. **Scribes** used around 1000 different symbols to write messages. Their writing has survived in stone carvings, plaster, wood, pottery and in paper books such as the Dresden Codex (see Source 4).

The Mayan writing system combined symbols representing whole words with symbols that represented syllables. The **hieroglyphics** were translated in the 20th century and now most of the writing can be understood.

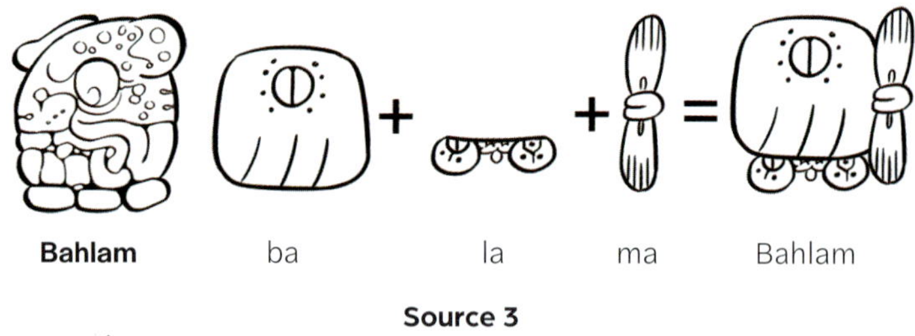

Source 3

The Mayan word for 'jaguar' was *bahlam*. Using the Maya writing system, 'jaguar' could be written as a pictorial symbol that looks like a jaguar's head. It could also be written as a cluster of three syllables: ba-la-m(a).

Source 4

The Dresden Codex is the oldest and best preserved of only four Mayan books that still exist today. The codex includes hieroglyphs from the Mayan writing system, which contain religious rituals and movements of the Sun, moon and Venus. [The Dresden Codex (c. 1200–1250), chalk-coated fibre, Saxon State Library, Dresden, Germany.]

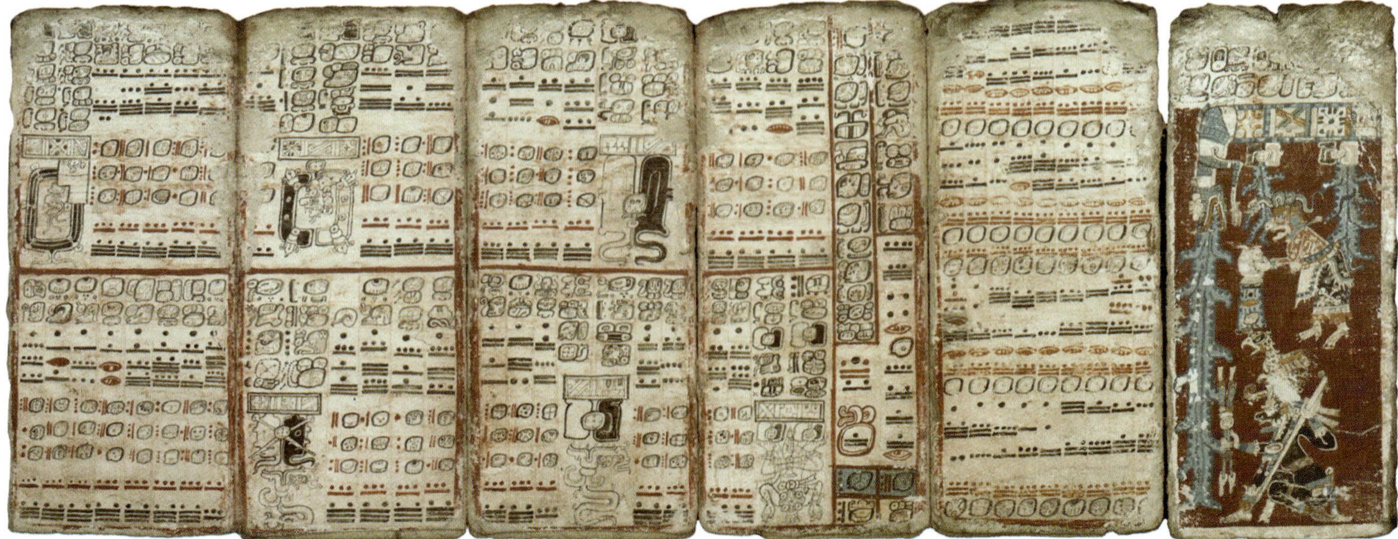

Source 5

This jade funerary mask was made around 683 CE for Pakal, king of the city-state of Palenque. Mayans fashioned funerary masks from jade stone to look like the deceased person. Funerary masks covered the face of dead royalty to stamp them as a god and give them eternal life. [*Funerary mask of jade and funerary offerings of Pakal King of Palenque* (c. 683 CE), jade, National Museum of Anthropology, Mexico City, Mexico.]

Science

The Maya were very advanced in astronomy and mathematics. Astronomers plotted the position of the Sun, the moon and Venus. The Maya created their own solar calendar that divided the 365 days of the year into 18 months of 20 days each, plus the extra month of Wayeb, with five extra unlucky days.

The mysterious end of the Maya civilisation

The Maya Empire ended quite suddenly around 950 CE, but historians cannot agree why. Something unknown forced the Maya people to leave their city-states. One by one they were abandoned and left for the jungle to reclaim.

Some of the theories put forward are:

- they exhausted all the resources and could no longer support their growing population
- continual warfare between competing city-states may have forced cities to be abandoned
- a major catastrophe such as a drought could have forced people to leave.

By the time the Spanish invaders arrived in the 1500s, most Maya were living in agricultural villages and their amazing cities had been reclaimed by the rainforest, until they were excavated by archaeologists several centuries later (see pages 170–71).

Learning ladder H5.3

Show what you know

1 What natural resources did the Maya get from the rainforest?

2 How do we know about the ancient Mayan writing system?

3 What were pyramids for and how were they built?

4 Why would it have been difficult to live in the dense rainforest?

5 Look at Source 4. What kinds of things might it tell us about the Maya? What things do you think it cannot tell us?

Cause and effect

Step 1: I can recognise a cause and an effect

6 Which of these caused the Maya to mine limestone?
   - a They needed material to build with.
   - b There was lots of limestone around.
   - c There was no other stone around.
   - d They lived in a jungle.

Step 2: I can determine causes and effects

7 What caused the Mayan city-states to be in conflict with each other?

8 What were the effects of Mayans farming in the forest?

Step 3: I can explain why something is a cause or an effect

9 Which of the suggestions about why the Mayan cities were abandoned do you think is most believable? Give evidence for your response.

Step 4: I can analyse cause and effect

10 How do you think it would have been different if the Maya were one unified civilisation, rather than a collection of individual city-states? What benefits and disadvantages would there be?

Cause and effect, page 203

.4

how it works

How is laser technology used in archaeology?

New lidar laser technology is producing images of Mayan cities hidden beneath the thick jungles of Central America. Lidar stands for Light Detection and Ranging.

Uncovering Mayan cities

Archaeologists are using advanced technology called **lidar** to uncover hidden Mayan cities in the jungles of Central America. Lidar lights up an area with laser light and then measures the differences in laser return times to make a 3D image of the area.

Source 1

Using lidar laser technology a huge network of buildings and structures at Tikal is revealed, producing a map for archaeologists to explore on the ground.

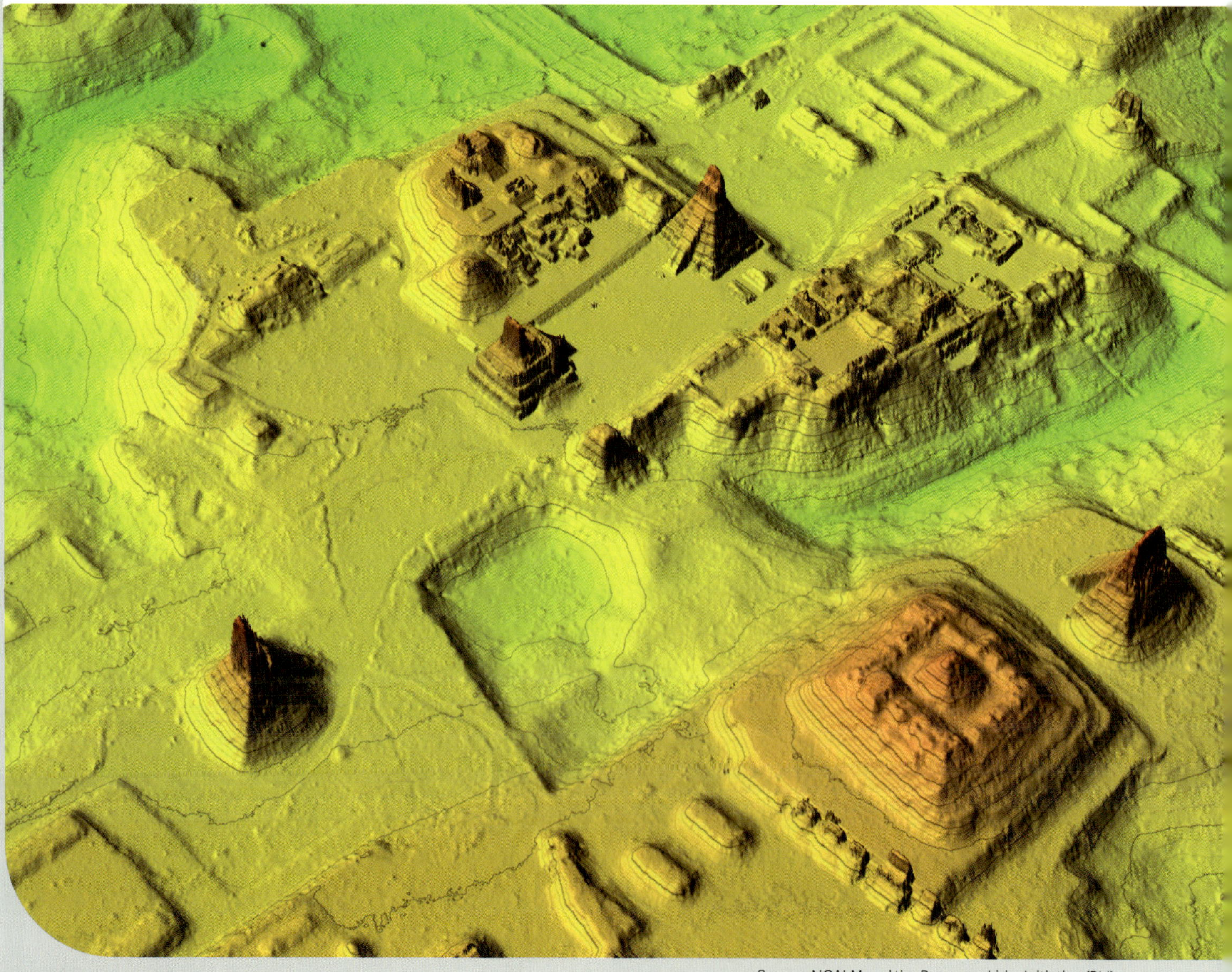

Source: NCALM and the Pacunam Lidar Initiative (PLI)

Source 2

To the naked eye, dense rainforest covers the Mayan city of Tikal in Guatemala.

In the thick jungles of Guatemala in Central America, lidar imagery is saving archaeologists years of searching. The new technology looks through the thick forest to reveal the structures built underneath.

New information provided by lidar technology is causing archaeologists to rethink the size of the Mayan population and how their cities were organised. At its peak about 1500 years ago, the Maya civilisation was estimated to have a population of five million people. The new data provided by lidar imagery suggests the population might have been double or triple this.

Archaeologists have been surprised by features such as walls, fortresses and moats that showed they did more to defend themselves than was previously thought. Raised wide highways linking all of the cities in the area have also been revealed. These might have allowed travel and trade even during the rainy season.

Learning ladder H5.4

Show what you know

1 What is lidar and how does it work?

2 How does laser technology save archaeologists time and effort?

3 What new things has lidar taught historians about the Mayans?

4 What year CE was the Mayan civilisation at its peak?

5 How might Mayans have travelled during rainy periods?

6 Look at Source 1. What buildings and structures are suggested by the lidar scan?

7 Do you think lidar would be useful for all historical work? Explain your answer.

Source analysis

Step 1: I can determine the origin of a source

8 How was the image in Source 1 made?

Step 2: I can list specific features of a source

9 Describe in detail the features shown in the lidar image in Source 1.

Step 3: I can find themes in a source

10 What have archaeologists deduced about Mayan life from using lidar technology?

Step 4: I can use my outside knowledge to help explain a source

11 What do the discovery of walls, fortresses and moats suggest about life for Mayan people in Central America?

Source analysis, page 197

What was life like for the Aztecs?

The Aztecs developed a sophisticated culture and built what was, at the time, the grandest city in the world. They supported themselves by farming from floating gardens and by demanding tributes of food and goods from surrounding tribes.

The mighty Aztec Empire

Before the arrival of the Spanish to the region, the powerful **Aztec** Empire dominated Mexico. The Aztecs arrived in Mexico around 1250 CE and established their capital city of Tenochtitlan in 1325 on five swampy islands in the middle of Lake Texcoco. By the 1400s, the Aztecs had taken control of the land surrounding the city. They made allies with or demanded control of local tribes to establish their dominance over much of Mexico.

Source 1

Eagle knight statue from the Templo Mayor, the main temple in Tenochtitlan. [*Knight-eagle* (c. 1480), stucco and terracotta statue, 170 × 118 × 55 cm, Museo del Templo Mayor, Mexico City, Mexico.]

Source 2

Daily life in the Aztec capital of Tenochtitlan

Source 3

A model of the floating gardens known as *chinampas* on Lake Texcoco

'The city is as big as Seville or Cordoba. The main streets are very wide and very straight; some of these are on the land, but the rest and all the smaller ones are half on land, half canals where they paddle their canoes ... [it had a great marketplace where] sixty thousand people come each day to buy and sell ... [Its merchandise included] ornaments of gold and silver, lead, brass, copper, tin, stones, shells, bones and feathers.'

Source 4

Hernán Cortés, quoted in Mary Wiesner-Hanks, *An Age of Voyages: 1350–1600*, Oxford University Press, 2005.

Emperor Moctezuma I ruled the Aztec Empire for 28 years from 1440. During his reign the Aztecs expanded their empire across central Mexico. Moctezuma I's grandson, Emperor Ahuitzoti (1486–1502), doubled the size of the **empire** and ruled over four million people. As a great military leader, Ahuitzoti conquered many other tribes. He was followed by his nephew, Moctezuma II, who reigned from 1502 to 1520.

Courageous and talented warriors received rewards from the **emperor** each year such as headdresses, clothing made from animal skins or feathers and jewellery. The best warriors became jaguar or **eagle knights** who commanded army regiments.

The amazing Tenochtitlan

Emperor Ahuitzoti also oversaw the reconstruction of Tenochtitlan to restore its grandeur. The island city was home to 200 000 people and was connected to the mainland by three causeways. Its temples were huge stepped pyramids up to 45 metres high.

In a letter written to King Charles I in 1520, Hernán Cortés (discussed on pages 178–79) described the city that he was about to attack (Source 4).

Around 1000 men were employed to clean the city streets, and boats sailed through the city canals to collect the waste. Private homes had their own toilets, so the roads didn't have the stench of many European cities.

Lake Texcoco surrounded Tenochtitlan. It provided good protection from attack, but its water was too salty to drink. Aztec engineers built a canal to divert water from freshwater springs to the city. Further canals were built within the city to supply all areas of the city with fresh water.

There was little land around Tenochtitlan to farm, so the Aztecs built floating gardens called ***chinampas***. These artificial islands were built by weaving reed platforms under the water to support layers of soil and vegetation until the top layer of soil was above the surface of the lake. *Chinampas* were used to grow crops of corn, beans, squash, tomatoes and chilli peppers.

The large population needed more resources than could be supplied, so the people relied on trade and **tributes** from villages and tribes outside the city. Tributes were taxes paid by surrounding societies to the Aztecs in the form of goods such as corn, grain, cotton, chocolate, tobacco, feathers and precious metals.

Aztec society

The Aztecs were organised into a strict social hierarchy where each person had their place. At the top of the hierarchy was the emperor, followed by nobles, warriors, commoners and **slaves**. The emperor and nobles wore cotton clothing and feathered headdresses and jewellery made of gold, silver and precious stones.

Commoners paid tributes of food and goods to the nobles who owned the land. Commoners were not allowed to wear cotton clothes. Education was compulsory for all children and schooling was for all – the society did not discriminate against the poor or women, which was different to many other cultures at the time. Laws were very strict, with death being the common punishment for those who broke them.

‘... they sacrifice their own persons, some hacking the body with knives; and they offer up to their idols all the blood which flows, sprinkling it on all sides of those mosques, at other times throwing it up towards the heavens ... They have another custom, horrible, and abominable, and deserving punishment ... they take many boys or girls, and even grown men and women, and in the presence of those idols they open their breasts, while they are alive, and take out the hearts and entrails, and burn the said entrails and hearts before the idols, offering that smoke in sacrifice to them. Some of us who have seen this say that it is the most terrible and frightful thing to behold ...’

Source 5

Hernán Cortés, *Letters and relations of Hernán Cortés to King Charles V*, (c. 1519; translated 1866), Library of Congress, Washington DC, USA.

Religion

The Aztecs worshipped many gods that they believed brought them good fortune. Some of the most important Aztec gods were:

- *Tlaloc* – the god of rain, water and farming
- *Huitzilopochtli* – the god of Sun and war
- *Quetzalcoatl* – the god of knowledge and creator of the world
- *Tezcatlipoca* – god of life on Earth.

Important gods had their own pyramid temples, many of which were used for human sacrifices to the gods. The Aztecs believed that sacrificing animals and humans paid a blood debt to the gods to ensure good fortune. Captured prisoners from surrounding tribes were sacrificed at an altar on the top of the temple, watched by a crowd below.

The Spanish invaders considered the practice barbaric, and it was another reason why they believed that they had to convert the Aztecs to **Christianity** (see page 188). Hernán Cortés explained the practice of human sacrifice in a letter to Emperor Charles V, the Spanish King (Source 5).

Source 6

Unfortunately, the Spanish destroyed most original sources from the Aztecs. What survives are documents from after the Spanish colonisation of the Americas. This illustration shows a priest performing a human sacrifice and offering the heart to the war god Huitzilopochtli. [*Codex Magliabechiano* (c. mid-16th century), paper, Biblioteca Nazionale Centrale, Florence, Italy.]

Ulama

Ulama was a ball game similar to volleyball that was played throughout Mesoamerica. It is the oldest known team sport in the world and was played by all levels in society. Nearly 1500 *ulama* courts have been found throughout Mesoamerica.

Ulama developed around 1600 BCE. The game had slight variations in rules for different regions and eras but, in general, the ball needed to be kept in the air using forearms, hips and thighs. The hard rubber ball had to cross a line or pass through a stone ring. Serious injury was common on the stone courts and members of the losing team could be sacrificed to a god.

Source 7

Ulama is a Mesoamerican ball game that is still played today. It is the oldest continuously played sport in the world. It is also the oldest known game that used a rubber ball. Evidence of *ulama* or similar games have been found throughout Mesoamerica. Archaeologists have uncovered nearly 1500 ball courts, figurines of players dating from 1200 BCE and rubber balls dating from 1600 BCE.

Learning ladder H5.5

Show what you know

1 How did the Aztecs come to dominate central Mexico?

2 How were great warriors rewarded?

3 How did Aztecs living in Tenochtitlan get food and water?

4 What was education like for the Aztecs?

5 Draw a diagram showing the Aztec social hierarchy. You could draw it in the style of a stepped pyramid.

6 Read Source 5. What language in this source suggests the author was biased? For each piece of language you identify, briefly explain why you think it might be biased.

7 Look at Source 3. What are the advantages or disadvantages in farming in this way compared to farming on land?

Chronology

Step 1: I can read a timeline

8 How long had the Aztecs been in Mexico before they founded Tenochtitlan?

Step 2: I can place events on a timeline

9 Write down six of the events discussed in this section in order from earliest to most recent.

Step 3: I can create a timeline using historical conventions

10 Using the historical conventions discussed on page 195, create a timeline that features the events from question 9.

Step 4: I can distinguish causes, effects, continuity and change from looking at timelines

11 Look at the timeline on pages 162–63. Suggest events that represent change for the Aztecs. Rank your changes from most important to least important and justify your selections.

Chronology, page 194

How did the Spanish find the Americas?

At the end of the 15th century, European powers looked to the Americas to expand their territories, find new sources of wealth and convert more people to Christianity.

In what is now called the Age of Discovery, the Portuguese and Spanish set off on great voyages to find wealth in new territories and convert local people to the Catholic religion. Explorers went in search of gold, silver and spices in southern Asia, known to the Europeans as the **Indies**.

Accidental discovery of the Americas

King Ferdinand and Queen Isabella of Spain financed the Italian explorer Christopher Columbus to find a new route to the Indies. In 1492, he arrived in what he believed was Cipango (modern-day Japan). The fleet had actually arrived in the Caribbean Islands off the coast of America. During four separate trips between 1492 and 1504, Columbus explored various Caribbean Islands. Still convinced he had discovered the Indies, Columbus referred to the local people as Indians.

Italian explorer Amerigo Vespucci's exploration of South America's east coast confirmed that Columbus had stumbled across a new continent. Vespucci named it after his first name – America.

Conquistador explorers

Columbus was not the first European to reach the Americas. **Viking** explorer Leif Eriksson had explored and settled part of North America 500 years earlier (see pages 46–47). However, Christopher Columbus' discovery was important because it began a period of conquest – where the Spanish King encouraged many more expeditions to explore and conquer the Americas.

Expeditions were led by the **conquistadors**. They were soldier explorers who came in search of fortune.

Source 1

Christopher Columbus' ships, named the Niña, Pinta and Santa Maria, set sail from the Spanish port of Palos in 1492 in search of a trade passage to the Indies (southern Asia).

Source 2

Christopher Columbus sets foot on what he thought, based on his readings from Marco Polo (see page 123), was the island of Cipango (Japan). In fact, it was Cuba, in the Caribbean.

Voyages of discovery to the Americas

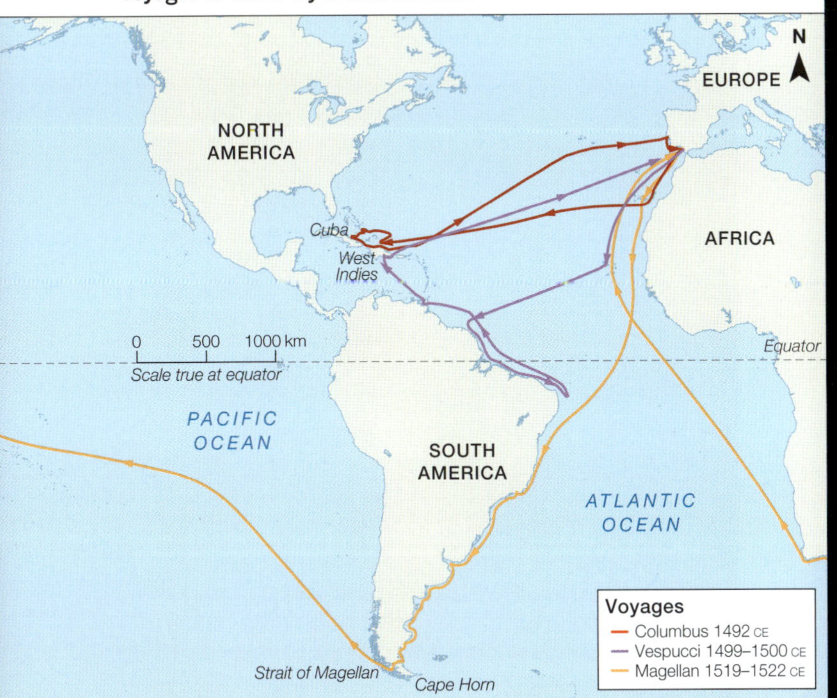

Source: Matilda Education Australia

Source 3

Voyages of discovery to the Americas

Learning ladder H5.6

Show what you know

1 Why did the Portuguese and Spanish want to explore new lands?

2 Why were native Americans called 'Indians'?

Historical significance

Step 1: I can recognise historical significance

3 Which of the following is the most historically significant?
- a Expeditions were led by conquistadors.
- b Conquistadors searched for treasure.
- c Explorers converted locals to Catholicism.

Step 2: I can explain historical significance

4 What do you think the locals pictured in Source 2 thought when they saw the Spanish coming off their ships? Justify your answer.

Step 3: I can apply a theory of significance

5 What period did Columbus' discovery of the Americas begin? Who was affected? Is Columbus' arrival in the Americas still important today?

Step 4: I can analyse historical significance

6 There have been three major arrivals in the Americas that historians can be sure of: Indigenous Americans around 15 000 BCE, Vikings in about 1000 CE, and Columbus in 1492. Rank these in order of significance. What factors should you consider? What other factors could also have been important to determine historical significance?

HOW TO

Historical significance, page 207

key individual

How did Cortés conquer Mexico?

Hernán Cortés was a Spanish conquistador whose small force of well-armed soldiers fought their way across Mexico to conquer the Aztec civilisation.

What was Cortés doing in the Americas?

Hernán Cortés was a Spanish **conquistador**, a soldier explorer who sailed from Spain and Portugal to the Americas (North, Central and South America), Oceania, Africa and Asia. Cortés had heard of Christopher Columbus' discoveries in the Americas and wanted to see these new lands. He was clever and ambitious, sailing to the Americas to conquer lands for Spain, convert the local people to **Catholicism** and steal their treasures.

In 1511, Hernán Cortés joined the conquistador Diego Velazquez de Cuellar in conquering Cuba. Velazquez became lieutenant governor of the Spanish territory in the region. In 1518, Velazquez decided to send Cortés on an expedition to investigate the Mexican coastline, make contact with **Indigenous** tribes and report back.

Cortés quickly pulled together ships, men, weapons, food and horses. Velazquez became concerned that Cortés was arming himself for a military conquest and decided to replace him. Cortés heard of these plans and bribed city officials to sign paperwork so he could set sail in February 1519 before Velazquez could stop him.

Conquest of Mexico

Cortés arrived on the mainland with a small force of 11 ships, 110 sailors, 553 soldiers, 10 heavy guns, four small cannons and 16 horses. He immediately set out to conquer the mighty Aztec Empire (see pages 172–75). Cortés was a commander of great ability and the weapons of his conquistadors were superior to those of the Aztecs (see pages 182–83).

Cortés and his troops destroyed one of the most spectacular cities in the world and wiped out much of Aztec culture. A new settlement with Spanish colonists called Mexico City was built on top of the ruins of Tenochtitlan, with Cortés in control.

Source 1

In 1521, Cortés and his conquistadors were joined by 8000 native people to launch a second attack on the Aztecs in their capital city of Tenochtitlan. The army stormed across the causeways to attack the island city. In a battle lasting 80 days, the Aztec empire fell to the Spanish with 10 000 Aztecs slain (see page 180). Cortés ordered his troops to destroy the city and they left the world's most spectacular city in ruins. [*The Conquest of Tenochtitlan* (c. 17th century CE), oil on panel.]

Cortés' expedition in Mexico 1519–21

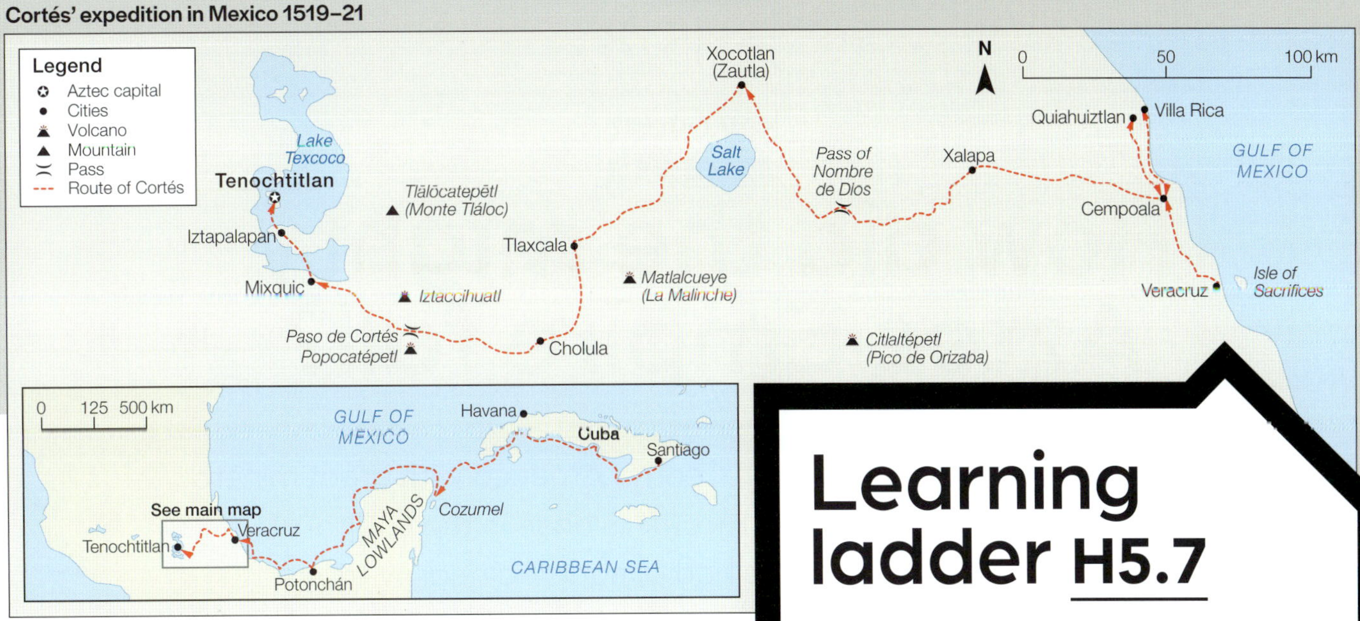

Source: Matilda Education Australia

Source 2

Cortés and his small army arrived on the Mexican coast in February 1519. By November 1519, Cortés and his conquistadors had fought their way inland to the Aztec capital of Tenochtitlan, the largest city in the world. Along the way he grew his army with native tribes who had a score to settle with the Aztecs.

Learning ladder H5.7

Show what you know

1 Why did Cortés want to go to Mexico?

2 What technological advantage did Cortés have over the Aztecs?

3 Why do you think Cortés ordered his soldiers to destroy Tenochtitlan?

4 What has the artist of Source 1 done to show Cortés' dominance over the Aztecs? Give examples from the painting. Why do you think this was painted?

Chronology

Step 1: I can read a timeline

5 Look at the timeline on pages 162–63. In what year did Cortés conquer the Aztecs?

Step 2: I can place events on a timeline

6 Source 2: How far did Cortés and his invading party travel?

February to July 1519 Santiago to take Veracruz

July to September 1519 Veracruz to Tlaxcala to persuade them to join his army

September to October 1519 Tlaxcala to attack Cholula

October to November 1519 Cholula to invade Tenochtitlan

Step 3: I can create a timeline using historical conventions

7 Use the information in question 6 to create a timeline of Cortes' march to Tenochtitlan.

Step 4: I can distinguish causes, effects, continuity and change from looking at timelines

8 From your timeline and Source 2, explain how Cortés arrived in Tenochtitlan with a much larger army than he arrived with on the Mexican mainland.

HOW TO Chronology, page 194

H5.8

key event

How did the mighty Aztec Empire fall?

The Aztecs controlled a large territory in Central America with great mineral wealth. They were ruled by Moctezuma II from his base in the capital of Tenochtitlan (see page 173). The Aztecs had conquered nearby tribes and ruled over 10 million people. Moctezuma II used a network of spies to watch over these tribes and therefore knew that Hernán Cortés and his Spanish conquistadors were on their way to Tenochtitlan.

Cortés and his men landed at the Yucatan Peninsula in Mexico in 1519 with the aim to conquer the **Aztecs** and claim their land, gold and other treasures for Spain. Cortés said: 'We Spaniards suffer from a disease of the heart, which only gold can cure'. The Spanish were greatly outnumbered by the Aztecs, but Cortés was far better armed. His armoured **conquistadors** would use their cannons, rifles, guns and crossbows against the spears and arrows of the Aztecs.

As the Spanish troops made their way inland to Tenochtitlan, they met Indigenous tribes who were unhappy paying taxes to the Aztecs and having their people taken as human sacrifices to the Aztec gods. Cortés persuaded many of these tribes to join him in fighting Moctezuma and the Aztecs. By the time he reached Tenochtitlan, Cortés had added up to 30 000 Indigenous allies to his army.

Knowing the Spanish lust for treasure, Moctezuma sent an ambassador with gifts of gold to meet Cortés. Moctezuma offered the gifts in exchange for Cortés to stop his march to Tenochtitlan. The gifts only made Cortés more excited about the treasures and natural resources the Aztecs owned. He told the ambassador that he planned to visit Tenochtitlan to pay tribute to Moctezuma – a lie to cover his true motive.

Moctezuma guessed the true intent of Cortés and knew he had a problem. To make matters worse, there had been signs from the gods that something bad was going to happen. The volcano Popocatépetl had erupted, there were fires in two temples and a comet had been sighted. Quetzalcoatl – the dragon god of knowledge who created the world and its people – was believed to return soon to rule over the land. In his human form, Quetzalcoatl could be bearded and white-skinned. Could Cortés be Quetzalcoatl coming back to rule?

Source 1

On May 22, 1520 the celebration of the Feast of Toxcatl was taking place in the Templo Mayor.

The conquistadors attacked the unarmed Aztecs, killing all those celebrating in the temple. The Spanish claimed they were intervening to stop a human sacrifice and the Aztecs claimed the Spanish were trying to steal the gold they were wearing at the special event.

When Cortés returned to Tenochtitlan from the coast, he was met by a full-scale uprising against the Spanish and was forced to retreat from the city, losing many soldiers in the process, who were killed in battle or captured and sacrificed.

[Emanuel Leutze, *The conquest of the Teocalli temple by Cortés and his troops* (1849), oil on canvas, 215 × 250 cm, Wadsworth Atheneum Museum of Art, Hartford, USA.]

The Spanish reach Tenochtitlan

Moctezuma welcomed Cortés to Tenochtitlan, either believing he was a god or trying to arrange a peace settlement with the invaders. The visitors were given treasures and invited into the palace. According to a Spanish source, the Florentine Codex, Moctezuma told Cortés:

> 'My lord ... to the land you have arrived. You have come to your city ... here you have come to sit on your place, on your throne. Oh, it has been reserved to you by a small time, it was conserved by those who have gone, your substitutes ... Come to the land, come and rest: take possession of your royal houses ...'

Source 2

Bernardino de Sahagun, *Florentine codex, book XII, chapter XVI* (c. 16th century; translated by Angel Maria Garibay c. 1960).

Cortés responded by imprisoning Moctezuma, destroying Aztec idols, stopping human sacrifices, turning temples into places of Catholic worship and raiding the city for treasure.

Cortés was forced to leave Tenochtitlan to fight off a rival Spanish force that had been sent to arrest him. While Cortés was absent, his deputy, Pedro de Alvarado, slaughtered hundreds of Aztec nobles and priests. Cortés arrived back to find the capital in chaos.

Cortés and his men were holed up in the palace, surrounded by thousands of angry Aztecs. Cortés sent out Moctezuma to calm to crowd. They rejected their leader as a traitor and stoned him to death.

The Spaniards loaded up as much treasure as they could and tried to sneak out of the capital at night. They were spotted on a causeway and surrounded by Aztecs in canoes. Many conquistadors jumped in the water, but sank from the weight of the gold in their pockets.

Source 3

An Aztec warrior

Source 4

A Spanish conquistador

Heavy steel helmet to protect against full-force impact. It also provided protection for the neck and the eyes.

Steel armour protected the upper body and forearms.

One-metre-long steel swords with very sharp edges.

Circular metal shield was convex to help deflect blows.

Other conquistador weapons included:

- light, transportable cannons called falconets, with a range of nearly two kilometres
- rifles known as arquebuses
- handheld pistols known as hackbuts
- metal crossbows with a range of 300 metres
- halberds – two-metre-long combinations of spears and axes.

Source 5

Hernán Cortés kneels before Moctezuma II, leader of the Aztecs in 1519. [*Reception of Hernando Cortez by the Emporer Montezuma* (1878), colour lithograph, Library of Congress, Washington DC, USA.]

Hundreds of Spanish and Aztecs were killed in the skirmish and Cortés was lucky to escape with his life. The Aztecs had driven Hernán Cortés and his Spanish conquistadors from Tenochtitlan.

The fall of the Aztecs

The conquistadors left behind something far more deadly than the weapons they had used to invade Tenochtitlan. In 1520, a **smallpox** epidemic struck the Aztec capital city, killing large numbers of people. Moctezuma's successor died from smallpox so his nephew, Cuauhtemoc, became leader.

The next year Cortés and the conquistadors returned with a huge army to again invade Tenochtitlan. The battle lasted for 80 days until the Aztecs, weakened by smallpox and fighting, were finally defeated. Cortés had Cuauhtemoc tortured and drained the lake in search of treasure he thought they were hiding from him.

Learning ladder H5.8

Show what you know

1 List the factors that made the Aztec a powerful force, as well as the factors that made the Spanish under Cortés a strong military force.

2 What did Moctezuma II do to try to stop Cortés defeating him?

3 Why were the Spanish able to defeat the Aztecs?

4 Look at Source 1. Do you think this is an accurate representation of the slaughter led by Pedro de Alvarado? Explain your answer with reference to the painting and your historical knowledge.

5 Describe the similarities and differences between the equipment and clothing of the Spanish and Aztec warriors in Sources 3 and 4 using a Venn diagram. Why do you think they looked so different?

Source analysis

Step 1: I can determine the origin of a source

6 Is Source 1 a primary or a secondary source? Explain.

Step 2: I can list specific features of a source

7 Source 5 shows Cortés kneeling before Moctezuma II. Examine the Aztecs in this image closely. What is the artist trying to show about the Aztecs? Give evidence from the painting.

Step 3: I can find themes in a source

8 Describe the artistic style in Source 1. Include mention of colour, movement, focus, foreground/middle ground/background and realism. Why do you think the artist painted it in that style? What was the artist trying to say about the event?

Step 4: I can use my outside knowledge to help explain a source

9 Look at Sources 3 and 4. Are these primary or secondary sources? What makes you say that? What do you know about the technology of the time in Spain and Mexico that explains what they are wearing? How was the level of technology different in the two parts of the world?

HOW TO

Source analysis, page 197

How did the Spanish overpower the Incas?

European diseases and a civil war between brothers weakened the Inca Empire just as Spanish conquistadors arrived on a mission to conquer the empire.

The Incan Empire

By 1493, the **Incas** commanded the largest empire of its time. Even though they lacked horses, wheeled vehicles, steel tools and weapons, and a written language, they had created cities, roads, **aqueducts** and terraced farms in the steep Andes Mountains of South America. The Inca **emperor** ruled over 10 million people in an **empire** that stretched for 4000 kilometres from Quito in Ecuador to Santiago in Chile.

Incan society was ruled by the emperor and noble class. To run such a large empire efficiently, all common people had to work directly for the empire. In return, the government paid the people with food, clothes and other goods.

The Spanish attack

Following Cortés' success in conquering the **Aztecs** (see pages 180–83), the Spanish King sent more expeditions to explore and conquer lands in the Americas. After a number of explorations in 1524, Francisco Pizarro and a small army of **conquistadors** set out in 1529 to conquer the Inca Empire and their wealthy land, rumoured to be rich in silver and gold.

Pizarro's task seemed impossible. He was leading a small army of just 177 conquistadors to meet Inca emperor Atahualpa and his army of 80 000 men in Cajamarca. However, many things went Pizarro's way:

- an epidemic of **smallpox** swept through the Inca, killing large numbers of people
- Incan emperor is killed by smallpox and two brothers fight for control
- there was unrest among the Inca people after years of paying **tributes** to Inca nobles
- the Inca army had never seen the horses or superior weapons of the Spanish soldiers (see pages 182–83).

Pizarro's exploration and conquest of the Inca Empire

Source: Matilda Education Australia

Source 1

Pizarro's exploration and conquest of the Inca Empire

Source 2

Spanish conquistadors under the leadership of Francisco Pizarro capture Incan Emperor Atahualpa. [Juan Lepiani, *La captura de Atahualpa* (c. 1920–1927), 60 x 85 cm, Museo de Arte de Lima, Peru.]

The Battle of Cajamarca

In November 1532, Pizarro arranged for an audience with Emperor Atahualpa in Cajamarca's town square. Pizarro positioned cannons on a rooftop and hid soldiers in buildings nearby.

The town square was quickly filled with Pizarro's hidden army and the sound of cannon fire. The Inca weapons were no match for the Spanish, and hundreds of Incas were killed. Emperor Atahualpa was taken hostage and had to offer to fill a large room with gold and two large rooms with silver for his freedom. The Spanish also demanded that Atahualpa be baptised as Catholic.

The Incans believed their emperor was a living god and did not attack the Spanish while he was their prisoner. With Atahualpa captured there was no leader to challenge the invaders. This allowed the Spanish time to call for more reinforcements.

In 1533, Emperor Atahualpa was killed and the Spanish installed a **puppet emperor** on the throne. The better-armed conquistador army attacked and took the important cities of Cuzco and Quito. The Spanish eventually took control of the empire.

Learning ladder H5.9

Show what you know

1 What technology did the Inca have?

2 How was Incan society run?

3 What things went in Pizarro's favour when he tried to conquer the Incas?

4 What mistakes did the Inca make before and during the battle of Cajamarca?

5 What played a bigger part in the conquest of the Incas, luck or good strategy? Back up your answer with historical evidence.

6 Look at Source 2. Who are the Spanish and who are the Incas? How do you know? What is the difference between the facial expressions of the Spanish and the Incas? Why are they different?

Cause and effect

Step 1: I can recognise a cause and an effect

7 How would horses have helped the Spanish conquer the Incas?

Step 2: I can determine causes and effects

8 Did smallpox help the Spanish conquer the Incas?

Step 3: I can explain why something is a cause or an effect

9 Create a flowchart, outlining all the decisions made by the Incas when confronted with the Spanish. Make up some alternatives that the Incas could have chosen and suggest their effects.

Step 4: I can analyse cause and effect

10 Copy and complete the diagram below:

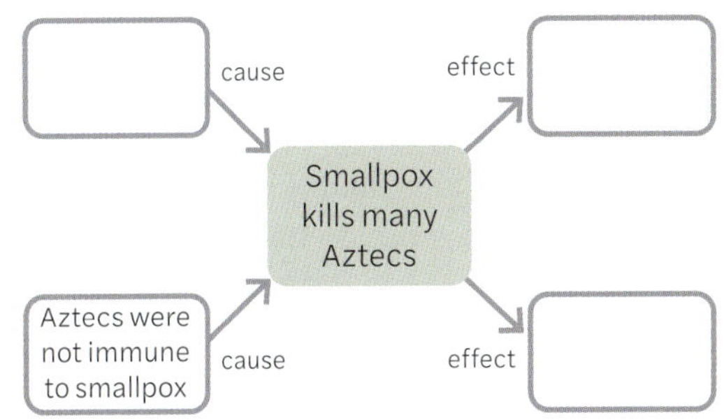

HOW TO Cause and effect, page 203

How do governments influence markets?

There are many ways to solve economic problems and come up with the best solution for a business or even a country. The Incans came up with a unique solution to successfully run their civilisation without money or trading for goods.

Running the huge Inca Empire

The Inca Empire was the largest that South America has ever known. The Incans engineered amazing monuments, built terraced fields on steep mountains and had plenty of food, textiles and gold. However, the Inca Empire is possibly the only advanced civilisation in history to not have merchants, money or any form of commerce.

How did they solve this **economic problem**? The Inca government controlled the production and distribution of all goods. All citizens in the Inca Empire worked for the government and in return were issued the food, clothing, tools and other goods they required from government storehouses. There were no markets to trade or purchase goods and therefore there was no need for money.

Australia's market economy

Australia's economy is quite different to the centrally controlled economy of the Incans. Australia has what is referred to as a **market economy**, where the availability and pricing of goods and services is driven by demand from consumers in Australia and overseas. There is little government intervention, as market economies work on the assumption that the forces of **supply and demand** best determine the health of the economy.

Source 1

Terraced fields at Machu Picchu on the steep slopes of the Andes Mountains in Peru surround government storehouses. The Inca Empire did not use commerce or money – people worked for food and other goods supplied from the storehouses by the government.

Markets refer to the relationship between the buyers and sellers. Markets are formed when **producers** make and sell goods and services to **consumers**. Producers depend on consumers to buy goods and services from them. Consumers depend on producers to provide the goods and services that they *need* and *want*.

We are all consumers and everyone in the workforce is also a producer. When we work, we provide labour to make goods or services for other consumers to buy. Workers earn money to buy goods and services from businesses to satisfy their needs and wants. When we spend money, we are consumers.

Source 2

The term 'market' refers to an exchange between producers and consumers, such as this retail market at Fremantle. The Australian economy has many markets in operation, such as the retail market, the labour market, financial markets and the stock market.

Government intervention

Modern market economies such as Australia do have some government intervention to provide economic stability. Individuals and businesses pay money to the government in the form of **taxes** and **rates**. The government uses these payments to provide welfare payments such as **unemployment benefits** and **pensions**, or services such as education and health.

Governments can stimulate an economy by spending money on building schools, hospitals and roads. Money is provided to businesses to build the services, helping those businesses grow and employ more workers. Governments can also slow an economy by increasing the levels of taxation, leaving consumers with less money to spend on goods and services, and businesses with less money to expand and employ workers.

Learning ladder H5.10

Economics and business

Step 1: I can recognise economic information

1 How did the Inca solve the economic problem of running a large empire with no form of money or commerce?

Step 2: I can describe economic issues

2 What is a market economy and how is it different to the economy of the Inca Empire?

Step 3: I can explain issues in economics

3 Look at the image in Source 2. From the detail shown, identify what goods producers are selling to customers.

4 Source 1: What issues did farmers face as producers in the Incan economy? How were they rewarded for their efforts?

Step 4: I can integrate different economic topics

5 Compare government intervention in the Incan economy and in modern market economies such as Australia.

Step 5: I can evaluate alternatives

6 Do you think it would be a good idea for modern Australia to do away with money? Justify your response with examples from each economy.

How did Spanish colonisation affect the Indigenous people?

The Indigenous populations in the Americas were devastated by conflict with the Spanish as well as by the diseases, famine and enslavement that were caused by Spanish settlement.

Colonisation and conversion

Following the defeat of the **Aztecs**, the area now called New Spain became part of the Spanish Empire. The capital was Mexico City, built on the ruins of Tenochtitlan. The Spanish viewed the local people as savages who worshipped devils. In 1513, King Ferdinand of Spain instructed the local populations to convert to **Catholicism** or:

> 'With the help of God we shall use force against you, declaring war upon you from all sides and with all possible means, and we shall bind you to the yoke of the Church and Their Highnesses; we shall enslave your persons, wives, and sons, sell you or dispose of you as the King sees fit; we shall seize your possessions and harm you as much as we can as disobedient and resisting vassals.'

Source 1

King Ferdinand of Spain, 1513

Source 2

European colonisation of the Americas

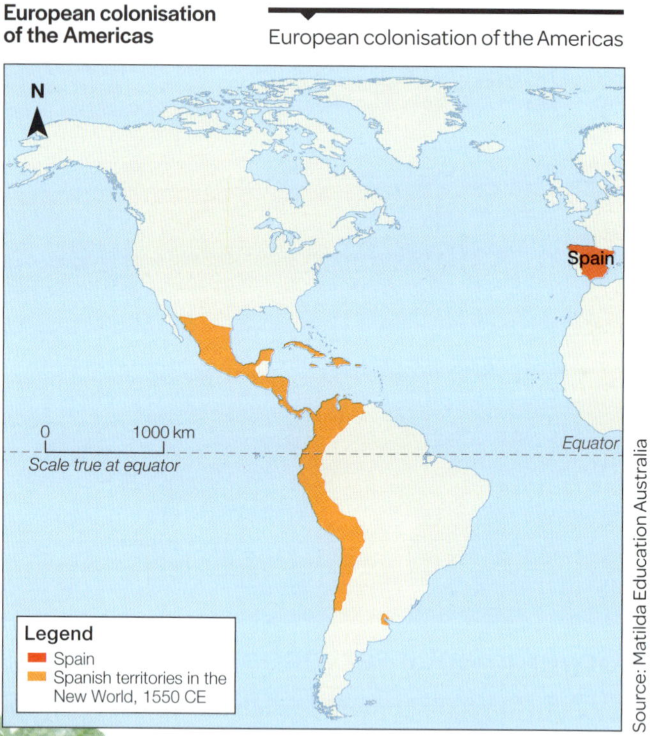

Source: Matilda Education Australia

Source 3

Spanish conquistadors march into an Incan village.

Indigenous workers

Conquistadors and settlers were given grants of land that included the Indigenous people who lived there. In return, the settlers would 'civilise' the local people and convert them to the Catholic faith. Under this ***encomienda*** system the Indigenous people were considered free, but really they were **slaves** and the property of settlers.

The Spanish monk Bartolome de Las Casas wrote about the treatment of the Indigenous population on the Island of Hispaniola in 1542:

> 'As if those Christians who were as a rule foolish and cruel and greedy and vicious could be caretakers of souls! And the care they took was to send the men to the mines to dig for gold, which is intolerable labour, and to send the women into the fields of the big ranches to hoe and till the land, work suitable for strong men ... And the men died in the mines and the women died on the ranches from the same causes, exhaustion and hunger.'

Source 4

Bartoleme de Las Casas, *Brief Account of the Devastation of the Indies* (1542)

Declining population

The populations of the Aztec, Maya, Inca and other Indigenous groups of New Spain plummeted after the arrival of the Spanish. Fighting, killing and new diseases caused the local population of Mexico to decline by 90 per cent by the early 1600s.

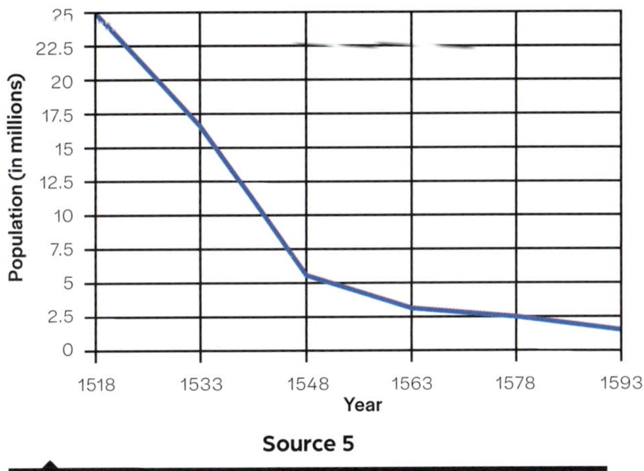

Source 5

Estimated Indigenous American population of Mexico, 1518–93

Learning ladder H5.11

Show what you know

1 Source 2: Using an atlas, work out which modern countries were a part of 'New Spain'.

2 By what percentage did the Indigenous American population decline by from 1518–93?

3 How did the Spanish benefit from the *encomienda* system?

4 Why did the Spanish want to convert the Indigenous Americans to Christianity; specifically, Catholicism?

5 Source 4: How did Bartolome de Las Casas criticise the Spanish settlers in his text?

Continuity and change

Step 1: I can recognise continuity and change

6 Source 1: Read King Ferdinand's instructions. In your own words, what does the king threaten if the Indigenous Americans refuse to convert to Christianity?

Step 2: I can describe continuity and change

7 Other than loss of life and religion, what other things would the local people have lost from Spanish colonisation?

Step 3: I can explain why something did or did not change

8 Source 5: Why did the Indigenous population of the Americas decline after Spanish colonisation?

Step 4: I can analyse patterns of continuity and change

9 When settlers from outside come to a new land, such as the Americas or Australia, the local population often declines. Why is this? Is this an accident or intentional? What of the local culture remains after this loss of population?

Continuity and change, page 200

HOW TO

Masterclass

Learning Ladder

Work at the level that is right for you or level-up for a learning challenge!

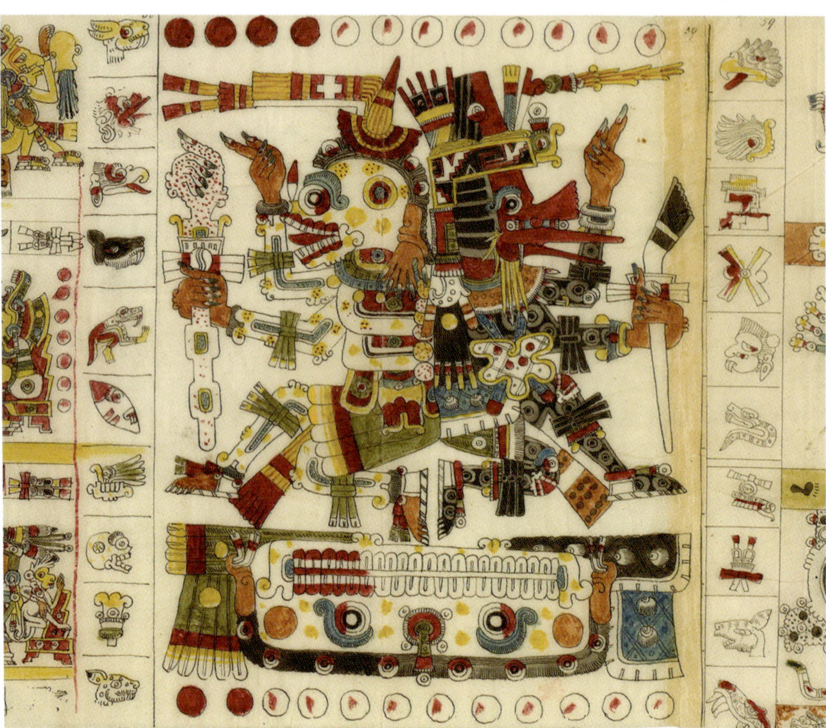

Source 1

Picture drawn on paper, Mexico, c. 1500 [*Codex Borgia* (facsimile) (original c. 1500; facsimile 1825–1831), paper, The British Museum, London, UK.]

Step 1

a **I can read a timeline**

Look at the timeline on pages 162–63.

i How long did the Inca Empire last?

ii How long was it between the development of writing by the Mayans and them building their first pyramids?

iii How long after Columbus arrived were the Aztecs defeated?

b **I can determine the origin of a source**

Where was Source 1 drawn? How long ago was it drawn?

c **I can recognise continuity and change**

Which of these things have changed, and which haven't changed from the past to the present?

i Boats travelling from Europe to the Americas

ii People believing in Christianity in Mexico

iii People farming in South America

iv Smallpox killing many people

v Worshipping at pyramids

d **I can recognise a cause and an effect**

Match the cause and effects in the following:

Aztecs had many enemy tribes.	Cortés was able to ally with them.
Columbus wanted to find a new route to Asia.	He set sail west on the Atlantic Ocean.
Mayans used up all their natural resources.	Many people died.
Indigenous Americans were not immune to smallpox.	People walked from one continent to the other.
There was a land bridge from Asia to North America.	Their civilisation declined.

e **I can recognise historical significance**

Which section in this chapter represents the most significant development in the Spanish conquest of the Americas?

Step 2

a **I can place events on a timeline**

Place these events in order from earliest to most recent:

- 65 years after settling there, the Aztec build the Templo Mayor in Tenochtitlan.
- After Huitzilihuitzli is king, Chimalpopoca is king, reigning for 10 years.
- Five years after the building of the Templo Mayor, Huitzilihuitzli is king. He reigns for 22 years.
- The Aztec have a three-year war with the Tepaneca, starting the year that Chimalpopoca's reign ends.
- The Aztecs settle in Tenochtitlan in 1325.

b **I can list specific features of a source**

Describe in detail all the figures down the left-hand side of Source 1. Include shape, colour and what they resemble.

c **I can describe continuity and change**

Describe how the lives of Mesoamerican people changed after the arrival of the Spanish.

d **I can determine causes and effects**

i What made the Mayan civilisation decline?

ii What happened after Pizarro asked the Inca for gold?

e **I can explain historical significance**

Why was the desire of Europeans to trade with Asia historically important?

Step 3

a **I can create a timeline using historical conventions**

Take the events in Step 2a and place them on a timeline, using correct historical conventions, which are outlined in the History How-to section on page 195. Note that there are two events and three spans of time in the list above.

b **I can find themes in a source**

Source 1 is likely to be related to religion. What religious symbols can you see in the image?

c **I can explain why something did or did not change**

Aztec religion is not as popular as it was in 1400 CE. Why?

d **I can explain why something is a cause or an effect**

The Incan Empire made sure everyone was fed and clothed. How did they do this?

e **I can apply a theory of significance**

What are some theories as to why the Mayan empire disappeared? Is the demise of the Mayans still important today?

Step 4

a **I can distinguish causes, effects, continuity and change from looking at timelines**

Look at the timeline on pages 162–63. Over the period 1150–1572, what things changed the most? What things in the Americas may have stayed the same? Give historical evidence in your answer.

b **I can use my outside knowledge to help explain a source**

Using your historical knowledge, what figures or images can you recognise in Source 1?

c **I can analyse patterns of continuity and change**

When the Spanish first arrived in the Americas, they arrived as conquerors of the local peoples. Now people live more equally in Central and South America. Why do you think this is?

d **I can analyse cause and effect**

Copy and complete the following table:

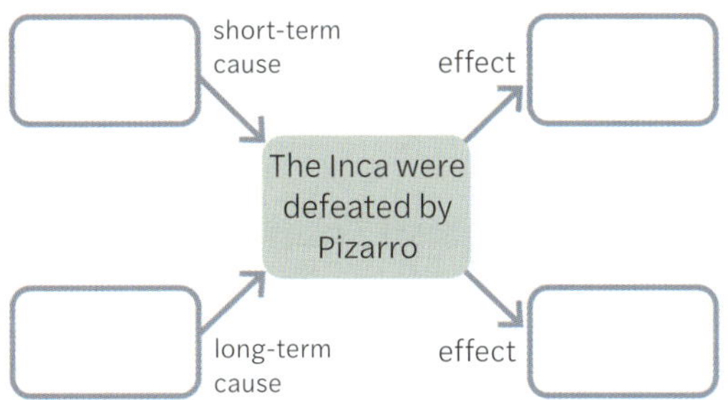

Masterclass

e **I can analyse historical significance**

List three reasons for and against the statement: 'Columbus's voyage to the Americas was the most important thing to happen in the Americas for centuries'.

Step 5

a **I can describe patterns of change**

Look at the timeline on pages 162–63. The Mayan civilisation lasted much longer than the Aztec or Incan civilisation. Using the timeline and your historical knowledge, explain this difference and suggest why this is the case.

b **I can use the origin of a source to explain its creator's purpose**

Knowing when Source 1 was created, which people created it? Knowing that it may be concerned with religion, why do you think it was created? Is it similar or different to religious art you know of from another civilisation?

c **I can evaluate patterns of continuity and change**

Do you think the Spanish settlement of the Americas was a good or bad thing? Use historical evidence to back up your answer.

d **I can evaluate cause and effect**

Was the settlement of the Spanish in the Americas really the main cause in the decline of their culture? Would it have declined anyway? Give reasons for your answer.

e **I can evaluate historical significance**

Was the conversion of Indigenous Americans from traditional religious beliefs to Christianity/Catholicism significant? Explain your answer with reasons.

Historical writing

1 Structure

Imagine you are given the essay topic, 'Evaluate the impact of the Spanish colonisation of the Americas on the Aztecs'. Write an essay plan for this topic. Include at least three main paragraphs.

2 Draft

Using the drafting and vocabulary suggestions on page 214, draft a 400–600 word essay responding to the topic.

3 Edit and proofread

Use the editing and proofreading tips on page 215 to help you edit and proofread your draft.

Historical research

4 Organise and present information

Imagine you are doing a large research project: 'A comparison of the Aztec, Inca and Maya people'. Write a contents page for this project. There should be an introduction, conclusion, at least four main sections and many subsections. Number your chapters.

Capstone

How can I understand Spanish conquest of the Americas?

In this chapter, you have learnt a lot about Spanish conquest of the Americas. Now you can put your new knowledge and understanding together for the capstone project to show what you know and what you think.

In the world of building, a capstone is an element that finishes off an arch, or tops off a building or wall. That is what the capstone project will offer you, too: a chance to top off and bring together your learning in interesting, critical and creative ways. You can complete this project yourself, or your teacher can make it a class task or a homework task.

mea.digital/GHV8_H5

Scan this QR code to find the capstone project online.

H6

History

How-To

History has its own set of skills to help us analyse and understand societies in the past and the key ideas, people and changes that shape the world we live in today. Historical skills are based around interpreting sources of evidence from the past, promoting debate and encouraging investigation.

Chronology

Chronology is the arrangement of events into the order they happened. Historians put events in order, from earliest to most recent. They do this to:

- identify continuity and change
- determine cause and effect (both short-term and long-term)
- provide a narrative showing how history unfolded.

I can read a timeline

When answering questions, include part of the question in your answer. This way, your answer is a full sentence by itself, and you could use it in a piece of history writing. If a question asks, 'When did the Normans conquer England?' you should answer with, 'The Normans conquered England in 1066 CE', not just '1066 CE'.

Working out the years between dates of BCE and CE is the same as when you work out the difference between positive and negative numbers in maths.

For example, how many years were there between 490 BCE and 150 CE?

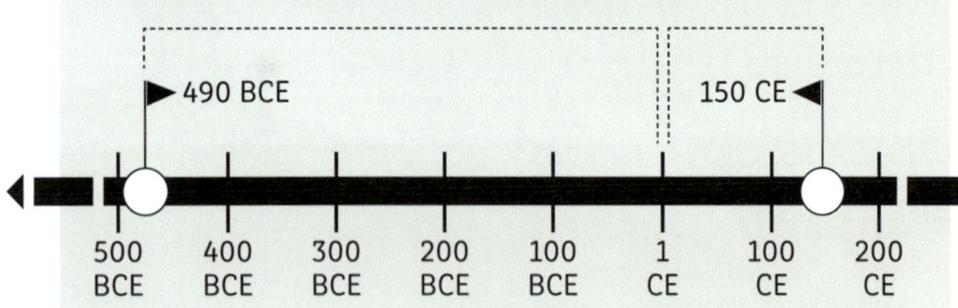

490 BCE is the same as –490

So, the difference in years between 490 BCE and 150 CE would be:

490 plus 150

= 490 + 150 = 640 years

When years are noted as 'BP', this means 'before the present'. You need to subtract that number of years from the current year to get a date that is in CE or BCE.

For example, 10 000 BP is 10 000 years before the current year. Say the current year is 2020.

10 000 – 2020 = 7980

So 10 000 BP = 7980 BCE.

I can place events on a timeline

Placing events on a timeline means putting them in order from earliest to most recent. The order goes like this:

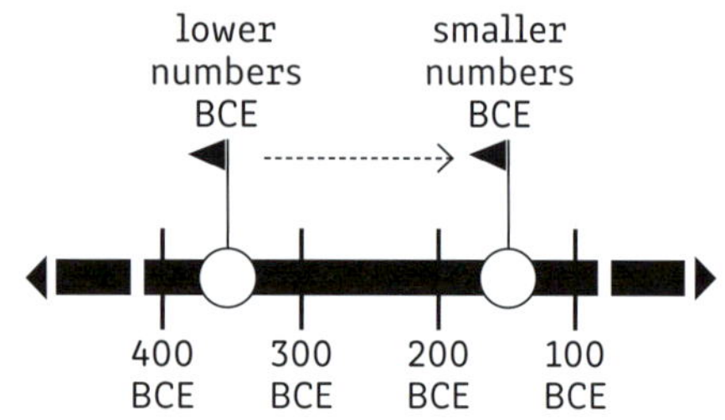

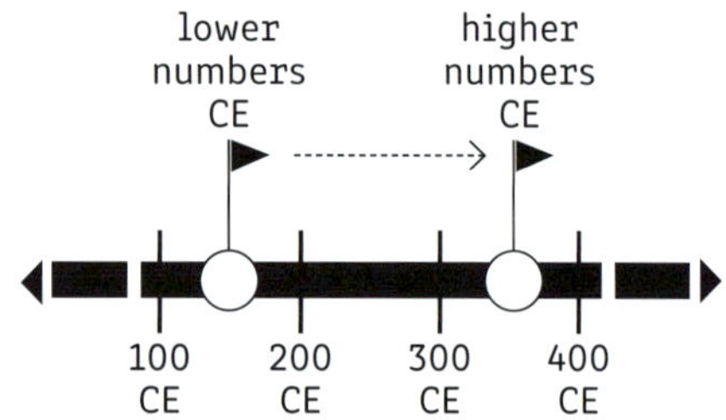

Just like maths, with positive and negative numbers, an event that lasts for 10 years BCE will start with a higher number and go *down*. An event that lasted for 10 years in CE will start with a lower number and go *up*.

For example:

700–690 BCE = an event taking 10 years

700–710 CE = an event taking 10 years

I can create a timeline using historical conventions

Follow these steps to make a timeline.

1 Select events to include on your timeline.
2 Put them in order, from earliest to latest.
3 Work out the earliest and latest dates you want to include.
4 Pick a span of years that your timeline will cover, so that the earliest and latest dates will fit.
5 Choose which unit of time you want to use: years, decades or centuries.
6 Work out how many segments your timeline will need. Figure this out based on how much space you have on your sheet of paper.
7 Draw a straight line and divide it up into segments.
8 Number the segments into the units you selected in Step 5.
9 Label the events on your timeline.

If an event goes over many years, add it as a span rather than a line.

If there is a large gap between the events on your timeline, add a zigzag line to show there has been a jump in time.

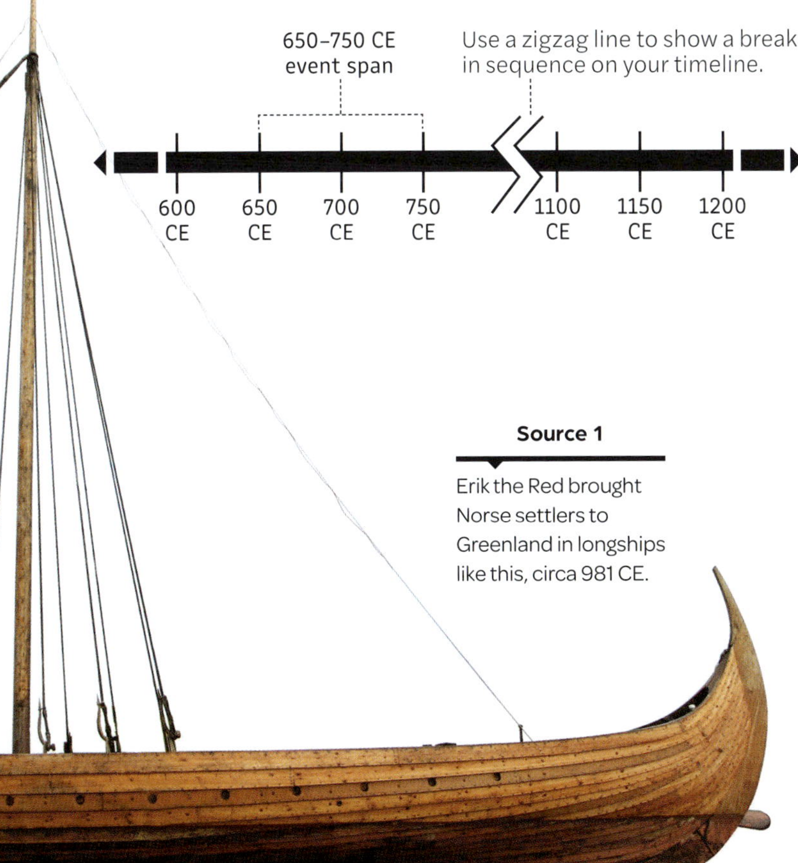

Source 1

Erik the Red brought Norse settlers to Greenland in longships like this, circa 981 CE.

Creating a timeline – an example

1 Choose some events to include on your timeline:
981 CE Erik the Red discovers Greenland, 1028 CE King Canute ruler of England, 793 CE Viking raid on Lindisfarne

2 Put them in order from earliest to latest.
793 CE Viking raid on Lindisfarne, 981 CE Erik the Red discovers Greenland, 1015 CE Vikings abandon Vinland settlement in North America

3 Work out the earliest and latest dates you want to include.
793 CE is the earliest, 1015 CE is the latest

4 Pick a span of years that your timeline will cover, so that the earliest and latest dates will fit.
775–1025 CE (250 years)

5 Choose the unit of time you want to use: years, decades or centuries.
Segments of 25 years

6 Work out how many segments your timeline will need. Figure this out based on how much space you have on your sheet of paper.
Ten blocks of 25 years. I'm using an A4 exercise book. I have about 20 cm to draw in. Ten blocks at 2 cm per block would give me 20 cm.

7 Draw a straight line and divide it up into segments. Note that if you need 10 blocks of space, you will need 11 lines.

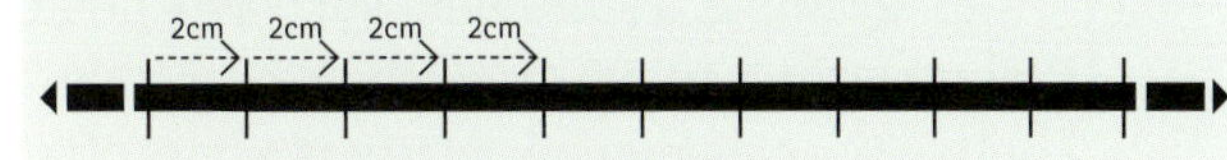

Ten blocks at 2 cm per block

8 Number the segments into the units you selected in Step 5.

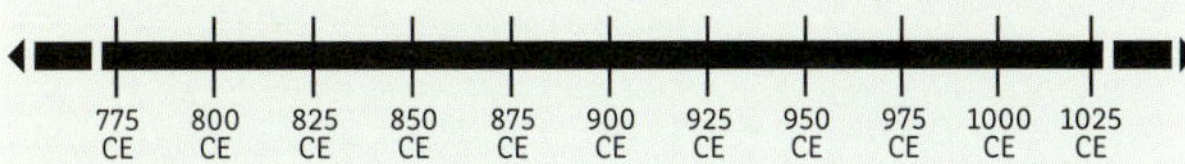

9 Add the events you want to include on your timeline.

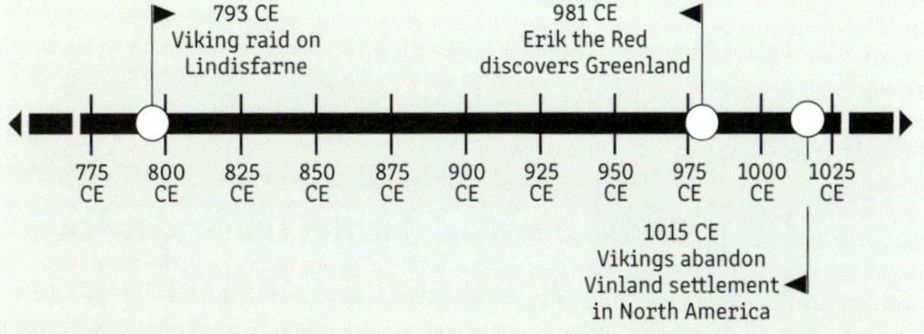

step 4 I can distinguish causes, effects, continuity and change from looking at timelines

Causes and effects

Distinguishing cause and effect on a timeline involves looking at the timeline for events and developments that are linked.

In looking for events that are linked, ask yourself:

- are both events about the same thing? Perhaps two or more events are about religion, war, trade, social upheaval or changes in thinking.
- are both events about different things, but you suspect there is a cause–effect link between them? For example, there is often a boom in artistic output in times of strong government, so perhaps one event causes the other.
- are more than two things linked? For example, does one event have two effects? Or are there two causes of one effect?

If you have found links, try to confirm your belief. See if you can find other information to back up your belief.

Continuity and change

To be able to distinguish continuity and change in a timeline, you need to classify the events or developments in the timeline into things that can continue or change.

For example, a statement such as 'Christopher Columbus sails to the Americas' doesn't show continuity or change because it is a one-off event. You need to work out what *type* of event it is. In this case, it could be considered as the discovery of a new continent or an impressive sea voyage.

You can then look at other parts of the timeline, or compare it to the present, and ask yourself, do we still see discoveries of new continents? Do we still see impressive sea voyages?

Look for continuity and change within the timeline, or find it by comparing timeline events with later times, like the present.

It is easiest to compare timeline events with the present, because it is the time we know most about. For example, seeing 'Mongols improve trade networks' on a timeline could show continuity with the present, because trade networks are still being improved today. But if we saw 'Normans conquer England', we might say this is a change. In modern times, we do not generally see one civilisation conquering another.

step 5 I can describe patterns of change

Describing patterns of change from looking at a timeline involves bringing lots of events together. You should find a *group* of events that are linked.

Things you need to remember about patterns of change include:

- You can show a pattern of change from within the timeline, or from the timeline to a later period, such as the present.
- Patterns of change won't necessarily be all about the same thing. A pattern of change in history is like finding a theme in a book.
- What is changing? Examples could include trade, society, life-expectancy or technology.
- How does the change occur? Is it quick improvement, slow evolutionary improvement, quick decline, slow and steady decline, or rises and falls.
- Typical things that could be considered patterns of change in history include:

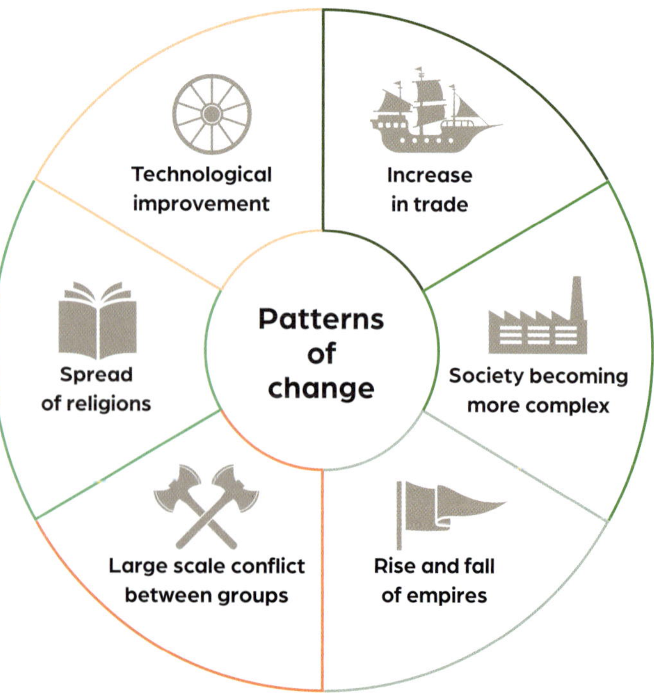

Source analysis

Source analysis asks us to look at evidence and ask, 'How do we know what we know about the past?' A good source analysis interprets and makes meaning of the source:

- Who created it?
- When was it created?
- What was the author or creator's purpose?
- What is the historical context of the source?

I can determine the origin of a source

To figure out who produced a source and when it was produced, look at the clues in the source. You can often find this information in the caption. If not, you can research it online or at the library.

I can list specific features of a source

Listing detailed features usually means writing more. Listing general features of a source is like giving a summarised version. Listing detailed features would be the long version, describing as much as you can, as shown in the table below.

General features	Detailed or specific features
The most obvious features	Obvious and minor details
The most important things in the source	Everything in the source, whether or not you think it is 'important'
Using vague words: 'big', 'small', 'very', 'good' …	Using specific words and phrases, such as: 'three times bigger than …', 'small/big when compared to …', 'in the background/ foreground', 'useful for …'

I can find themes in a source

Often, a source contains a theme. The theme can help you uncover more meaning in a source. A theme is something that you might notice after recognising specific features.

Some examples of themes and how they might be shown are listed in this table.

Theme	How this might be shown in a source
Beauty	A statue of a handsome person with a muscular body and symmetrical facial features
Faith	A decoration on a vase showing people offering sacrifices to a god
Good vs bad	A statue of two figures fighting – one that looks like an angel and one that looks like a demon
Hierarchy	An image going from top to bottom, with gods on the top, then people, then animals, then rocks
Humanity vs nature	A building placed in a natural place, dominating it; e.g. a temple on a mountain
Technology in society – good or bad	Good: a statue of a person using a new farm implement looking happy. Bad: a painting of people happily working in the fields, while a person using a machine to work looks unhappy
War or conflict	A decoration on a building depicting a large-scale war or conflict

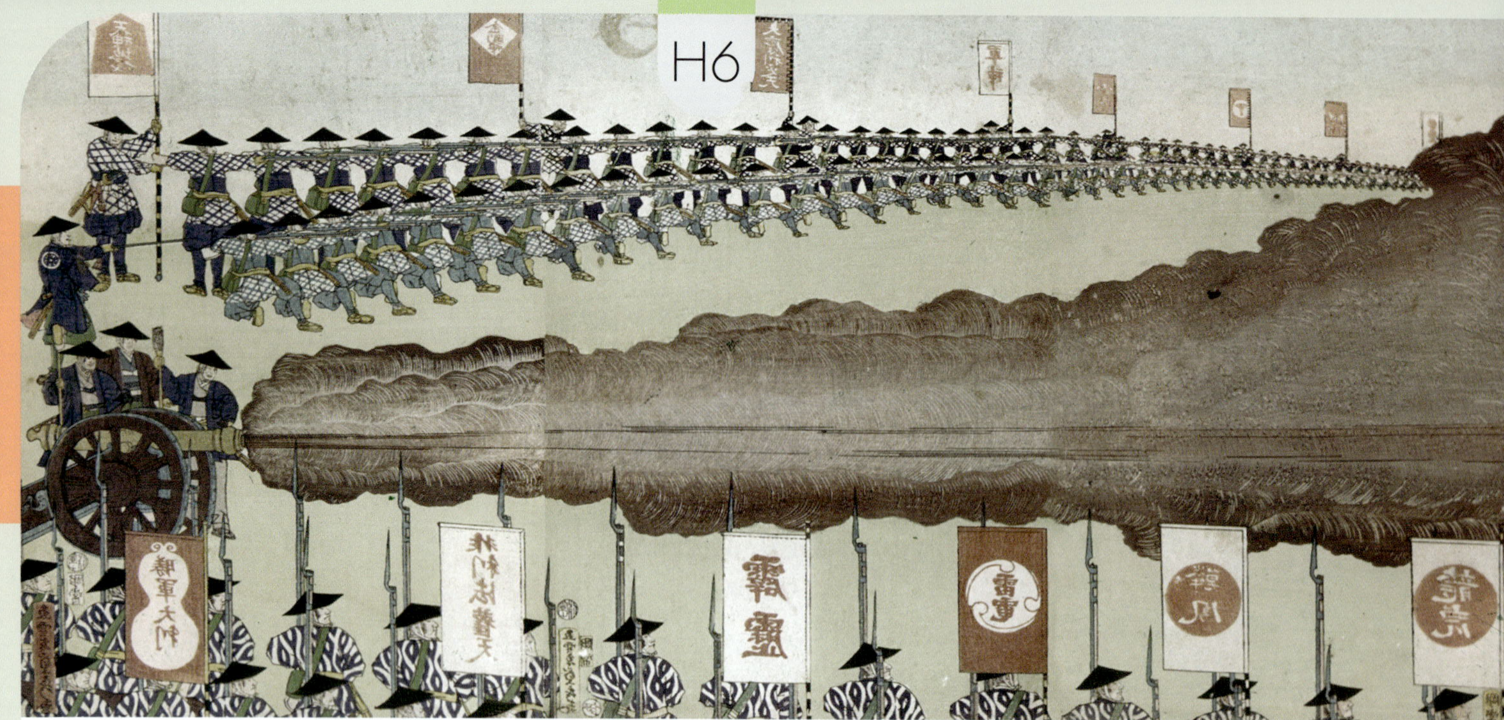

Source 2

Soldiers of the Tokugawa Shogunate shoot a cannon in this late-period woodblock. What themes do we see here? [Gountei Teishu, *Japan Military Training* (c. 1867), woodblock print, triptych.]

If you think there is an underlying theme, after looking at all the details in a source, always give evidence *from the source* to back up your answer.

Not all sources have themes. If the source is a simple piece of technology, like a sword, it might not have a 'theme' like the painting shown in Source 3. However, you could still comment on the type of society that produced it. For example, what materials did they use? How advanced were they? Who did they fight with?

step 4 I can use my outside knowledge to help explain a source

A visual source can tell us a lot about a time in history. However, we can also use information that we know about a period to help understand a visual source.

These are two different skills. They both help us to understand, but they should not be confused.

Evidence from visual source ▷ helps to understand ▷ time in the past

Outside knowledge about the time ▷ helps to understand ▷ the visual source

Before using outside knowledge to help understand a source, you first need knowledge about the period. Then, think what knowledge is relevant to the source, as outlined in this table:

What's in the visual source	Outside knowledge that might help to understand it
People coming ashore	The many new lands Vikings explored; e.g. Greenland, Iceland and North America
A scene from the Bible	Knowledge of the structure of the Christian Church and its power in medieval Europe
A person on horseback with bow and arrow	Mongol war technology and cavalry tactics
A Japanese castle	The nature of warfare that took place between *daimyo* in feudal Japan
A person standing atop a square pyramid	Aztec religious rituals

Both evidence from the visual source and outside knowledge about the era can help us gain understanding.

I can use the origin of a source to explain its creator's purpose

Here are three ways you could explain a creator's purpose. You could:

1 use *who* created it to explain *why* it was made
2 use *when* it was created to explain *why* it was made
3 use both when it was created *and* who created it to explain why it was made.

The third explanation is the best. Here are some questions to ask of the source, that will help you explain the creator's purpose.

- What do you know about who created the source?
  - How old were they?
  - What gender were they?
  - What job did they do?
  - What position did they have in society? For example, were they powerful or powerless?
  - What beliefs did they have?
- What was going on at the time the source was made?
  - Were there any important events taking place?
  - What was going on politically?
  - What biases might people have had?
  - What was normal behaviour in society?
- Why was it produced?

The purpose of a source refers to what the source was originally made for. Don't get confused and think about what *we*, as historians, might use it for. Sources aren't usually created to leave records for historians.

Try to get into the head of the creator *at the time they were making the object*. For example, were they trying to:

- influence people?
- sell something?
- tell their version of events?
- make art? If so, who would enjoy it?
- make something practical, such as a tool?

The table below provides two examples of using the origin of a source to explain its purpose, using a Spanish manuscript and a Viking runestone.

Source	**Origin: Who created the source**	**Origin: When the source was created**	**Why you think the source was created (its purpose)**	**Using the origin of the source to explain its purpose**
Runestone	Swedish Vikings	400–550 CE	As a gravestone	Swedish Vikings created this runestone sometime between 400–500 CE. The markings and the design of the stone suggest that it is a gravestone for a deceased Viking. It looks similar to gravestones that we see in modern times as well.
The *Florentine Codex* (a research study that includes lots of Aztec primary sources)	A Spanish monk called Bernadino de Sahagun	1545–90 CE	To study and record Aztec culture	The *Florentine Codex* was made in the 16th century by the Spanish monk Bernadino de Sahagun. He included many primary sources from the Aztec culture. Because the Spanish conquerors and European diseases were killing many Aztecs at this time, he collated the codex to record Aztec culture, so that it would not be lost.

Continuity and change

Scale of change from least (on the left) to most (on the right)

One reason we study history is to see how life was different in the past. We can also see how an idea or a piece of technology evolved over time.

Continuity and change are on a scale, between 'no change' and 'completely different', as shown below.

Continuity and change can exist at the same time: some things stay the same, while others change. Change can be fast or slow, happen gradually or in a burst.

No change at all (e.g. human DNA)	A bit different (e.g. food)	Quite different (e.g. attitudes to race, gender and sexuality)	Completely different (e.g. transportation technology)

step 1

I can recognise continuity and change

The easiest kind of continuity or change to notice is from a historical period to the present, because we know a lot about the present. This table looks at continuity and change in medieval Europe, during Mongol expansion, and in Japan under the *shoguns*.

Civilisation	Continuity from then until now	Change from then until now
Medieval Europe	Europeans speak different languages, such as English, French and German	Kings and queens have no real power
Mongols	People farm cows and ride horses	There is more equality between men and women
Feudal Japan	Geisha are still female entertainers	People are not allowed to carry swords around

When you are trying to recognise continuity and change, ask:

1 Did it exist in the past and *still exists* now? That is continuity. You can say: 'This situation still exists today and represents continuity from the past to the present.'

2 Did it exist in the past but *doesn't exist* now? That is change. You can say: 'This situation is different today, and represents a change from the past to the present.'

Source 3

The Florentine Codex includes information on the Aztec gods, such as Quetzalcoatl, also depicted in this stone statue on display at the National Museum of Anthropology, Mexico City, Mexico.

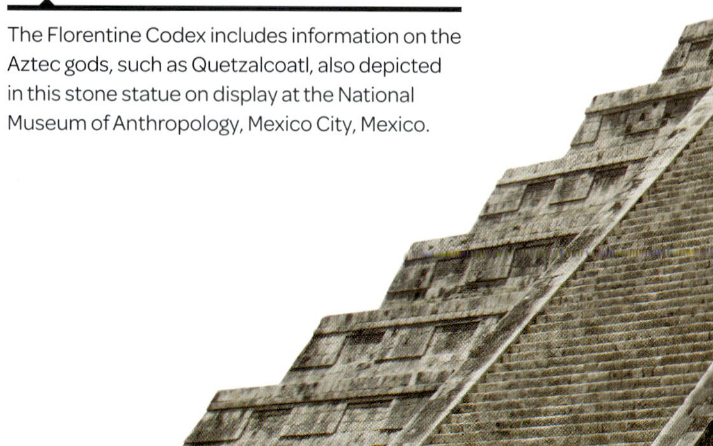

I can describe continuity and change

Describing continuity and change is more difficult because you have to recognise it yourself (without help) and describe it, rather than just noticing it.

Civilisation	Earlier time	Later time	Continuity between these times	Change between these times
Vikings	800 CE: Vikings begin raiding European coasts	1050 CE: Vikings settling around Europe	Many vikings have red hair and long beards	Viking descendants live in harmony with other Europeans rather than attacking them
Aztec	1325: Aztecs first settle Tenochtitlan	1521: Capital Tenochtitlan conquered by Spanish	Aztecs practice traditional religion	Military and political power in Mexico is in the hands of Spanish descendants, not the Aztecs

After recognising continuity or change, you must then write descriptive sentences. These sentences should use historical evidence to back up your claim.

I can explain why something did or did not change

Explanations require you to answer the question *why?* When explaining continuity and change, you need to:

- recognise it
- describe it
- know what caused it, or what effect it had.

I can analyse patterns of continuity and change

Examples of patterns of continuity include:

- close family bonds
- the importance of food
- the impact of disease
- natural disasters.

Examples of patterns of change include:

- the improvement in technology
- the rise in population
- the spread of new ideas.

Source 4

El Castillo, the Temple of Kukulcan, a Mesoamerican step pyramid at Chichen Itza, Mexico

step 5

I can evaluate patterns of continuity and change

Evaluation is a higher-order thinking skill. It involves remembering what you have learned and being able to use that information to make sense of something or make a judgement.

Evaluation can involve:

- assessing whether a theory or belief is true; for example, 'Is it true that all empires rise and fall?'
- comparing different ideas; for example, 'Which is a better life for humans: being a nomad or being a farmer?'
- judging between different things; for example, 'Some people think Mongol expansion was more influential than feudal Japan: who is right?'
- asking 'So what?' for example, 'Was it a good or bad thing? Or was it both good and bad?' A historian might include both positive and negative aspects to form a balanced view, as shown in this table. (See the key in the next column for an explanation of the colours.)

Each balanced view in the table contains four elements:

1 A statement about whether the situation was a continuity or a change.

2 A statement showing the positive aspect of it.

3 A statement showing the negative aspect of it.

4 A statement summarising the balance.

There is no 'right' answer when evaluating. Having a balanced view is not always the best answer, either. For example, it would be difficult to argue for the benefits of Hitler's policy to exterminate specific groups of people. However, some answers are better or worse than others. Better answers:

- use more historical evidence as examples in their evaluation
- use more logical reasoning – they show directly how beneficial the patterns of change were.

Situation	Positive thing	Negative thing	Balanced view
Continuity: Mongolian nomads live outdoors in yurts (large circular tents)	Mongolians enjoy a high quality of life and live in nature	They can experience harsh weather such as rain, wind and cold temperatures	Many Mongolian people continue to live as nomads, staying in yurts. These people benefit from a strong connection to nature that helps their wellbeing. However, they are also subject to harsh weather such as rain and cold. Overall, Mongolian nomads lead rewarding yet sometimes tough lives outdoors.
Change: Vikings spread across Europe and the North Atlantic Ocean	Vikings improved trade networks between Europeans	Vikings attacked and killed many when they raided	Vikings originally raided and attacked European coastal areas, killing many people. As time wore on, they settled and intermingled with the locals. This greatly improved trade networks in Europe, to the benefit of all. On balance, the Viking colonisation of Europe benefited trade in the long-run, but caused violence in the short term.

Source 5

The helmet of a Norman soldier, worn during the conquest of England

Cause and effect

Cause and effect is visible every day. For example, when we open the fridge, the light comes on; when we wave our hand, the bus stops for us.

There are short-term and long-term causes. The short-term cause of the fridge light coming on is opening the door. The long-term cause is because we are hungry. Equally, there are short-term and long-term effects. After I wave my hand, the short-term effect is that the bus stops. The long-term effect is that I arrive at school on time.

Source 6

Statue of Christopher Columbus, erected in Barcelona, Spain

Cause and effect requires understanding which events are linked and why. When we say things are linked by cause and effect, we say they have a *causal link*. This means that one thing *caused* the other.

Most things that happen have multiple causes and effects – some of which are more important than others. Two main types of causes are:

- historical actors: the individuals or groups involved; for example, Charlemagne, Christopher Columbus, Moctezuma II, Genghis Khan and Erik the Red
- historical conditions: social, political, economic, cultural and environmental; for example, mass unemployment, drought, world religions and bad harvests.

However, just because one event happens after another, it doesn't always mean that the first event caused the second. For example, when a rooster crows in the morning, we don't think it makes the sun rise. You also need to be able to tell a believable story about why something caused its effect.

Events in history are not inevitable. When we study cause and effect, it can seem like things were always going to work out in a certain way. Yet, change a few conditions and things could have happened differently. If Genghis Khan had died from an illness as a boy, would the trade networks in Asia be so significant today? It is easy, with hindsight, to think our cause and effect explanations are perfect. But we need to be cautious when we make claims about cause and effect, as many events happen at random.

I can recognise a cause and an effect

Recognising cause and effect means correctly choosing from a list of possibilities. For example, which of these is most likely to be the cause of the Norman conquest of England, when a group of ex-Vikings from northern France won the Battle of Hastings in 1066 and took over England?

A The Normans introduced feudalism to England.

B The Normans had superior military tactics and lots of archers.

C Vikings spread throughout the Northern Atlantic.

D Normans spoke French.

E There was another battle at Stamford Bridge in 1066 as well.

The only one of these options that is linked to the conflict is B. Option A refers to a time *after* the wars, so it can't be the cause. Option C refers to Vikings, not Normans, so it is not relevant. Option D and Option E just tell us facts that wouldn't necessarily lead to conflict.

For events to be causally linked:

- one event must come before the other
- you must be able to tell a believable story about why one event caused the other
- if possible, you should have some historical evidence that one event caused the other.

There is not necessarily a right or wrong answer, but there are better or worse answers. Better answers use more historical evidence, and have more logical reasoning.

step 2

I can determine causes and effects

Determining cause and effect means deciding what the cause or effect of something might be. Knowledge of the period will help with this.

Examples of historical causes include:

- conditions:
  - social
  - political
  - economic
  - cultural
  - environmental.
- actors:
  - individuals
  - groups.

If you suspect that two things are linked, see if one of the above items is the cause.

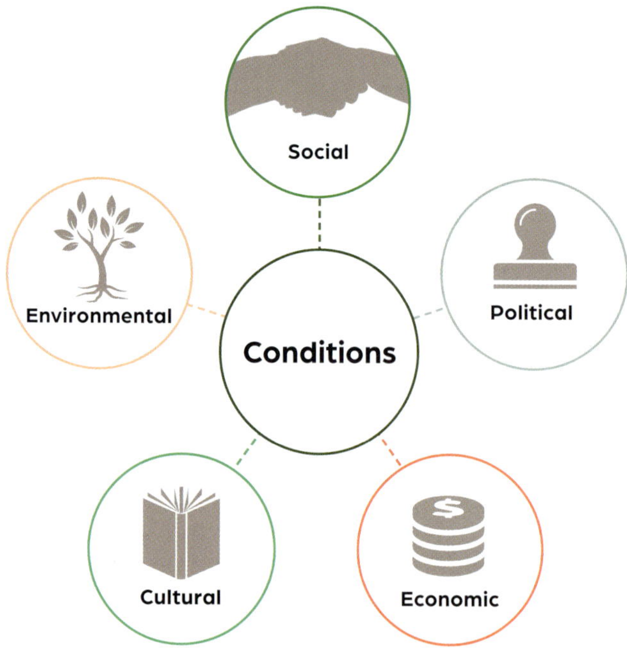

Type of cause	Example cause	Example effect
Social conditions	Peasants were treated badly in medieval England	The Peasant Revolt in 1381 CE
Political conditions	In 1532 the Inca were in a civil war	The Spanish conqueror Pizarro more easily defeated the Inca Empire
Economic conditions	Population growth in Scandinavia meant fewer resources per person	Vikings began to settle in other parts of Europe
Cultural conditions	Japan's social hierarchy was set in stone	The feudal system remained in Japan for many centuries
Environmental conditions	Central Asian soil was not good for farming	Mongolians were nomadic
Individual actor	Hernán Cortés wanted riches and fame	He attacked the Aztec Empire
Group actor	French-speaking Norman nobles began ruling England after 1066	French words for meat were introduced into the English language

I can explain why something is a cause or an effect

Explaining cause and effect involves stating *how* or *why* a cause led to an effect.

Cause	Effect	*Explaining how* the cause led to the effect
Genghis Khan conquered the largest land empire ever	Trade networks from Asia, the Middle East and Europe were strengthened	Before the Mongol Emperor Genghis Khan conquered a huge area of land in Asia, trade was patchy. After he united a region spanning from Korea in the east to Russia in the west, the Mongols enforced peace within their empire. This peace allowed trade to increase. It strengthened the Silk Road and saw an increase in trade across Asia and between Asia and Europe.
Admiral Perry visited Japan in 1853 demanding trade	Japan opened up to the world and westernised	Before 1853, Japan isolated itself from the world. When Admiral Perry from the USA came into Tokyo Bay with huge warships and demanded Japan trade with them, it forced Japan to open up to the rest of the world. The Japanese realised how behind other civilisations they were and begun a modernisation program known as the Meiji Restoration.

Source 7

The kimono was a traditional garment of the Japanese people, but was worn less during and after the Meiji Restoration.

I can analyse cause and effect

Analysing means the ability to break down something into its parts. If you can identify these different parts, and explain how together they make up the whole, you are analysing. If you can explain the rules or theories that show how these parts are organised, you are analysing.

The first step is being able to break something down into its parts. For example, what caused the Aztecs to fall to the Spanish?

We can break the cause into four parts:

1 The Spanish had superior military technology.
2 The Spanish used better tactics.
3 Rival Central American tribes formed an alliance with the Spanish.
4 Hernán Cortés was a skilled commander.

Breaking down the cause is the first part of the analysis. Now we can try to explain how these combined causes would have led to the Spanish conquering the Aztecs. For example:

> The Spanish had two major advantages over the Aztecs: they had superior military technology such as guns, steel swords and heavy armour. They also used better tactics than the Aztecs, led by their skilful commander Hernán Cortés. Moreover, the Spanish forces were supported by many other tribes, who were against the Aztecs.

Here we have linked Cause 1 and Cause 2 together.

Another way to analyse cause and effect is to look at how strong the causal link is. Causes can have different strengths.

Small causal link	Large causal link
Gradual, minor, almost no part, short-lived, partly, partial, to some extent, small extent	Radical, powerfully, significant, important, to a great extent, considerable, main, crucial

Once you have decided how strong a cause was, here are some words you can use to describe it.

- If something had only a minor effect you could say:

  A few places in England have Viking place names because Vikings ruled England for a few years during the Middle Ages.

- If something had a major effect you could say:

  The Mongols' ability to fight from horseback while charging at speed made them superior to other warriors in Asia, which allowed them to conquer huge areas of land.

Source 8

Mongol warriors carried quivers of arrows and used bows designed to be fired from horseback.

step 5 I can evaluate cause and effect

Evaluation is a higher-order thinking skill. Evaluation can be:

- assessing whether a theory or belief is true or not; for example, some historians think Mongols increased peace in Asia after initially conquering many peoples. Is this true?
- comparing different ideas; for example, which is a more important effect of the destruction of Tenochtitlan by the Spanish: the loss of Aztec culture or the start of a Spanish Empire in the Americas?
- judging between different things; for example some people think Japan would have westernised eventually anyway, others think the Americans were vital in opening up Japan to the world. Who is right?
- asking, 'So what?' Were the causes or effects good or bad, or perhaps a bit of both? A historian may include both positives and negatives to form a balanced view.

Look at Step 5 in 'Historical significance' (page 210) for more tips on evaluating.

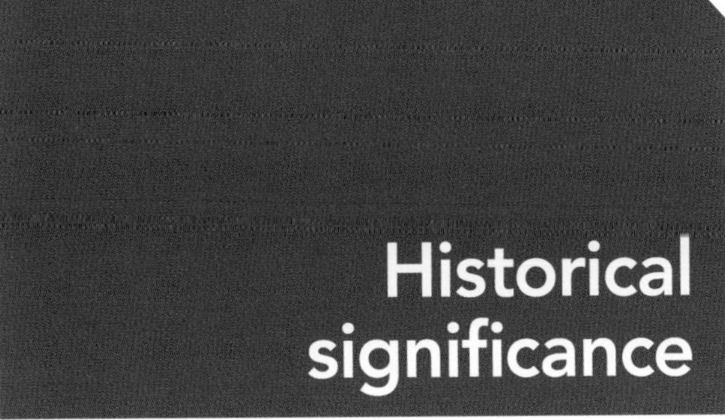

Historical significance

How do we decide what is important to learn about from the past? How do we decide which events or time periods have historical significance?

We can use a model or theory to help us decide. A useful model is Geoffrey Partington's model of significance.

Partington's model states that you can determine historical significance by asking the following questions:

1. Importance: How important was it to people living at the time?
2. Depth: How deeply were people's lives affected?
3. Number: How many people were affected?
4. Time: For how long were they affected?
5. Relevance: How relevant is it to the present?

The tables below show some examples using Partington's model of significance.

The Meiji Restoration (when Japan westernised)	
Importance	It originally affected mostly upper class and wealthy and/or urban people.
Depth	To start with, it only affected outward appearance but, over time, most of society was changed.
Number	Affected most Japanese but perhaps the peasants to a lesser extent; total population was 33 million, so perhaps 10 million people?
Time	For about 150 years; it started in 1868 but the westernisation program has continued since.
Relevance	Very relevant because Japan is an important economic leader in the world, based on its Western-influenced business model.

The use of *chinampas* (floating gardens) by Aztecs at Tenochtitlan	
Importance	It was helpful to residents of Tenochtitlan because they could grow food close to where they lived.
Depth	The depth was significant – this was people's main source of vegetables and many people were agricultural workers.
Number	About 200 000 people lived in Tenochtitlan at the time.
Time	They were used from about 1325 in central America. Their use declined after Spanish settlement, starting in about 1525.
Relevance	Not very relevant. Very few still practice this form of farming.

I can recognise historical significance

Recognising historical significance means looking at a list of events or developments and deciding how important they are. (Significant means important; something worth noting.)

You should have some way of determining significance. You might ask: Was it important back then? Were people deeply affected? Did it affect a lot of people for a long time? Is it still relevant to modern times? The more you answer 'Yes' to these questions, the more significant the event was.

Historical significance is not a black and white issue, as it can be shown on a significance scale like the one below:

Least significant			Most significant
Name of Kublai Khan's mother	The date Henry VIII of England died	Use of bow and arrow from horseback by Mongols	Invention of the printing press

Source 9

The invention of the printing press was very historically significant.

I can explain historical significance

Explaining historical significance means asking *how* or *why* something is important.

Here are some examples of significant and less significant events, based on Partington's model.

	More significant	Less significant
Importance	The destruction of Tenochtitlan by Spanish conquerors is really important because it greatly affected the 200 000 Aztecs living there. Many were killed by violence or disease, and the rest had to live under Spanish rule, which denied them many freedoms.	The carvings of large heads by the Olmec people are fascinating but not that important to history. They did not have a major impact to people at the time, and had even less importance for Aztecs. Today they are tourist attractions and historical curiosities but little else.
Depth	The banning of weapons in feudal Japan by the shogun affected people deeply. Most people would have had a sword in their family that they then were not allowed to use. It also made times more peaceful, reducing violent deaths, which is important.	The writing of the first novel, *The Tale of Genji*, while interesting, is not historically significant because it didn't affect anyone really deeply at the time. It was 'merely' an artwork.
Number	The sheer number of people that were within the borders of the Mongol Empire at its peak makes this a highly significant period in history.	Marco Polo being the special envoy of Kublai Khan isn't that significant because it is only really relevant to Marco Polo and the few people he met. His stories are useful for historians but aren't important in their own right.
Time	The settlement of Vikings in Europe is important, because they settled permanently; meaning Viking DNA is still in modern Europeans from all the places they settled.	Viking religion is not very significant because it did not last long after Christianity took hold of Europe. After about 1000 CE, almost all Europeans were Christians.
Relevance	The invention of the printing press is very significant because printing is still widespread and an important way we spread knowledge, even today. The ideas spread by the printing press, such as philosophy and science, have a big impact on modern life.	The use of heraldry isn't that significant because it has not continued into the modern day.

Good explanations of historical significance will discuss more than one of these elements.

Partington's model of significance is just an aid to your thinking. There are other things that could explain whether a historical event was significant. For example, a person might be important if they changed other people's ideas, or provided a good or bad example of how to live. An event might be important if it reveals underlying themes or patterns in history.

step 3 I can apply a theory of significance

Applying Partington's theory would mean looking at a set of events and ranking them against his categories. When applying his theory, uses phrases such as those in the table below:

Importance	Issue A was more significant than Issue B because it was more important to people *at the time*.
Depth	Issue A is more important in history, because it affected people more *deeply* at the time than Issue B did.
Number	Issue A deserves the status of historical importance more than Issue B. Put simply, it affected more people.
Time	Issue B has been shown to be less important historically than Issue A, because it didn't last as long.
Relevance	Issue A is a more significant event than Issue B because it is still relevant to the present. It helps us to understand the modern world, whereas Issue B doesn't help as much.

step 4 I can analyse historical significance

Analysing means the ability to break down something into its parts. If you can identify these different parts, and explain how together they make up the whole, you are analysing. If you can explain the rules or theories that show how these parts are organised, you are analysing.

In the next paragraph, we will analyse the significance of the Magna Carta . Three main points are made rather than just one, and an example of how each leads to significance is added. (See below the paragraph for a colour key.)

> There are many reasons the Magna Carta, signed in 1215, was an important historical document in history. First, it stated that the monarch was not above the law, and required monarchs to take into account the voices of others (although only nobles at first). This began the move from absolute rule by one person to the more democratic systems we see today. Secondly, it enshrined in law important legal rules such as habeas corpus. This was very influential in future laws passed in England, making legal protections more fair and equal for all. Thirdly, it has a lot of relevance today because many of its ideas were incorporated into the Constitution of the United States of America. Basic equality before the law in the constitution was originally from the Magna Carta. Many other countries have been inspired by the US constitution to pass similar laws and enshrine similar protections. Therefore, the Magna Carta is significant because of its rejection of absolute power by the ruler, its support for justice in law and its influence on other legal documents.

Key:

Main point

Claim

Evidence backing up claim

Summary statement

Look at Step 4 in 'Cause and effect' (page 205) for more tips on analysing.

Source 10

The Magna Carta is a historically significant document. This picture shows an original parchment copy of the Magna Carta (November 1217), kept at the History Museum in Speyer, Germany.

I can evaluate historical significance

Evaluating can mean asking, 'So what?'
In terms of evaluating historical significance, evaluating could be:

- questioning Partington's model of significance:
  - Perhaps the model suggests that Event A is more important than Event B, but you don't agree. Evaluating would involve you explaining why you think the model of significance doesn't give the right result in this instance.
  - Maybe you think some of the questions about significance are more important than others. Perhaps you think that relevance to today is more important than how significant it was to people at the time.
- making a judgement call about the worth of important events:
  - Were these events 'good' or 'bad' in some way? When making an evaluation like this, make sure you define what 'good' or 'bad' means in this context.

Questioning the model of significance	Some historians think relevance to today is really crucial when determining historical significance. However, perhaps it isn't as simple as that. The Aztec people are not very significant today, but that is because their culture and people were practically wiped out by European violence and disease. We need to imagine how important Aztec people might have been had this not happened, to really know how significant they could have been.
Judging the worth of an important event	How important was the death of Genghis Khan? The death of great leaders is often seen as extremely important, but this may not be true in this instance. The Mongol Empire continued for almost a century after Genghis Khan's death. Moreover, the Mongol influence on the Chinese (via the Yuan Dynasty) lasted even longer. We too easily think that one person is highly influential, when often it is the slow spread of ideas that make a bigger difference.

Look at Step 5 in 'Cause and effect' (page 206) for more tips on evaluating.

Source 11

Statue of Francisco Pizarro in Trujillo, Spain

Historical  writing

I can identify the writing purpose

If you are given a writing task, it will usually involve certain 'task words', such as *analyse, argue* and *compare*. These task words are explained below.

- *analyse*: look at the features of something, showing the relationships between the parts, how they're related and why they're important
- *argue*: make a case for or against something
- *compare*: discuss two things, emphasising what is the same and what is different between them
- *contrast*: discuss two things, emphasising what is different between them
- *describe*: write a detailed description of something, showing what something looks like, what it is for and how it works. Don't judge.
- *discuss*: write about something, talking about the arguments for and against and issue. Provide a balanced description, but make a judgement at the end.
- *evaluate*: make a judgment about something, but back it up with lots of evidence.
- *explain*: answer the question 'why?' about something. Go into detail about the reasons for it, causes of it and effects of it.
- *justify*: provide reasons why a decision was or should be made, or why a conclusion was reached.
- *summarise*: briefly state the main points. Leave out the details.

After you know your purpose, figure out:

- *what kind of information you need to gather.* This relates to Stage 2 of the history-writing process: gathering information. Gather the right kind of information – but avoid gathering lots of irrelevant material.
- *how that information should be organised.* You will eventually need to write up any information you have gathered. How you do that – and what structure your writing takes – should be determined by the purpose of the writing.

Here is an example history-writing question, and how it can be tackled: 'Discuss how important Francisco Pizarro was in the decline of Incan civilisation.'

This is a 'discuss' type question, so it is asking us to write about the topic, discussing arguments for and against it. You should discuss both sides but also make a judgement.

Information needed to answer the question:

- details about the conquering activities of Pizarro (but *not* details about every aspect of his life)
- details about the decline of Incan civilisation. Gather information about the history of the Incan civilisation, especially the last 20 years when it declined.

How should that information should be organised? A graphic organiser like the one below is a great way to structure your note-taking.

	General information	Evidence he helped the decline of the Incan Empire	Evidence he did *not* help the decline of the Incan Empire
Pizarro's conquering activities	○ ○ ○	○ ○ ○	○ ○ ○
Information about the decline of the Incan Empire	○ ○ ○	○ ○ ○	○ ○ ○

I can gather information

Good history writing involves providing lots of evidence. The more *relevant* information you use, the better. Relevant means the information is closely connected to what is being studied.

Gathering information will involve taking notes from historical sources. Academic historians look at many primary and secondary sources. For most school projects, you are likely to rely on secondary sources. Secondary sources provide a wide range of easily accessible information for young people. Textbooks and reference books are easy to obtain, relatively cheap, easy to read and contain pictures, facts, explanations and examples.

Follow these steps when taking notes for your history writing:

1. Purpose: *why* am I taking notes?
2. Organise: use a graphic organiser or codes
3. Skim-read the source. This is so:
   - you can look for topics, headings and so on.
   - you *don't* have to read the entire source.
4. Find the *most* important information *for your purpose*:
   - rewrite it in your own words
   - write as briefly as possible
   - include keywords, and definitions of any words you don't know.

Source 12

Genghis Khan's horde of warriors led the expansion of the Mongol Empire. This photograph shows a re-enactments of the unification of the Mongolian tribes for Genghis Khan's 800th anniversary.

I can organise information

Two ways to organise your information are to use a graphic organiser or use codes.

A graphic organiser is best for:

- when you know in advance the kind of information you will be taking notes about
- when the question you are answering has obvious parts to it that you can divide information into; for example, for and against.

Using codes is a process that involves:

- taking notes
- reading through your notes several times and seeing what patterns, themes or categories emerge
- making up a code for each pattern, theme or category; for example, 'W' for *war*; 'I' for *individual*; 'ME' for *Mayan Empire*
- going through your notes and writing the code beside each point
- rewriting your notes in the code categories. (This is much easier if you have taken notes electronically, because you can change their order without having to rewrite them.)
- using your notes in their coded categories to form the basis of your essay structure.

With either of these methods, don't forget to ask yourself which notes you should *not* use. You will always take notes that you thought were important but later realise don't actually matter. Get rid of them. Remember: the final written piece is what is most important, not your notes. Don't worry that you spent time writing those notes in the first place, because only your final piece of writing matters. Next time, try to take fewer irrelevant notes.

Here is an example of the process, with all the steps from note-taking to organising your notes.

Essay question: 'Was the legacy of Mongol expansion positive or negative?'

1 Original notes

- Mongols killed whole Jin army in 1211
- in 1215 Genghis Khan starved then slaughtered all the inhabitants of the city of Zhongdu
- in the Mongol army, if one person from a unit fled, the entire unit would be executed
- Mongol troops always looted places they defeated, often burning them as well
- Mongol troops tortured people
- Mongols killed prisoners and people who surrendered
- the Mongols built profitable trading routes
- people under the Mongol Empire could practice their own religion
- Genghis Khan appointed skilled administrators to important government positions
- people gained promotions under Genghis Khan because of talent, not because they were nobles
- the Mongols established a vast communication network in the empire.

2 Put into a graphic organiser, the notes look like this:

	Mongol expansion was negative	Mongol expansion was positive
General comments	○ Mongols killed whole Jin army in 1211 ○ in 1215 Genghis Khan starved then slaughtered all the inhabitants of the city of Zhongdu	○ the Mongols built profitable trading routes ○ the Mongols established a vast communication network in the empire
Government actions	○ Mongol troops always looted places they defeated, often burning them as well ○ Mongol troops tortured people ○ Mongols killed prisoners and people who surrendered	○ Genghis Khan appointed skilled administrators to important government positions ○ people gained promotions under Genghis Khan because of talent, not because they were nobles

3 Notes not needed:

- people under the Mongol Empire could practice their own religion
- in the Mongol army, if one person from a unit fled, the entire unit would be executed

4 You could then put your notes into paragraphs:

- Mongol expansion was negative: general comments, actions of soldiers
- Mongol expansion was positive: general comments, improvements in government administration

step 4 I can structure a piece of writing

History essays should have an introduction, several body paragraphs and a conclusion.

When you are starting out writing essays, a paragraph structure that is easy to learn is **TEEL**. TEEL is an acronym for:

- Topic sentence
- Evidence
- Explanation
- Link.

These words are explained below. Every paragraph should have TEEL.

Introduction

The introduction should:

- show you understand what the question is asking
- say your overall response to the question
- introduce the main points.

Paragraphs using TEEL

Paragraphs using TEEL should include:

- Topic sentence: one sentence that summarises the whole paragraph
- Evidence: use *specific* examples, not general examples
- Explanation: how evidence supports your claim
- Link: at the end of the whole paragraph, link the main point in the paragraph back to the main question.

Conclusion

In your conclusion, make sure to:

- summarise your main points
- restate your response to the question.

step 5

I can write a draft

Drafting tips

- Focus on answering the question; don't just write everything you know about the subject.
- Don't worry about making mistakes when drafting – you will fix this later.
- Don't worry too much about punctuation, grammar or spelling when drafting.
- Start with the paragraphs. Draft the introduction and conclusion last.
- If you can, use a computer, as it makes it easier to edit and proofread your work later.
- Write the first draft quickly. Then edit and proofread slowly.

Sentence starters

Here are some sentence starters for introductions:

- This essay will discuss ...
- This essay will focus on ...
- The issue being focused on is ...

Other words you could use in sentence starters in place of *focused* are:

- described
- analysed
- evaluated
- explained
- explored
- justified
- outlined.

For conclusions, some starters include:

- In conclusion, ...
- In summary, ...
- It has been shown/demonstrated that ...
- Therefore/Thus/Hence, ...
- To summarise, ...

For comparing within your answer: (when things are the same)

- By comparison, ...
- In the same way, ...
- Likewise, ...
- Similarly, ...

For comparing: (when things are different)

- However, ...
- In contrast, ...
- On the other hand, ...
- Then again, ...

For adding more:

- Additionally, ...
- Also, ...
- First, ... Second, ... Third, ... Finally, ...
- Furthermore, ...
- In addition, ...
- Moreover, ...

For giving examples:

- For example, ...
- For instance, ...
- An illustration of this is ...
- As an example, ...
- ... such as ...

For showing effects:

- As a result, ...
- For this reason ...
- It can be seen that ...
- The evidence suggests ...
- The result of this is that ...
- These factors contribute to ...

Different ways to say 'caused':

- resulted in
- created
- lead to
- determined
- is attributed to
- meant that
- is dependent on
- forced
- made.

step 6

I can edit and proofread

The point of writing is to communicate – using words to pass ideas from you to another person. So, keep your writing clear and simple.

Editing means checking for meaning, to make sure your text answers the question and meets the task requirements. For example, does your writing need a bibliography? Does it need labelled pictures? Proofreading means checking the grammar, spelling and punctuation of your work. Always edit first, then proofread.

Editing tips

- Always use headings, unless told otherwise.
- Delete any words, sentences or paragraphs that do not help the piece of writing overall.
- Check what you have written against the requirements of the task. Ask yourself:
  - does my writing answer the question asked?
  - is it clear that I have done the full task?
  - is there an assessment schedule or rubric my writing will be marked against? Mark yourself against these criteria. Is there time to improve at least one aspect of what you have written?
- What is your worst paragraph? Why? What would it take to make it your best paragraph?

Proofreading tips

Proofreading is going back over your finished work looking for errors.

- Don't try to fix every problem at once. Pick one thing to correct each time you proofread. For example, first look at spelling, then look at punctuation, then look at confused words, then look at making your vocabulary more interesting.
- Read your work aloud. Even better, record it, then play it back to yourself a bit later.
- Ask someone else to read your work aloud.
- Read sentences backwards to check for mistakes. This will help you pick up more errors.
- If you know you are a bad speller, don't trust your instincts. Check words you are unsure about in the dictionary.
- Spellcheckers are not perfect. A word can be spelled correctly but still be the wrong word, so don't rely on a spellchecker!
- Read a printed copy of your work, rather than reading on screen.

Common errors

Following is a list of common errors:

- only use apostrophes for shortening words and ownership, not for plurals
- write shorter sentences, preferably less than 25 words
- only use capital letters at the start of sentences and for proper nouns
- confusing 'your' and 'you're'
- confusing 'there', 'their' and 'they're'
- writing informally: don't use 'I' (unless told to), '&', 'etc.', 'e.g.', 'i.e.', 'wanna', 'heaps', 'stuff'
- could of / would of / should of are incorrect; replace with could have / would have / should have
- confusing 'to/ too/ two'
- confusing 'much/many': much = for an item that can't be counted (e.g. water), many = individual item that can be counted
- then = something happening after something else; than = comparing
- subject–verb agreement. If the subject is a plural, the verb must be too, for example 'towels *are* in the closet'
- too many commas. Instead of writing long sentences with lots of commas, write shorter sentences
- be careful about starting a sentence with 'and', 'but' or 'because'
- a full sentence should have a subject (doer), verb (action) and an object (the thing the verb is happening to)
- use the same tense (future, present, past) in the whole text
- avoid using boring words: very, good, bad, amazing, interesting, crazy, mad, extremely
- avoid 'passive' sentences. Instead of 'The army was led by Genghis Khan', write 'Genghis Khan led the army'.

Historical research

I can define the problem

To define the subject to be researched, get some background information and build up a list of keywords.

Start by reading a simple Wikipedia page about your subject.

Get keywords for your topic. Think of different ways of saying your topic, or google 'synonyms of ...' and insert your search term.

I can decide what information to find and where to find it

What type of evidence do you want?

Include these kinds of words in your search:

- facts, examples, definitions, quotes, artefacts, images, data, statistics
- primary and secondary sources
- databases, links, archives, collections, references, research, museums, journals, graphs, tables, letters.

Where is your evidence?

There are many different types of websites to look at: scholarly works, databases, archives, reference sources and information pages.

How credible is the evidence?

Ask yourself the following questions:

- Is the content *relevant*? Is it useful for my purpose? Does it contain links to other relevant sources? Is it at an appropriate reading level?
- Is the source *believable*? What type of source is it? (Published or official sources are better.) Who is the author? (Experts are better.) When was it published? (Newer is usually better.) Is the source biased?
- Is the source *true*? Is it backed up by other sources? Does it *sound* right? Does it fit in with other things you know?
- Does the source state where its information comes from? This means it is more likely to be credible (able to be believed).

I can find information

Online search strategies

Following are some search strategies:

- After you type in a search term, scan through the first page of results. If they are not relevant, change your search.
- Start with a wide search, then get more specific.
- Learn *from* your search. Change what you are searching for based on what you learn after you start searching.
- Be ready to stop a search if it is taking you in the wrong direction.

Tips for searching with Google

- Every word matters
- The order of the words matters
- Capitalisation doesn't matter
- Punctuation doesn't matter
- Specific search terms are better

Source 13

Searching using Google

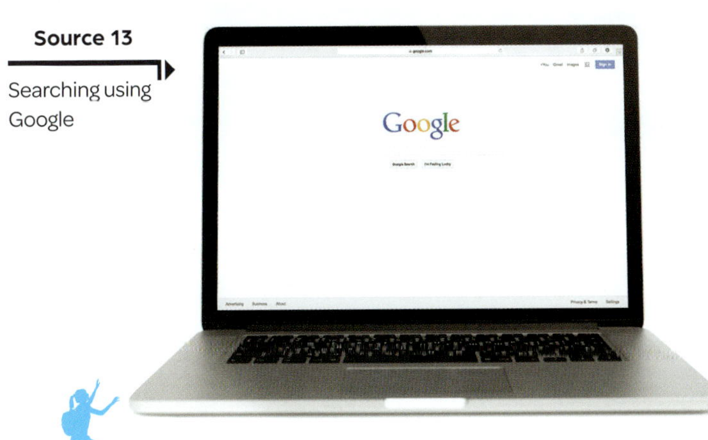

- Use these capitalised terms to narrow your search: AND, OR, NOT
- A search with 'filetype:' will find specific files. For example, 'Mayan Empire filetype:ppt' will find PowerPoint files about the Mayan Empire.
- A search with 'site:' will find things *within* a website. For example: 'Vikings site: britishmuseum.org' will find Viking-related material from the British Museum website
- Use the tabs along the top for different types of results such as images, news, videos, maps and books.
- Use a hyphen to exclude words and narrow your search. For example, 'Danish -pastry' will find information about Denmark, not Danish pastries.
- Search for a range of numbers using two full stops between speech marks: '..' For example:
  - '2001..2004' searches between 2001 and 2004
  - '..2004' searches before 2004.
  - '2004..' searches after 2004.
- An asterisk acts as a wildcard. So, for example, 'teen*' will return results with any of the words *teen, teens, teenager* in them.
- Use exact phrase searching by putting speech marks around a search to find exact text.

step 4 I can extract information

This note-taking stage is the same as Step 2 in the Historical writing section. Read that section on page 212.

step 5 I can organise and present information

This stage is very similar to Step 3 in the Writing Process. Read that section on page 212.

Research will be presented in a number of different ways, and will usually include some history writing. History writing is generally presented as text with perhaps some supporting pictures.

You should also edit and proofread your research, just as you do with your writing. Read Step 6 from the Historical writing section on page 215.

step 6 I can evaluate information

You can improve every time you conduct research by asking yourself these questions after you finish:

- What worked? What didn't work?
- How could I work smarter next time?
- Can I apply what I've learnt to other situations?
- How could I have improved:
  - the project?
  - the way I worked on my project?
  - the way I managed my time?

Glossary

Althing an annual event where new Viking laws were made – the whole population could express opinions on important topics such as taxes, treaties and even selecting a king; the *Althing* was also a religious festival and gave scattered people a chance to trade

Anglo-Saxons a cultural group who inhabited England from the 5th century CE; a merging of the Angle and Saxon people

aqueduct an artificial water channel built to move water from place to place; has to be built with precision because the water has to run downhill over large distances

arban a 10-soldier military unit in the Mongol army

archaeology the study of human history through the excavation of sites and the analysis of artefacts and physical remains found there; people undertaking archaeology are known as archaeologists

artefact an object made by a person, such as a tool or implement, usually of historical interest

artisan a skilled craftsperson who makes things by hand

Aztecs a civilisation that ruled a large empire in central Mexico between 1250 and 1521 CE

bailey a protected courtyard within the outer castle wall of a medieval castle

barbarian a term used to describe different non-Latin and Greek speaking groups such as the Celts, Franks, Saxons, Angles, Visigoths and Huns who moved into western Europe in the Early Middle Ages; also used to describe people in general thought to be unsophisticated

barbican a towered wall to protect the main gate of a medieval castle

basqaq a Mongolian military governor put in charge of conquered territory

battlement the top of a stone castle wall that contained gaps called crenels to allow defenders to shoot arrows or throw spears at the enemy while sheltering behind the higher parts of the wall (the merlon)

beqlare-beq a Mongolian prime minister and chief advisor to the Great Khan

berserker an elite Viking soldier who could enter a trance-like 'berserker' state where they charged forward with no fear, hacking at their foes; the modern term 'berserk' comes from the berserkers

bias to be for or against an idea, a person or a group, especially in a way that could be thought of as unfair

Black Death the bubonic plague, a pandemic that swept through Europe in the middle of the 14th century, killing one-third of the population

blacksmith a craftsperson who creates objects from iron or steel by heating the metal and shaping it

blót a Viking sacrifice to the gods where pigs or horses were killed and the blood was sprinkled on statues of the gods and on the participants themselves

Buddhism a belief that after people die, they are reincarnated – reborn into another life; living a proper life would stop the cycle of rebirth, and this stage, called nirvana, could be reached by leading a good life and no longer wanting things

Bushido the ethical code of the samurai that stressed loyalty, respect and courage at all times

carbon dating estimating the age of historical material (such as archaeological or paleontological items) by calculating the item's content of carbon-14

Catholicism the faith and practice of the Roman Catholic Church

cavalry soldiers or warriors mounted on horseback

census an official count of statistics relating to a population

chinampas floating gardens used in the Aztec city of Tenochtitlan

Christianity a religion based on the teachings of Jesus Christ; Catholic and Protestant religions both preach Christianity

chronology the process of organising events into the order they happened to bring structure and order to events in time

city-state an independent city (and surrounding lands) with its own government; it is linked to other cities by trade, language and culture, but is not under a centralised government

civil war war between citizens in the same country

coat of arms a shield with symbols and colours that represent a family, group or organisation through the process of heraldry

Confucianism a philosophy taught by Confucius (551–479 BCE) that taught that people should be honest, brave and knowledgeable and never violent or arrogant; the path to happiness lay in obeying the law, doing one's duty and respecting older people – especially parents and grandparents

conquistador an armed and armoured fortune-hunting soldier from Spain

conscription compulsory military service

consumer a person who purchases goods and services

crusade a religious war to regain control of the Holy Land for Christians

daimyo the noble class in medieval Japanese society who were granted land by the emperor in return for military protection

Dark Ages the period following the fall of the Roman Empire between 476–800 CE, when there was no emperor in western Europe and barbarian peoples such as the Goths and Franks moved into Europe; historians now more commonly use the term Early Middle Ages to describe this period

democracy rule by the citizens, who elect officials and leaders

dohyo the sacred ring in which sumo contests take place

Domesday Book a book that outlined how much land there was in England, what it was used for and who owned it before 1066, who lived there and how much it was worth

donjon a fortified tower inside a castle wall in medieval Japan; it served as the daimyo's residence and the last place of refuge when defending the castle

duelling to decide on a dispute by fighting; the rich with swords and the poor with staffs or fists

dynasty where one family maintains power over many years by handing on the throne to an heir, usually the oldest son

eagle knight the most feared and skilled type of warrior in the Aztec army; a soldier of noble birth who had taken many prisoners in battle; also known as a jaguar knight

ecomienda a grant by the Spanish king to colonists in America to give them the right to demand tributes and forced labour from the Indigenous people of an area

economic problem the problem of how to satisfy people's needs and wants with limited resources

economy how money is made and used in a country or region, including the production and consumption of goods and services

emperor a ruler or monarch of an empire; i.e. in 800 CE Charlemagne became emperor of western Europe

empire a number of nations or states ruled by a single leader such as an emperor, king, queen or an oligarchy; i.e. the Aztec Empire

entrepreneur a person who sets up a business, taking on risks to explore an opportunity that will earn a profit or satisfy a personal goal

excommunicate to cut off a Christian from the church and guarantee they would be sent to hell after they died

export where goods or services produced in one country are shipped to another country for sale or trade; exports are crucially important for a country's economy to add to its gross national income

fealty formal acknowledgement of a vassal's loyalty to a lord

feudalism a system of rights and obligations throughout society, where lords in Europe and *daimyo* in Japan were granted land in exchange for military service; peasants worked on the lord or *daimyo*'s property in return for the chance to grow their own food and receive protection from the lord's knights or the samurai

fief a parcel of land given by a lord to a knight in exchange for military support and allegiance

fjord a long, narrow body of water with steep sides created by a glacier

flagellants groups of men who, during the Black Death In medieval Europe, travelled from town to town publicly whipping themselves in the hope that God would take pity on them and protect them from the illness

Frankish Empire the dominant Christian kingdom of early medieval western Europe, occupying present-day northern France, Belgium and western Germany

geisha an unmarried female entertainer and companion, highly trained in Japanese dance, poetry and playing musical instruments

guild a type of trade union established for each skilled craft to guarantee workmanship, oversee the training of apprentices and to settle disputes between employers and skilled workers; guilds were like modern trade unions except they also included employers

gyoji a Shinto ceremony to bless and purify the *dohyo* before a sumo contest

heraldry the system by which coats of arms are created, described and regulated; used to signify different noble families

heretic people whose beliefs did not align with those of the Catholic Church in Medieval Europe

hieroglyphics a writing system using pictures and symbols, such as the system used by Mayans

historian a person who specialises in the study of history by using evidence to answer questions about the past

history the study of past events, particularly relating to humans

holmgang a Viking duel used to settle disputes

homage an official ceremony where a vassal would swear allegiance to his lord

humanism a set of ethics or ideas about how people should live and act

Ice Age the period of time between 1.8 million and 12 000 years ago when much of the Earth's water was frozen at the northern and southern tips of the planet and the sea level was over 100 metres lower than it is today

import a good or service brought into one country from another country; along with exports, imports are the other key element of international trade

Inca the largest empire of its time that reached its peak in 1493 and stretched for 4000 kilometres from Quito in Ecuador to Santiago in Chile

Indies in medieval times, the area of southern Asia, as referred to by Europeans

Indigenous originating or naturally occurring in a certain place; used to describe peoples who live in, or have historical or cultural attachments to, a particular territory

Islam faith of Muslims taught by the Koran and based on the religion founded by the prophet Muhammad

jarl a wealthy Viking noble who could become rich and powerful enough to eventually become a king

jihad a fight or war against the enemies of Islam

joust a tournament where mounted knights used a lance to try to knock their opponent off their horse

Judaism a religion based on the Torah, with origins to 5781 BP; followers believe in one God, who is spiritual and eternal

kami in the Japanese Shinto religion, sacred ancestor spirits that took on forms in nature such as the Sun or rivers

kampaku the title given to a regent in medieval Japan who ruled as emperor until the young emperor came of age

karl an everyday Viking farmer, craftsperson, warrior or sailor

karve *a Viking cargo ship* used to transport heavy cargo around coastlines and up rivers

keep a fortified tower inside a medieval castle wall; the keep served as the lord's residence and the last place of refuge when defending the castle

kendo a safe form of sword training for samurai using bamboo sticks

khan a Mongolian tribal chief or leader

khanate a political region ruled by a khan; the large Mongol Empire split into four separate khanates after the death of Kublai Khan

knarr a broad Viking cargo ship designed for overseas trade

knight a heavily armoured soldier who rode on horseback; a member of a privileged class in medieval European society, who received a grant of land known as a fief from a lord in return for their protection

knighthood the official declaration that a person is a knight

konungr the chief, or king, of a Viking village; each Viking village or town had a chief or king because there was no central government

labour the amount of work used to produce goods and services in an economy; in return, workers receive a wage or salary to buy their own goods and services

lance a long spear used by a mounted warrior

lidar laser technology used by archaeologists to uncover objects covered by thick forest

longhouse the largest Viking building with no windows, a thatched roof and a large fireplace in the middle for heating, lighting and cooking. One end of the longhouse could be used as a barn to shelter livestock in the winter and the other end could be used as a workshop

longship a fast, versatile Viking ship with a very shallow draft, which was able to travel in shallow water

loot valuable goods taken in a battle, raid or attack

Magna Carta a document signed in 1215 in England that outlined basic rights for all people with the principle that no-one was above the law, including the king

maiko a young geisha in training from the age of six

mangonel a weapon that used the power of twisted rope to make a wooden arm spring up and hurl an object

manor a parcel of land given by a king to a lord in exchange for military support and allegiance

market when producers make and sell goods and services to consumers

market economy where the availability and pricing of goods and services is driven by demand from consumers

marketing action to promote the selling of a product or service

martial arts forms of self-defence or attack, such as kung fu, karate and *kendo* that originated in Asia

Maya/Mayans a Mesoamerican civilisation that lived in separate city-states in the dense jungles of Central America between 250 and 900 CE

mead an alcoholic drink made by fermenting honey and water

Medieval Period the period of time between the fall of the Roman Empire in 476 CE and the beginning of the Renaissance in the 14th century; also referred to as the Middle Ages

meritocracy a system of promoting those with talent, or rewarding those with skill, rather than those with a senior position in society

Mesoamerica the area from central Mexico down to Central America that was home to Olmecs, Mayans and Aztecs in ancient and medieval times

Middle Ages the period of time between the fall of the Roman Empire in 476 CE and the beginning of the Renaissance in the 14th century; also referred to as the Medieval Period

milpa a cleared field in the rainforest where the Mayans grew corn, beans and squash

Mongol a native or inhabitant of Mongolia; in medieval times, the term described the nomadic peoples united under the rule of Genghis Khan

Moors north-western African Muslim people who conquered parts of Spain in the 8th century; driven out of their last stronghold in Granada at the end of the 15th century

motte within a medieval castle, a raised hill with a wooden or stone keep (a fortified tower) on the top

Muslim a follower of the Islam religion

need something required to survive

ninja highly trained warriors who were specialist spies and assassins

ninjutsu the art of becoming a ninja and perfecting their specialist fighting and spying skills

nomad/nomadic people with no permanent home, who herd animals from place to place to find fresh pasture

Normans Viking settlers in the French region of Normandy who adopted the French language, customs and religion

noyan a Mongol army general, in charge of a *tumen*

Olmecs the first major civilisation in Mesoamerica; the first in the region to build stepped pyramids; best known for carving massive stone heads

outsource where businesses send some of their operations to areas of the world where they can take advantage of cheaper labour

pagan someone who doesn't follow the major religions of the world or believes in more than one god

page the son of a lord who, at age seven, was sent to another lord's family to undertake training as a knight

paiza the official metal pendant carried by riders in the *Yam* in the Mongol Empire, to show they were messengers of the Khan

palisade in a medieval castle, a defensive wall made of wooden stakes

peasant the lowest class in medieval European and Japanese society; peasants farmed the land for their lords or emperor in return for protection

pension a regular payment made by the government to people at the official retirement age and to some returned servicemen and disabled people

pilgrimage a spiritual journey to a place of religious importance

pillage a violent robbery where the attacker takes everything from the place and people they have conquered

prefecture a Japanese administrative region controlled by a governor

prehistory any time before 3500 BCE, when writing was invented in ancient Sumer (modern-day Iraq)

primary source a source that was created or existed at the time under study, such as books, letters, artefacts and buildings

producer a person who creates and supplies goods or services

puppet emperor (generally young) emperors in medieval Japan who carried the official title while the real power lay with the *shogun*

rate a fee charged against ownership of an asset such as property; e.g. people in Australia pay council rates based on the ownership of property to pay for rubbish removal and the upkeep of council parks and services

regent a person appointed to rule a country if the monarch is sick, or too young to rule themselves

ronin a samurai without a master who could hire himself out as a warrior

rule of law the principle that people should be ruled by law and should all obey it

rune Viking symbols used as part of their writing system

saga a medieval Viking story of achievements and historical events

sakoku a policy to shut off Japan to the outside world during the Tokugawa *shogunate*

samurai medieval Japan's warrior class, who worked for daimyo; samurai were similar to knights in medieval Europe

Saxon a member of a Germanic people who settled in England in the 5th century CE; they merged with the Angles to become the Anglo-Saxon people

scribe a well-educated Mayan person who could read and write complex hieroglyphics

secondary source a source created after the time under study, such as a textbooks, websites and documentaries

sensei the teacher of a martial art

seppuku ritual suicide practiced by samurai rather than facing the humiliation of defeat at the hands of the enemy

serf a member of the poorest of the peasant class who received food, shelter and protection from the lords and knights in return for their work on the land and rent payments

shah title given to the leader of Iran; the country also known as Persia

Shinto Japan's ancient religion based around sacred ancestor sprits known as *kami*, which took on forms in nature such as the Sun or rivers

shogun the military and administrative leader of Japan; shoguns ruled Japan for approximately 700 years from 1192

shogunate the government of the shogun in medieval Japan; i.e. the Tokugawa *shogunate* was the government controlled by shoguns during the Tokugawa dynasty; also known as *bakufu*

siege a military operation where enemy forces surround a town or building with the aim of cutting off supplies and forcing a surrender

siege tower a high wooden platform on wheels that was pushed against castle walls so the attackers could climb into the castle

Silk Road a 6500-kilometre trading network of routes over land and sea that connected China to Asia, Europe and the Mediterranean

slave a person who is owned by another person and forced to work for nothing; in Medieval society, slaves were at the lowest level of the societal hierarchy

smallpox a contagious viral disease introduced by the Spanish that wiped out large numbers of people in Mesoamerica and South America

sod turf made of grass and soil that was used by Vikings for housing materials

squire a stage of training on the way to becoming a knight, achieved at the age of 14; squires were treated as men and their training became far more dangerous

steppe a large area of flat unforested grassland

stereotype a widely held but generalised idea of a particular type of person or thing

sumo a martial art practiced by huge sumo wrestlers that is linked to Shinto traditions

sunstone a crystal rock used by navigators to follow the path of the Sun through clouds and fog

supply and demand the relationship between the amount of a product that producers wish to sell and the quantity that consumers wish to buy

Taoism a Chinese philosophy promoting humility and balance

tariff a tax imposed by a government on goods or services imported from another country

tax a compulsory payment to the government from a worker's income or business profits, or added to the cost of goods and services

Thing a local Viking meeting that took place over a few days every autumn and spring to decide guidelines for how the chief or king was to rule

thrall a Viking slave who had been abducted in a Viking raid, purchased overseas or had sold themselves into slavery to pay a debt

tithe a tax paid to the Church in medieval Europe, usually one-tenth of a person's yearly earnings paid in money or in goods (animals, seeds, grain)

tortilla a flat bread made by early Mesoamerican people

treason the crime of betraying your country or attempting to overthrow its government

trebuchet a tall wooden catapult with a swinging arm that could hurl projectiles high in the air at a target as far as 300 metres away

trial/trial by jury where a panel of regular citizens decides on the outcome of a legal case

tribute tax paid to the Aztecs in the form of food, feathers and precious metals

tumen a 10 000-soldier military unit in the Mongol army

uji Japanese clans made up of family groups related by blood or marriage

ulama a Mesoamerican ball game similar to volleyball; the oldest continuously played sport in the world

unemployment benefit a payment made by the government to an unemployed person

vassal in medieval Europe, someone who received land in return for providing military service and allegiance

Vikings Norse peoples from Scandinavia who raided and settled in Europe from around 800 CE to the 11th century; Vikings were great explorers, colonising Iceland and Greenland, and exploring parts of North America

vizier Mongolian government ministers

want something that people desire to have

Yam a system of relay stations every 30–50 kilometres across the Mongol empire, which enabled messages to travel 200–300 kilometres per day, as riders handed their message to a fresh rider, or chose a fresh horse and rode on

Yassa the code of laws used throughout the Mongol Empire

yurt a portable tent covered in felt and used by nomadic Mongolian herders

zuut a 100-soldier military unit in the Mongol army

Index

D

E

N

P

S

T

W

Acknowledgements

The author and publisher are grateful to the following for permission to reproduce copyright material:

PHOTOGRAPHS: adobestock/viktor, **92** (top); agefotostock /Photo12/Photosvintages, **4**; AKG Images, **180–1**, /Science Source, **183**; Alamy/Active Museum/ Le Pictorium, **57**, /Album, **68**, **80**, **106–7**, **77** (top right), /Allied Computer Graphics, Inc., **44–5** (longship), /Andrey Panchenko, **44–5** (warrior), /Art Collection, **2**, **3**, **70**, **168** (bottom), /Artokoloro Quint Lox Limited, **109** (top), **132**, /Derek Bayes/Lebrecht Music & Arts, **178–9**, /Chronicle, **121**, /Classic Image, **77** (top left), **137**, /COLUMBIA/GAUMONT/All Star Picture Library, **84–5**, /Concordia Discors, **134**, /FALKENSTEINFOTO, **3**, /Fine Art Images/Heritage Image Partnership Ltd, **i** (top right), **55**, **99**, **100** (left), **107** (bottom), /Tim Gainey, **31**, /James Gifford-Mead, **87**, /Diego Gomez, **i** (bottom right), **159**, /The Granger Collection, **153**, /Granger Historical Picture Archive, **21**, (top), **152**, **198**, /Bjorn Grotting, **38**, /Russ Heinl/All Canada Photos, **46**, /Historical Views/age fotostock, **115** (top), /INTERFOTO/History, **81**, /Lanmas, **73** (bottom), /Barry Lewis, **212**, /Jason Lindsey, **27** (top), /Tom Lovell/National Geographic Image Collection, **30**, /David Lyons, **22** (right), /Franck METOIS, **71**, /John Mitchell, **167**, /Manu Mielniezuk, **105** (top), /MARVEL STUDIOS/WALT DISNEY /AF Archive, **i** (top centre), **19**, /Mint Images, **141**, **142**, **143** (all bottom right), /J.Enrique Molina, **169**, /NETFLIX/AF Archive, **100** (right), **123** (top), **118**, /Niday Picture Library, **154–5**, **86** (top), /North Wind Picture Archives, **18**, /Photo Researchers /Science History Images, **174**, /The Picture Art Collection, **101** (top), **119**, **124**, **154**, **146–7**, /Picturehouse/Courtesy Everett Collection, **110–11**, /The Print Collector/Heritage Images, **83**, /Jorge Royan, **21** (bottom), /SONY PICTURES/AF Archive, **86** (bottom), /Bakhauaddin-bek Sopybekov, **110–11** (warrior), /SuperStock, **89**, /TCD/Prod DB © Marvel Studios, **36**, /Thor's Fight with the Giants by Mårten Eskil Winge, 1872/GL Archive, **37** (left), / Riccardo Croci Torti, **43** (top), /UtCon Collection, **141** (top), /Werner Forman Archive/Heritage Image Partnership Ltd, **37** (right), **145**, /Colin Weston/LatitudeStock, **51** (top), /World History Archive, **185**; Bridgeman Images/Bayeux Tapestry /Musée de la Tapisserie, Bayeux, France, **64–5**, /Leif Ericson sails past icebergs/Private Collection/Peter Newark Historical Pictures, **45** (top), /Battle of Shizugatake Pass 1583 (colour litho) /Peter Newark Military Pictures, **149** (top), /Toyotomi Hideyoshi (colour litho)/Peter Newark Military Pictures, **133** (bottom right), **149** (bottom), /Mongolia, China: A Mongol horseman with a composite bow, c. 13th century/Pictures from History, **108** (bottom), /Yuan dynasty banknote/Pictures from History, **123** (bottom), /Battle of Stamford Bridge, 1870 (oil on canvas) /Arbo, Peter Nicolai (1831-92)/Private Collection, **50–1**, /Dover Castle: the siege of 1216, Dunn, Peter (20th Century)/Private Collection/©Historic England, **76–7**, /Genghis Khan marched an army into China, Salinas, Alberto (1932–2004)/Private Collection/© Look and Learn, **112** (left), /William the Conqueror at the Battle of Hastings/Private Collection/© Look and Learn, **62–3**, /Leif Eriksson Discovers America (oil on canvas), Dahl, Hans (1849-1937)/Private Collection/Photo © O. Vaering, **23** (top), **47** (top), /Necklace, Gulf Coast, Mexico, c.1500-400 BC /Gift of Edward R. Roberts, **164** (top); British Museum Images /©The Trustees of the British Museum, **52** (left), **94**, **126**, **131**, **156**, **161**, **190**; Ally Crawford/Violet Sky Photography, **117** (top); DK Images/Dorling Kindersley, **39**, /Paolo Donati, **74–5**, /Steve Noon, **88**; Getty Images/Bettman, **59** (centre), /Stefano Bianchetti, **61**, /Bridgeman Art Library, **35**, /De Agostini/G. Dagli Orti, **172** (top), /DEA PICTURE LIBRARY, **85** (top), **144** (left), **163** (top), **172** (bottom), /DEA/A. DAGLI ORTI, **93**, **98**, **127**, /Peter Dennis, **176**, /Anthony Devlin - PA Images, **43** (centre), /DKart, **230**, /Dorling Kindersley, **74**, **78–9**, **163** (bottom), **162** (right), **175**, **188**, /duncan1890, **59** (bottom left), /Werner Forman, **25** (top), /David Goddard, **76** (left), /Heritage Images/Hulton Archive, **67**, /koya79, **162** (left), /David Madison, **151**, /Photo Josse /Leemage, **59** (top), /picture alliance, **209**, /RapidEye, **191**, /Mats Silvan, **72** (centre), /ManuelVelasco, **95**; iStockphoto, **196** (all), /RossellaApostoli, **142–3**, /bubaone, **204** (economic), /Michael Burrell, **211** (bottom), /channarongsds, **59** (bottom right), /clu, **58** (bottom), /dar woto, **204** (environmental), EAGiven, **187** (top), /Jan-Schneckenhaus, **208**, /KavalenkavaVolha, **i** (background), /Nikiteev_Konstantin, **204** (cultural), /Lorado, **28–9**, /MarinaMM, **5** (vintage car), /MR1805, **53**, **218–21**, /mura, **141** (bottom left), /Nastasic, **9**, /Nili, **186–7**, /Sean Pavone, **222–8**, /sx70, **204** (social), /Ross Tomei, **i** (top left), **1**, /UpdogDesigns, **52** (right), /xijian, **229**, /AlexanderYershov, **54**, **96**, **128**, **158**, **192** (arch icon); Library of Congress [LC-DIG-jpd-02051], **i** (centre right), **129**; The Metropolitan Museum of Art, New York/Gift of The Salgo Trust for Education, New York, in memory of Nicolas M. Salgo, 2010, **56**, /Purchase, Arthur Ochs Sulzberger Gift, 2005, **101**, /Gift of Margaret Wishard, in memory of her mother, Mrs. Luther D. Wishard, 1974, **132–3**, **157**; National Geographic Images /Fernando G. Baptista, **48–9**, /James L. Stanfield, **i** (centre left), **97**; National Centre for Airborne Laser Mapping/University of Houston and Pacunam Lidar Initiative (PLI), **170**; Shutterstock /AAR Studio, **200–1**, /Mark Agnor, **115** (bottom), /aphotostory, **112–13**, /AVIcon, **204** (political), /Alex James Bramwell, **i** (bottom right), **193**, /camilkuo, **104–5**, /Valeria Cantone, **133** (top), **150**, /cge2010, **16–17**, /Tony Craddock, **210**, /CW Pix, **7**, /Everett Historical, **15**, /Natalya Erofeeva, **165**, /funkyfrogstock, **202**, /J. Helgason, **195**, /Evgeniy Kalinovskiy, **8**, /Katiekk, **102**, /Kozlik, **194**, /m.o.arvas, **79** (top), /Mark Mortin, **171**, /naskami, **6**, /Nejron Photo, **22** (left), /Pajor Pawel, **205**, /Gary Perkin, **73** (top), /Leon Rafael, **200**, /Andrew Roland, **2**, /RPBaiao, **24–5**, /Sammy33, **147** (top), /Studio Driehoek, **138**, /Pises Tungittipokai, **108–9**, /Travel Stock, **135**, /Yeamake, **217**, /Yudhistira99, **206**; Swedish History Museum/Crister Ahlin/SHMM (CC BY 2.0), **23** (centre).

OTHER MATERIAL: Interview transcript reproduced by permission of the Australian Broadcasting Corporation - Library Sales. Richard Fidler © 2016 ABC, **87**; Extract from 'Viking Raiders' by Frances White, *All About History*, Future Publishing Ltd, 2015. Reproduced with permission of the Licensor through PLS clear, **28**; The Map Archive/Battle of Hastings map, **64**; Orpheus/q-files/illustrations © Orpheus Books, **24–5**, **26–7**, **40**, **166**, **177** (top); Engraving from *A Brief History of the United States* by Joel Dorman Steele and Esther Baker Steele, 1885, **90**; Aztec Chinampas model by Te Mahi, Photographer: Te Papa, © Te Papa, **173**; Map based on data from US Geological Society, Land Cover Institute, Broxton, P.D., Zeng, X., Sulla-Menashe, D., Troch, P.A., 2014a: A Global Land Cover Climatology Using MODIS Data. J. Appl. Meteor. Climatol, 53, 1593–1605. doi: http://dx.doi.org/10.1175/JAMC-D-13-0270.1, **103**.

While every care has been taken to trace and acknowledge copyright, the publisher tenders their apologies for any accidental infringement where copyright has proved untraceable. They would be pleased to come to a suitable arrangement with the rightful owner in each case.

good humanities

Flip your book over and begin your journey through **Geography**

contents

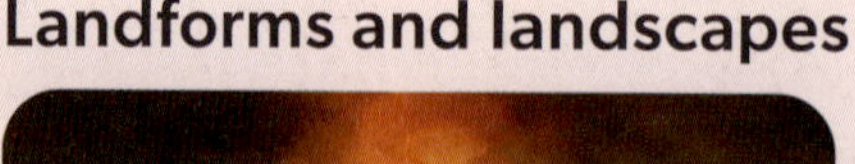

Landforms and landscapes

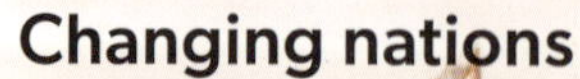

Changing nations

Geography concepts + skills

Good Humanities 8
Victorian Curriculum
1st edition
Ben Lawless
Danielle O'Leary
Peter van Noorden

Publisher: Catherine Charles-Brown
Publishing director: Olive McRae
Copy editor: Kate McGregor
Project editors: Kate McGregor, Isabelle Sinclair
Proofreader: Kelly Robinson
Indexer: Max McMaster
Text and cover designer: Jo Groud
Typesetters: Paul Ryan, Kerry Cooke
Illustrator: QBS Learning
Cartography: Wayne Murphy, MAPgraphics
Production controllers: Nicole Ackland, Sarah Blake
Digital production: Erin Dowling
Permissions researchers: Samantha Russell-Tulip, Hannah Tatton
Cover images photographed by:
Geography: Photo by Paul Tsui
History: Adobe Stock/ Patrycja
Spine: iStockphoto/valentinrussanov

First published in 2021 by Matilda Education Australia, an imprint of Meanwhile Education Pty Ltd
Melbourne, Australia
T: 1300 277 235
E: customersupport@matildaed.com.au
www.matildaeducation.com.au

National Library of Australia Cataloguing-in-Publication data
Author: Ben Lawless, Danielle O'Leary, Peter van Noorden
Title: Good Humanities 8 Victorian Curriculum
ISBN: 9781420247190

A catalogue record for this book is available from the National Library of Australia at www.nla.gov.au

Warning: It is recommended that Aboriginal and Torres Strait Islander peoples exercise caution when viewing this publication as it may contain names or images of deceased persons.

Printed in Malaysia by Vivar Printing
Jun-2025

G0

Introduction to Geography

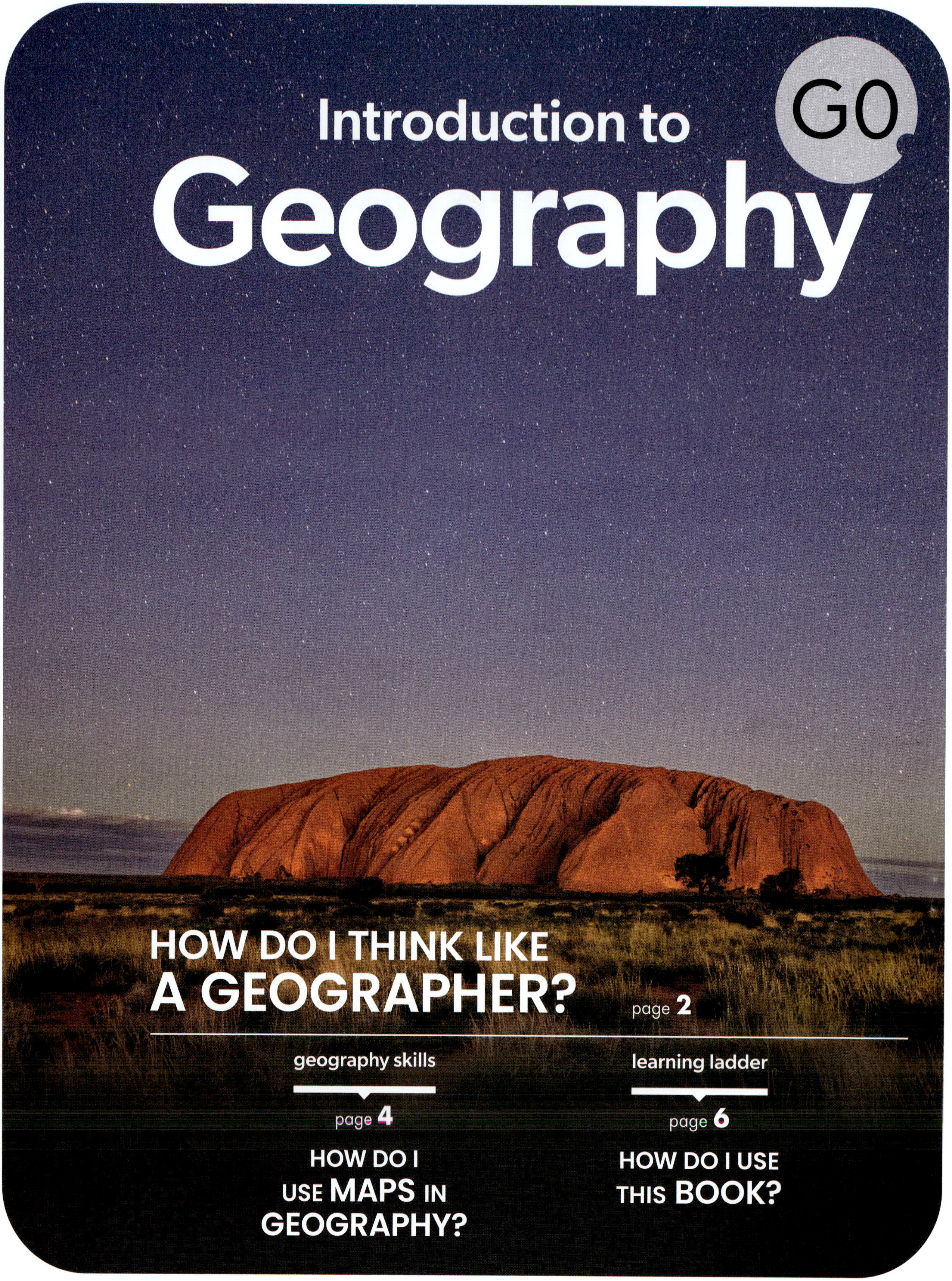

HOW DO I THINK LIKE A GEOGRAPHER?

page 2

geography skills

page 4

HOW DO I USE MAPS IN GEOGRAPHY?

learning ladder

page 6

HOW DO I USE THIS BOOK?

How do I think like a geographer?

Geography is the study of the world, its characteristics and patterns, and how it changes over time. In Geography we focus on two main factors: human activity and natural processes. We are interested in how humans influence the *space* in which they live and, in turn, how the natural *environment* influences how humans live. We use maps, images, graphs and other sources of data to explore spatial patterns and conduct fieldwork to answer research questions about different phenomena or *interconnections*. Geography is key to understanding the world around us, because 'without geography, you are nowhere!'.

Using geographic language: SPICESS

Using geographic language is important to help you write high-quality responses. Remembering the acronym SPICESS will help you improve your use of geographic terminology. Note: you do not have to use the terms from SPICESS directly in your responses; instead, try to use the concept to make your writing more geographical.

S Space (1)

In a geographical context, '*space*' does not refer to outer space. It refers to the way in which we use, *change* and distribute things on Earth's surface. For example, this *environment* can be broken up into two key *spaces*; the natural mountain *space* and human urbanised *space*.

P Place (2)

The concept of *place* allows humans to identify the location or position of something within a *space*. We can identify place through describing the **relative** or **absolute location** of that area. This *place* is Mt Fuji, Japan.

I Interconnection

3

Interconnection is the idea that two things or phenomena are related, interact or are linked in some way. For example, there is a strong *interconnection* between the Japanese people and Mt Fuji. Many people climb the mountain as a spiritual journey, stopping to pray at temples along the way.

C Change

4

Change refers to how a *place* is altered because of shifts in the environment or to meet the needs of humans. Change can be positive or negative and can occur over short or long periods. In this *place*, we can observe a change between the seasons through the vegetation. In spring, beautiful pink and white cherry blossoms bloom throughout Japan.

E Environment

5

An *environment* can be defined by its **geographic characteristics**. Some *environments*, such as coastlines, islands and forests, are largely natural and are untouched by humans. Other *environments*, such as cities and other urban areas, have undergone significant change and are largely unnatural. Within environments, we can observe processes, interconnections between **phenomena** and change over time. This is an example of both a natural and a human *environment*.

S Scale

6

Scale usually refers to the size of something. Scale can be literal, such as a scale on a map using quantitative data to show you how big something would be in real life. We can also use scale as a qualitative word, such as when describing a region. For example, patterns can exist on a local, regional, national or global *scale*. Mt Fuji is Japan's tallest peak at 3776 metres.

S Sustainability

7

Sustainability is the concept of maintaining and preserving resources and *environments* for future generations. This could be through using *sustainable*, renewable energy such as solar power or wind-generated electricity. Mt Fuji is a major tourist attraction in Japan and therefore this *place* is subject to a lot of foot traffic and rubbish dumping. A range of groups and organisations have been developed over time to help *sustain* the natural *environment* surrounding Mt Fuji to ensure it remains beautiful for future visitors. Mt Fuji was listed as a World Heritage Site in 2013.

Source 1

Mt Fuji, Japan

How do I use maps in Geography?

Maps are vital in Geography to visualise spatial patterns and identify *interconnections* between phenomena. There are many types of maps and each has a slightly different purpose.

World water use

WATER USE PER PERSON PER YEAR (LITRES)

More than 750 000

500 000 to 750 000

250 000 to 500 000

100 000 to 250 000

Less than 100 000

No data

N

0 3000 6000 km

Source: Matilda Education Australia

Source 1

A choropleth map

Contour maps

Contour maps show how mountainous a region is. Contour maps can be identified by having a series of wiggly lines with numbers along them, known as contour lines. The closer the contour lines are together, the steeper the hill is. The numbers indicate how far above sea level that particular point is. Source 2 shows Uluru in the Northern Territory. Source 3 is an aerial view of Uluru, with a map overlaid on the photo showing how contour lines can be used to indicate height and depth.

Choropleth maps

Choropleth maps use darker and lighter shades of colour so that the reader can instantly see a pattern. The choropleth map in source 1 uses darker shades to represent the areas with highest water use per person per year and the lighter shades represent the areas with lowest levels of use. It is best to use colours of the same shade, such as light to dark green to clearly show the pattern.

Source 2

This aerial image of Uluru shows its shape and the weathering it has endured.

Source 3

Uluru contour map

A contour map of Uluru, overlaid on an aerial photo

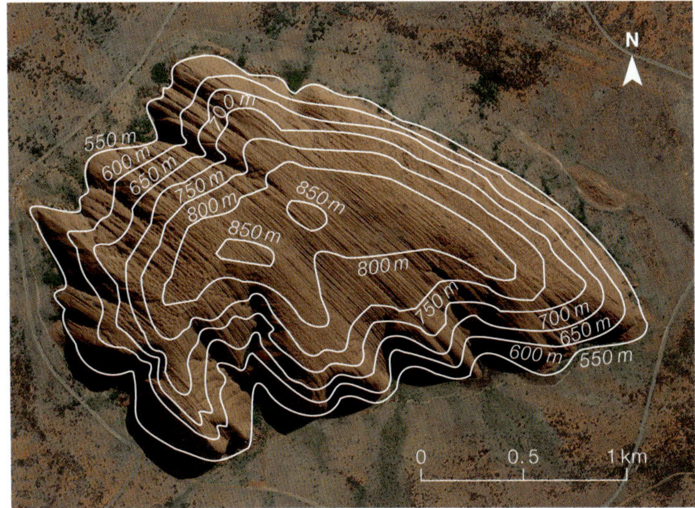

Source: contour map from Matilda Education Australia, Geoscience Australia

World political map

Source: Matilda Education Australia

Source 4

A political map of the world

Political maps

Political maps show the boundaries of different countries, determined by political governance or ownership. Over time, these boundaries change as wars or treaties affect the borders of different areas.

Australia rainfall variability

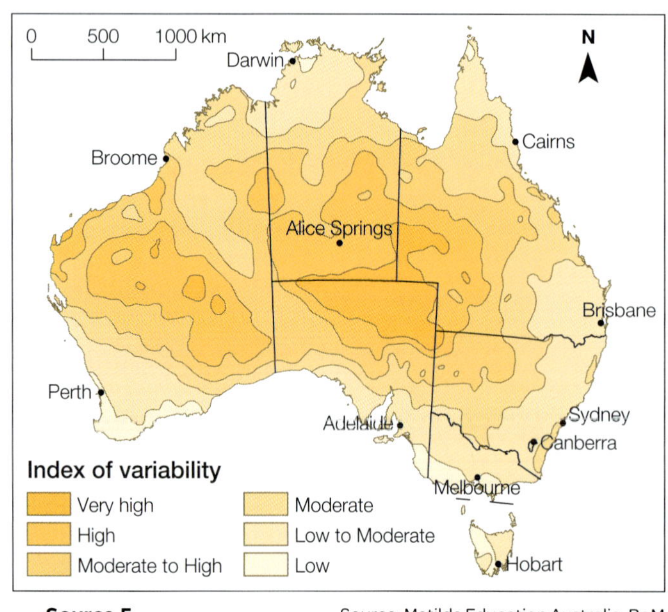

Source 5

Source: Matilda Education Australia, BoM

A climate map

Climate map

In Geography, **climate** is important for determining land cover and how liveable a place is. While weather is what we observe day to day through our classroom window, climate represents the average conditions (usually temperature and rainfall) within a region over a period. Climate is important when studying Geography, as it gives us an indication of expected rainfall. In the map in Source 5, we can clearly see a pattern of variability (changes) in rainfall in different parts of Australia.

Learning ladder G0.2

1 Note down which maps you are already familiar with. When have you used these maps?

2 In a small group, discuss the list of maps you have used. Are you all familiar with the different kinds of maps discussed in this section?

3 Why are maps vital in Geography? Use bullet points to list at least three reasons.

How do I use this book?

Good Humanities has been built to help you thrive as you work through the Level 8 Humanities curriculum and to demonstrate your progress in every single lesson. The Geography section of this book explores two geographical topics: Landscapes and landforms and Changing nations. You will also find a Fieldwork section and a Geo How-To skills section. The Geo How-To section is vital and you should refer to it often.

Climb the Learning Ladder

Geography is a skills-based subject; and learning content alone does not mean that you have a geographical understanding. In order to be a geographer, you need to be able to write geographically, read a variety of data, interpret data and conduct and communicate your own research.

Each chapter in this book begins with a Learning Ladder. This is your 'plan of attack' for the skills you will practise in each chapter. It lists the five geographical skills you will be learning, and five levels of progress for each of those skills. Read it from the bottom to the top. As you progress through the chapter, you will climb *up* the Learning Ladder.

Each skill described in the Learning Ladder is a higher progression than the one below it. To be able to accomplish the higher-level skills, you need to be able to master the lower ones. Practising activities at all the levels will help you to achieve 'higher-order' skills such as evaluating. This approach is called developmental learning and puts you in charge of your own learning progression!

Source 1 The Learning Ladder helps you to take charge of your own learning.

learning ladder

	Spatial characteristics	Interconnections	Geographical challenge	Record, collect and display data	Analyse data
step 5	I can analyse the impact of change on places	I can explore spatial association and interconnections	I can plan action to tackle a geographical challenge	I can evaluate data	I can draw conclusions by analysing collected data
step 4	I can predict changes in the characteristics of places	I can evaluate the implications of significant interconnections	I can evaluate alternatives for a geographical challenge	I can use data to support claims	I can analyse relationships between different data
step 3	I can explain processes influencing places	I can identify and explain the implications of interconnections	I can compare strategies for a geographical challenge	I can choose, collect and display appropriate data	I can explain the reasons behind a trend or spatial distribution
step 2	I can explain spatial characteristics	I can explain interconnections	I can compare responses to a geographical challenge	I can recognise and use different types of data	I can describe patterns and trends
step 1	I can identify and describe spatial characteristics	I can identify and describe interconnections	I can identify responses to a geographical challenge	I can record, collect and display data in simple forms	I can use geographic terminology to interpret data

Check your progress

Each chapter is divided into sections.
Each section is designed to cover one lesson, but your teacher may spend more or less time on a particular section. A section is usually two pages long, but some are four pages. At the end of every section, you will find different kinds of questions to help you check your progress.

1 Show what you know

These questions ask you to look back at the content you have read and to show your understanding of it by listing, describing and explaining.

Learning Ladder

These activities are linked to the Learning Ladder. You can complete one of the questions or several of them, depending on your progress. Throughout each chapter, you will complete at least one activity for each level of the Learning Ladder.

Case studies

Throughout every chapter, you will discover a variety of case studies that explore local, national and global places, issues and events. The local case studies are focused on Victorian contexts. The national case studies explore a case study from other parts of Australia and the global case studies describe places and events in other countries. Many of these link to interactive sources you can explore further using your eBook.

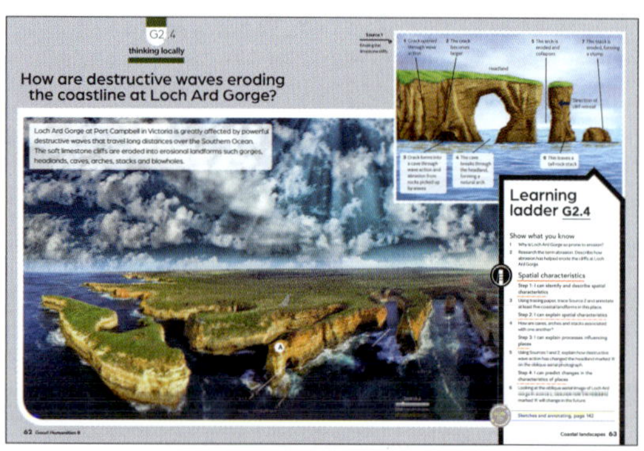
G2.4
thinking locally
How are destructive waves eroding the coastline at Loch Ard Gorge?
Learning ladder G2.4

Source 2

Case studies are an important part of your Geography course.

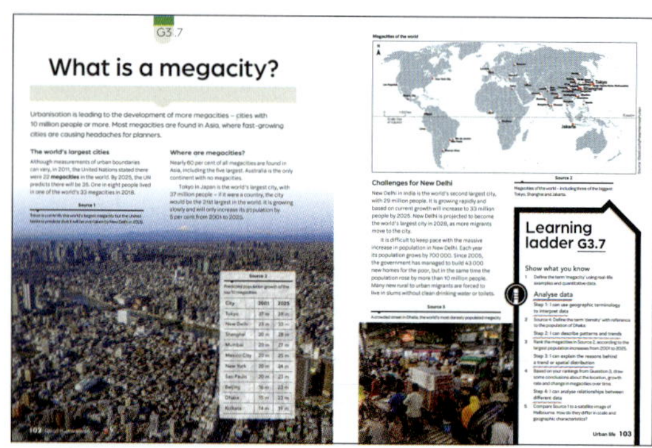
G3.7
What is a megacity?
Learning ladder G3.7

Source 3

Check your progress regularly. You can attempt one or all of the Learning Ladder questions.

civics+ citizenship **economics+ business**

The study of Humanities is more than just Geography and History – it is also the study of Civics and citizenship, and Economics and business.
In every chapter of this book, you will discover either a Civics and citizenship *or* an Economics and business lesson. School is busy and you have a lot to cover, so designing a textbook where the important Civics and citizenship and Economics and business content is placed meaningfully next to relevant Geography or History lessons will help you connect your learning to a wider context.

Geo How-To

In the middle of the book, you will find a guide to fieldwork and a skills section called Geo How-To. The How-To section features explanations about how to perform each skill. There are *many* examples. Refer to it often, especially when completing the Learning Ladder questions and Masterclass activities.

The Fieldwork section explains the skills you need for hands-on research as well giving you several suggested tasks.

Source 4

The Geo How-To section is your key to success – refer to it often!

Masterclass

At the end of each chapter is a review section – the Masterclass. The questions here are organised by the steps of the Learning Ladder. You can complete all the questions *or* your teacher may direct you to complete just some of them, depending on your progress.

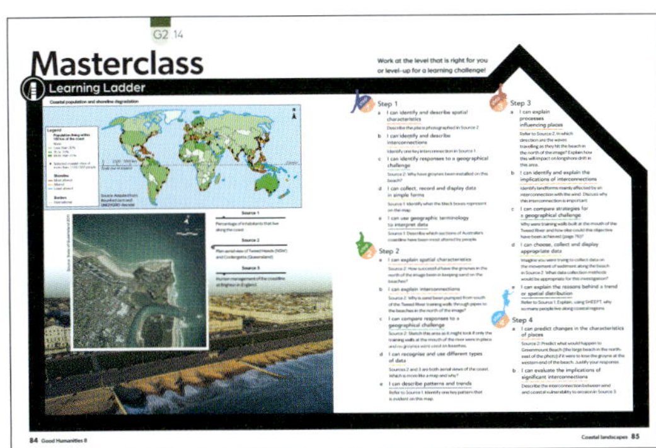

Source 5

You did it! We knew you could! The Masterclass is your opportunity to show your progress. Take charge of your own learning and see if you can extend yourself.

Capstone

After you complete a chapter, it's time to put your new knowledge and understanding together for the capstone project to show what you *know* and what you *think*. In the world of building, a capstone is an element that finishes off an arch, or tops off a building or wall. That is what the capstone project will offer you, too: a chance to top off and bring together your learning in interesting and creative ways. It will ask you to think critically, to use key concepts and to answer 'big picture' questions. The capstone project is accessible online; scan the QR code to find it quickly.

Source 6

The capstone project brings together the learning and understanding of each chapter. It provides an opportunity to engage in creative and critical thinking.

Learning ladder G0.3

1 How can you use the Learning Ladder to monitor your progress in Year 8 Geography?

2 In a small group, discuss the idea of 'monitoring your own progress'. Why is this important?

3 Read through the steps of the Geography Learning Ladder and consider where you may be up to for each skill, based on your prior learning.

4 Draw a logo for the Humanities learning area at your school. Make sure your logo includes elements from Geography, History, Economics and business, and Civics and citizenship.

G1

Landscapes and landforms

WHAT HAPPENS WHEN A VOLCANO ERUPTS?

page 28

geographical challenge

page 32

WHAT ARE GEOHAZARDS?

thinking nationally

page 40

HOW DID ULURU FORM?

economics + business

page 42

HOW DO LANDFORMS ATTRACT TOURISTS?

How can I understand landscapes and landforms?

Mountains, glaciers, valleys, plateaus and many other natural features that we see on the surface of Earth are called landforms. Landforms are produced by natural processes. Similar landforms are grouped together as landscapes.

learning ladder

	Spatial characteristics	Interconnections	Geographical challenge
step 5	**I can analyse the impact of change on places** I can analyse and evaluate the implications of changes such as deforestation at different scales and calculate the impact on people and environments.	**I can explore spatial association and interconnections** I can compare distribution patterns and the interconnections between them; e.g. mountain ranges and tectonic plate boundaries.	**I can plan action to tackle a geographical challenge** I can frame questions, evaluate findings, plan actions and predict outcomes to tackle a landform-based geographical challenge.
step 4	**I can predict changes in the characteristics of places** I can predict changes in the characteristics of places over time due to geomorphic processes.	**I can evaluate the implications of significant interconnections** I can identify, analyse and explain key geomorphic interconnections within and between places, and evaluate their implications over time and at different scales.	**I can evaluate alternatives for a geographical challenge** I can weigh up alternative views and strategies on a landform-based geographical challenge using environmental, social and economic criteria.
step 3	**I can explain processes influencing places** I can explain the series of actions leading to change in a place, such as the formation of block mountains.	**I can identify and explain the implications of interconnections** I can identify, analyse and explain geomorphic interconnections and explain their implications.	**I can compare strategies for a geographical challenge** I can compare strategies for a geographical challenge, taking into account a range of factors and predicting the likely outcomes.
step 2	**I can explain spatial characteristics** I can identify concepts of Space, Place, Interconnection, Change, Environment, Scale and Sustainability (SPICESS) when I read about landscapes and landforms.	**I can explain interconnections** I can describe and explain interconnections and their effects, such as frost wedging in rocks.	**I can compare responses to a geographical challenge** I can identify and compare responses to a geographical challenge and describe its impact on different groups.
step 1	**I can identify and describe spatial characteristics** I can talk about spatial characteristics at a range of scales; e.g. volcanoes of the world.	**I can identify and describe interconnections** I can identify and explain simple interconnections involved in phenomena such as shifting tectonic plates and earthquakes.	**I can identify responses to a geographical challenge** I can find responses to a geographical challenge such as a volcanic eruption and understand the expected effects.

Source 1

The basalt rock 'organ pipes' in the Organ Pipes National Park in Victoria formed about one million years ago from the lava flow of a nearby volcano. The lava filled a creek bed and cracked vertically as it cooled.

I can evaluate data
I can determine whether data presented about landscapes and landforms is reliable, and assess whether the methods I used in the field or classroom were helpful in answering a research question.

I can draw conclusions by analysing collected data
I can summarise findings and use collected data to support key patterns and trends I have identified for research.

I can use data to support claims
I can select or collect the most appropriate data and create specialist maps and information using ICT to support investigations into landscapes and landforms.

I can analyse relationships between different data
I can use multiple data sources, overlays and GIS to find relationships that exist in patterns of landscapes and landforms.

I can choose, collect and display appropriate data
I can select useful sources of data and represent them to conform with geographic conventions.

I can explain the reasons behind a trend or spatial distribution
I can identify Social, Historical, Economic, Environmental, Political and Technological (SHEEPT) factors to help me explain patterns in data.

I can recognise and use different types of data
I can define the terms primary, secondary, qualitative and quantitative data and represent data in more complex forms.

I can describe patterns and trends
I can identify Patterns, Quantify them and point out Exceptions (PQE) to describe the patterns I see.

I can collect, record and display data in simple forms
I can identify that maps and graphs use symbols, colours and other graphics to represent data.

I can use geographic terminology to interpret data
I can identify increases, decreases or other key trends on a map, graph or chart about landscapes and landforms.

Collect, record and display data

Analyse data

Spatial characteristics

1 Looking at Source 1, how would you describe the geographic characteristics of this *place*?

Interconnections

2 Source 1: The interconnection between which two phenomena formed the Organ Pipes?

Geographical challenge

3 What geographical challenge did the people of Hawaii face (see pages 30–31)?

Collect, record and display data

4 In the map of World mountain landscapes, Source 2 on page 17, is the data qualitative or quantitative?

Analyse data

5 Still looking at Source 2 on page 17, in which continent can the Himalayas be found?

What are landforms and landscapes?

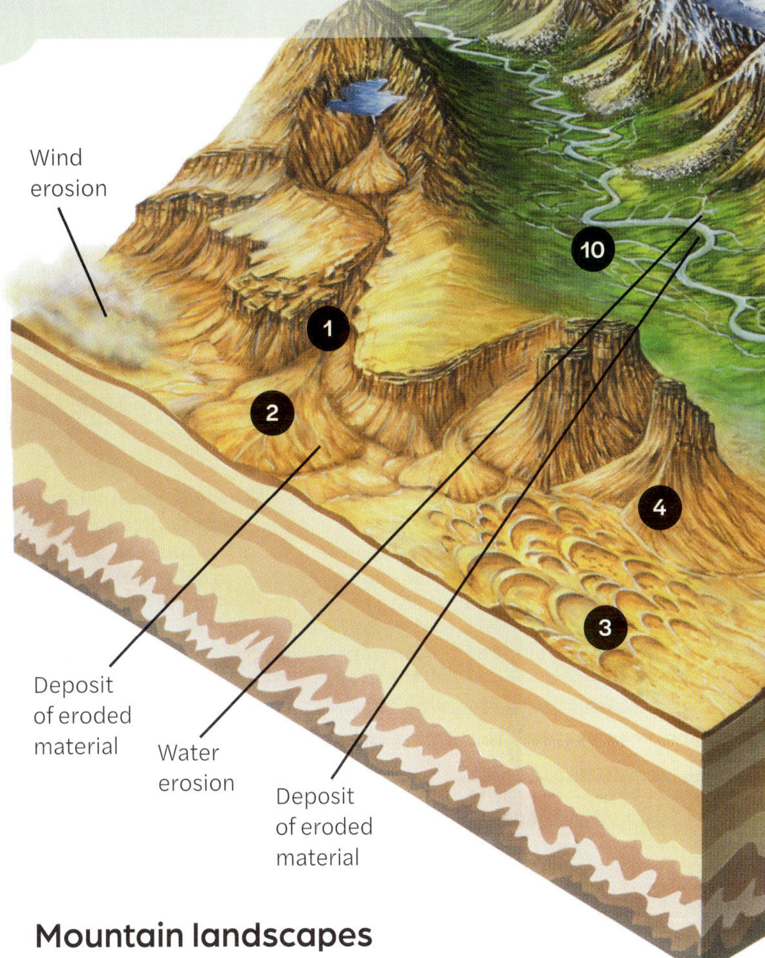

Earth contains many different types of natural **landscapes**: mountain, desert, coastal and riverine landscapes. Each landscape features unique **landforms** created by natural forces, such as the movement of **tectonic plates**.

Landforms are shaped over time by **weathering** and **erosion**. Weathering wears them down and breaks them apart and erosion removes the weathered material. The growth of human settlement and activities such as farming, mining, logging and tourism can change and sometimes degrade landscapes.

Desert landscapes

Deserts are dry environments. They can be hot or cold and can be rocky, sandy or even covered in ice. Rocky deserts can be weathered and then eroded by water and wind into amazing shapes. Use the numbers for each landform below to locate them in Source 1.

Desert landforms

1. **Wadi**: Dry watercourse in a deep, narrow valley that divides a **plateau**.
2. **Alluvial fan**: Semicircular build-up of eroded material deposited by water and wind at the end of a wadi.
3. **Barchan dune**: A crescent-shaped **dune** formed by the wind blowing from one direction. They are a common landform in sandy deserts.
4. **Butte:** Isolated, steep, flat-topped hill of resistant rock (rock that does not erode easily).

Mountain landscapes

Mountain landscapes sit high above the surrounding land. They are formed by movements in Earth's crust, which also cause earthquakes and volcanoes. Use the numbers for each landform below to locate them in Source 1.

Mountain landforms

5. **Mountain**: Steep-sided, lone peak rising above the surrounding land.
6. **Fold mountains**: Formed when two tectonic plates collide and the crust folds and is pushed upwards.
7. **Glacier**: River of ice that moves downhill in response to gravity.

Riverine landscapes

Riverine landscapes are formed by the natural movement of a network of rivers in a **drainage basin**. These landscapes attract farming because they contain fresh water and fertile land from deposited silt. Use the numbers for each landform below to locate them in Source 1.

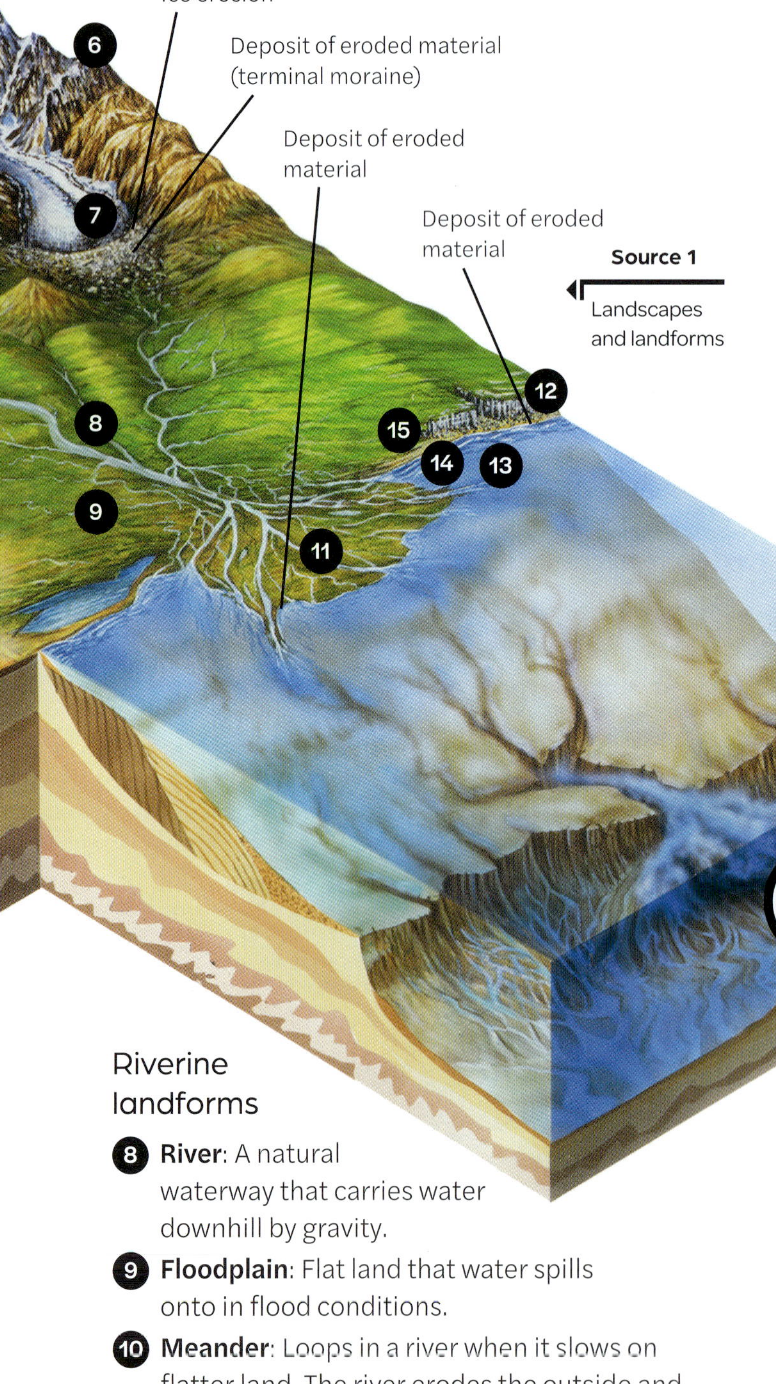

Source 1 Landscapes and landforms

Riverine landforms

- **8 River**: A natural waterway that carries water downhill by gravity.
- **9 Floodplain**: Flat land that water spills onto in flood conditions.
- **10 Meander**: Loops in a river when it slows on flatter land. The river erodes the outside and deposits on the inside of the meander.
- **11 Delta**: Deposited sediment at a river's mouth, which causes the river to block its own exit. The river splits into a number of finer channels that find their way to the sea.

Coastal landscapes

The coast is where the land meets the sea. Coastal landscapes are shaped by wind and waves that erode and deposit materials to create landforms that are constantly changing (see Chapter 2, Coastal landscapes). Use the numbers for each landform below to locate them in Source 1.

Coastal landforms

- **12 Headland**: Narrow, high land jutting out into the sea, formed of harder rock that resists erosion
- **13 Bay**: Curved indentation in the coastline.
- **14 Beach**: Deposited rock particles, such as sand or gravel, along the coastline.
- **15 Coastal dune**: Mound or ridge of wind-blown sand at the back of the beach.

Learning ladder G1.1

Show what you know

1 Name a landform that fits each of the following descriptions:
   a made of ice
   b made by erosion
   c created by eroded material being deposited
   d caused by movement of the Earth's crust
   e made by water.

Spatial characteristics

Step 1: I can identify and describe spatial characteristics

2 Identify different valley landforms from three different landscapes.

Step 2: I can explain spatial characteristics

3 Where do glaciers occur and why?

Step 3: I can explain processes influencing places

4 Explain the key process that drives the formation of desert landscapes.

Step 4: I can predict changes in the characteristics of places

5 Select one landscape or landform that will change the most dramatically due to global warming, where world temperatures are rising.

HOW TO Block diagrams, page 144

How are we connected to landscapes?

Landscapes and landforms around the world are valued in different ways. Some people may have a cultural, spiritual or aesthetic connection to a particular landscape, while others are more interested in the economic value of the place.

Aesthetic values

The **aesthetic value** of a landscape is linked to its uniqueness and its perceived beauty. People are drawn to a place for many reasons, including how it looks, how it makes them feel or through a personal connection with the place.

Mountains are among the most aesthetic landscapes (see pages 16–19). The beauty of mountains goes beyond an appreciation of the landscape and the **landforms** within it. It also includes concepts such as wilderness and adventure and emotions such as excitement, anticipation and even fear.

In Nepal, exploration and appreciation of the Mount Everest region is something that began when the peak was first climbed in 1953. Since then more than 5000 climbers have climbed to the peak of the world's tallest mountain (see page 22). Recently, mountaineers have complained that the crowds climbing Mount Everest are taking away from its aesthetic appeal.

Source 1

The aesthetic appeal of the Three Sisters landform in the Blue Mountains also provides an economic value, as the site attracts over 600 000 tourists every year. It is a place of cultural significance to the Gundangurra, Wiradjuri, Tharawal and Darug peoples.

Source 2

The aesthetic value of Mount Everest goes beyond the beauty of this mountain region. It includes its uniqueness and majesty as the world's highest mountain, along with the achievement of climbing it.

Spiritual and cultural values

Landscapes contain many landforms that are sites of spiritual importance, particularly for Indigenous peoples. In the Hawaiian religion, the ancient fire goddess Pele, who lives in the crater at the top of the Kīlauea volcano, created Hawaii's landscape. When Kīlauea erupts, lava is the physical embodiment of Pele. She is known to be unpredictable, so Hawaiians traditionally provide gifts and offerings to keep her happy. During the eruption of Kīlauea in 2018 (see page 30), Hawaiians left leaves in front of their homes and flowers in cracks caused by the volcano for good luck.

For Australia's First Nations Peoples, landscapes and landforms were formed by the Ancestors who continue to live in the land, water and sky. Indigenous Australians express the cultural value of the land to them through stories, song, dance and art.

The Three Sisters in the Blue Mountains (see Source 1) is significant to the Gundangurra, Wiradjuri, Tharawal and Darug peoples and was recognised as an Aboriginal place by the New South Wales Government in 2014.

One creation story tells the tale of three sisters from the Katoomba tribe who fell in love with three brothers from a neighbouring tribe. The law forbade the girls from marrying outside their own people. When the brothers decided to capture the girls, a major battle erupted. A clever man from the Katoomba tribe cast a spell to turn the sisters to stone to keep them safe. During the battle, the clever man was killed and unable to reverse his spell, so the sisters remain trapped in stone to this very day.

Learning ladder G1.2

Show what you know

1 What is your favourite landform or landscape? Create and deliver a short presentation on this place. Consider the following:
   a the location of the landform or landscape
   b how it was formed
   c why you have a connection to that place.

2 Conduct a survey of your class or a large group of students asking them the following questions. Record your data using a tally.
   a What is your favourite landform or landscape?
      i Mountains (including glaciers and volcanoes)
      ii Rivers
      iii Deserts
      iv Coastal (including beaches, bays and headlands)
      v Other
   b What is your main connection to that place?
      i Spiritual
      ii Visual
      iii Cultural
      iv Other

3 Create two bar graphs to display your findings from question 2. Discuss your findings as a class.

Interconnections

Step 1: I can identify and describe interconnections

4 Why do you think tourists are attracted to particular landscapes and landforms?

Step 2: I can explain interconnections

5 How do Indigenous Australians and Hawaiians show their connection to the land?

Step 3: I can identify and explain the implications of interconnections

6 Source 1: How does this place interconnect both spiritually and aesthetically with people?

Step 4: I can evaluate the implications of significant interconnections

7 Source 2: Evaluate why there is a strong interconnection between climbers and Mount Everest.

HOW TO

Graphing, page 146

How are mountains formed?

Mountains are landforms that rise high above the surrounding land. They are formed when tectonic plates in the Earth's crust move to push up the land. Sometimes this allows molten magma to reach the surface, which creates volcanic mountains.

Source 1

At 5895 metres above sea level, Mount Kilimanjaro is Africa's highest mountain and the tallest freestanding mountain in the world. It is a volcanic mountain formed over a hot spot in the Earth's crust. The top is covered in ice and snow even though the mountain is located near the equator.

World mountain landscapes

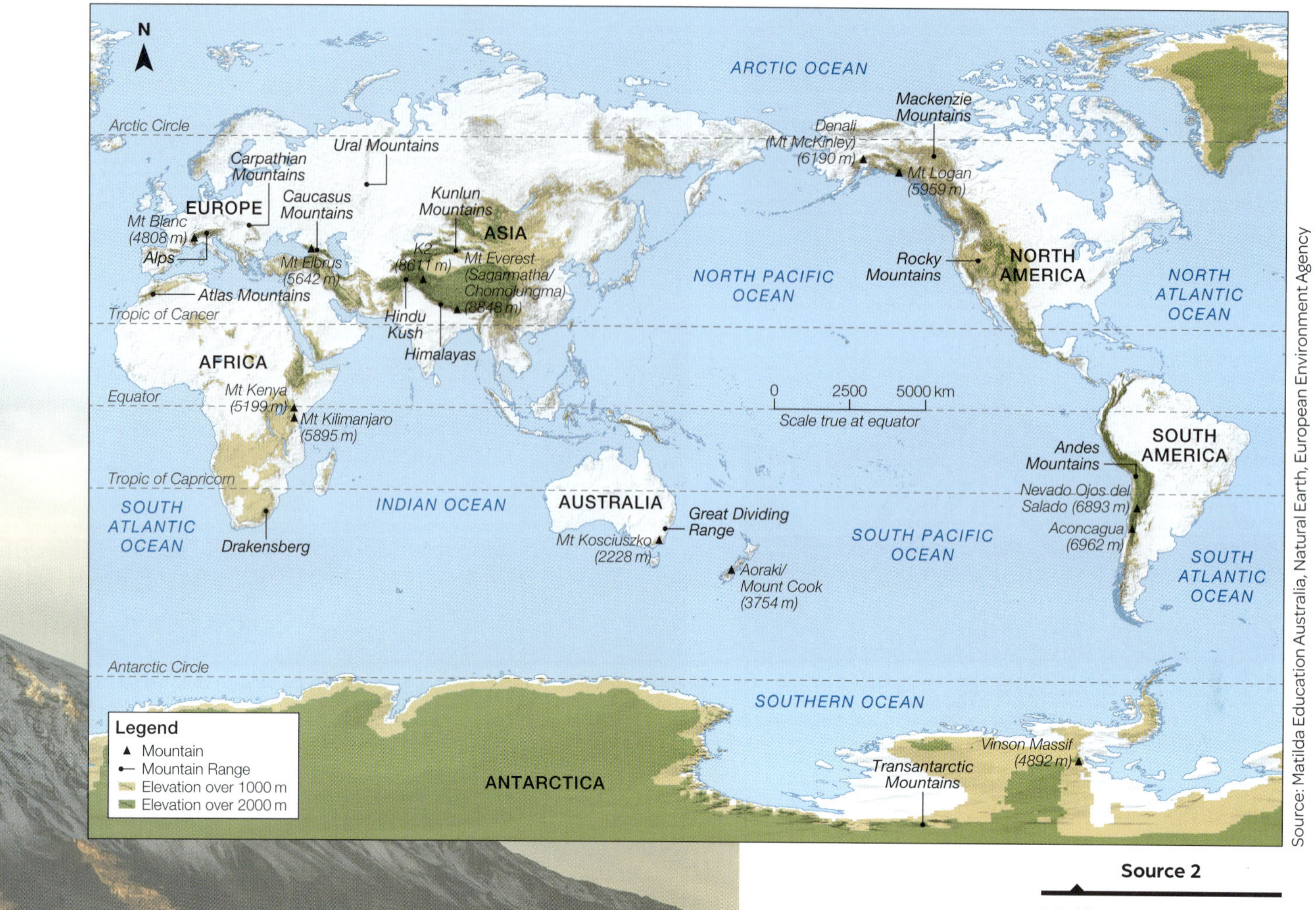

Source: Matilda Education Australia, Natural Earth, European Environment Agency

Source 2 World mountain landscapes

What are mountains?

A **mountain** is a large landform that sits above the surrounding land, usually in the form of a peak. A mountain is steeper than a hill. A few mountains, such as Mount Kilimanjaro in Africa, are isolated, but most are part of a series of mountains called a **mountain range**. Earth's highest mountain is Mount Everest in Asia's Himalayan mountain range. Its summit is 8848 metres above sea level. Australia's highest mountain, Mount Kosciuszko, is just 2228 metres tall.

The world's longest above-water mountain range is found in South America. Known as the Andes, this mountain range is 7000 kilometres long and travels through seven countries. Australia's longest mountain range is the Great Dividing Range, stretching more than 3500 kilometres through Victoria, New South Wales, the Australian Capital Territory and Queensland.

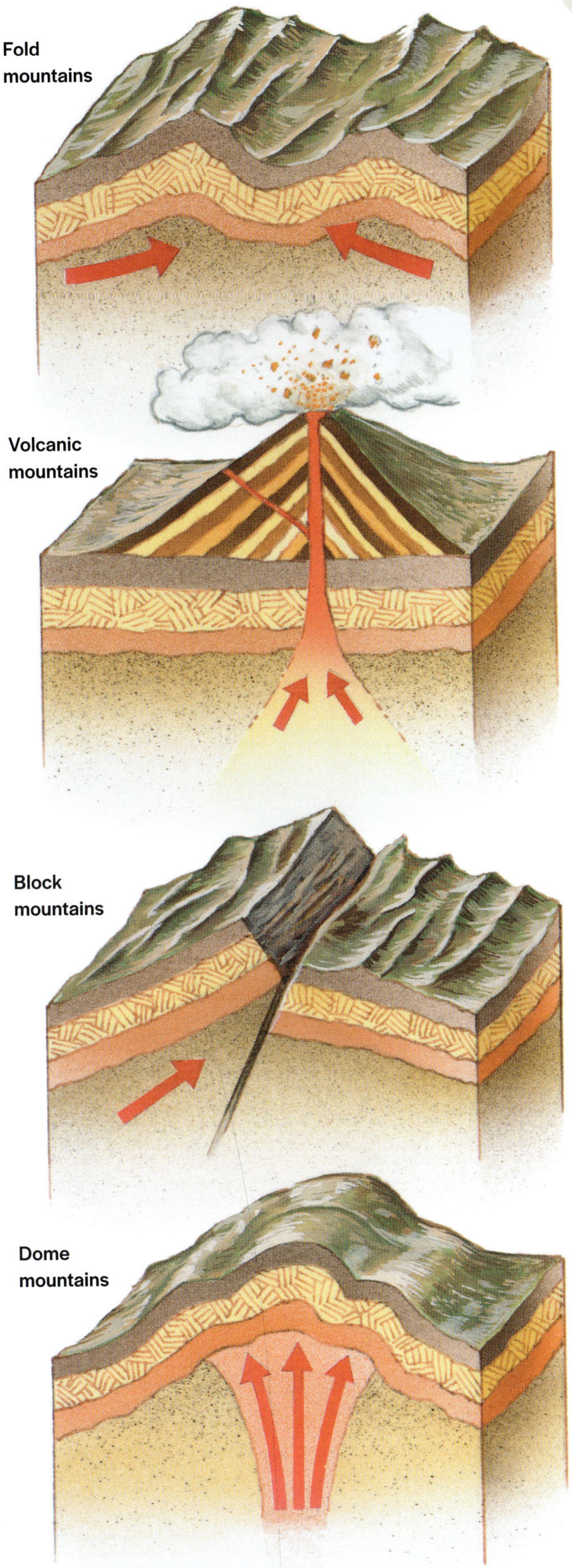

Fold mountains

Volcanic mountains

Block mountains

Dome mountains

Source 3

Different mountain formations: fold, volcanic, block and dome

How do mountains form?

Mountains are formed through earth movements or volcanic action. Strong forces inside Earth can push up the surface into long mountain chains or punch through the surface at **hot spots** (see page 25) to create **volcanoes** such as Kīlauea in Hawaii (see page 30) and Mount Kilimanjaro in Tanzania (see Source 1 on page 16).

Fold mountains

Fold mountains are pushed upwards where two continental **tectonic plates** (see pages 20–21) collide. Layers of rock are bent and pushed upwards to create the fold mountains. The world's largest fold mountains are found in the Himalayan mountain range (see pages 22–23), with over 50 peaks more than 7200 metres high.

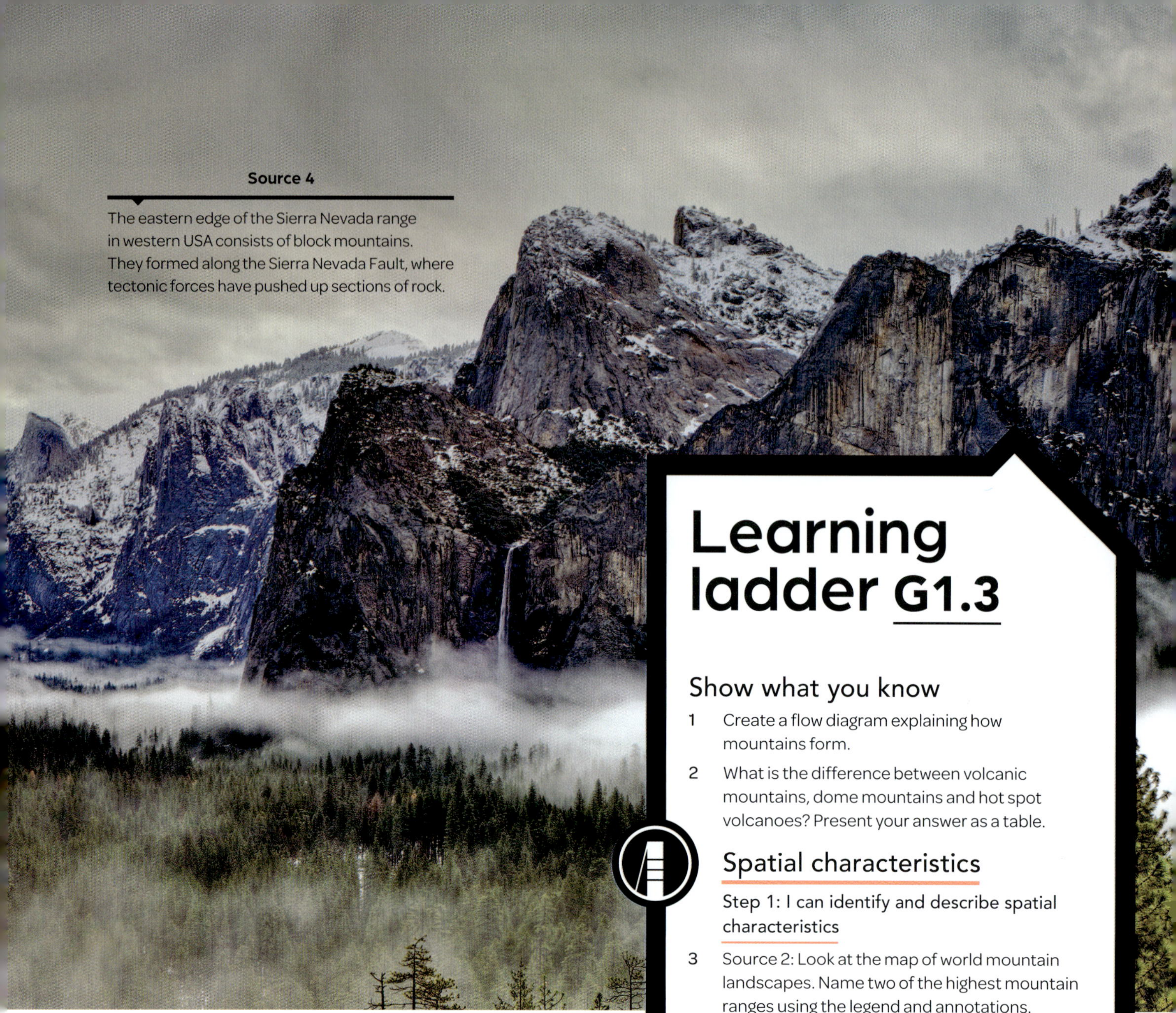

Source 4

The eastern edge of the Sierra Nevada range in western USA consists of block mountains. They formed along the Sierra Nevada Fault, where tectonic forces have pushed up sections of rock.

Volcanic and dome mountains

Volcanic mountains form when **magma** from below the Earth's surface is pushed through an opening in the Earth's crust. This material can build up to form a cone-shaped mountain known as a volcano (see pages 24–25). **Dome mountains** form when magma is injected between two layers of **sedimentary rock**, causing the top layer to bulge upwards forming the dome shape.

Block mountains

Block mountains form when blocks of rocks are forced upward or downward along faults or cracks in Earth's crust. The uplifted blocks are block mountains or **horsts**. Dropped blocks are called **graben**, which can form **rift valleys**.

Learning ladder G1.3

Show what you know

1 Create a flow diagram explaining how mountains form.

2 What is the difference between volcanic mountains, dome mountains and hot spot volcanoes? Present your answer as a table.

Spatial characteristics

Step 1: I can identify and describe spatial characteristics

3 Source 2: Look at the map of world mountain landscapes. Name two of the highest mountain ranges using the legend and annotations.

Step 2: I can explain spatial characteristics

4 Source 2: Which are the highest and flattest continents?

Step 3: I can explain processes influencing places

5 Simulate tectonic plates by placing two sheets of paper end to end on a flat surface (about two centimetres apart). Place four fingers on both sheets of paper and move the left sheet slowly towards the right.
   a Sketch what happens when the paper meets.
   b What type of mountains form like this?

Step 4: I can predict changes in the characteristics of places

6 Draw a labelled block diagram to explain how and where block mountains form.

Flow diagrams, page 144
Block diagrams, page 144

What is the theory of plate tectonics?

Earth's surface is made up of huge tectonic plates, much like a big jigsaw puzzle. These plates are always moving, but usually by only a few centimetres per year. As they collide, mountains form and volcanoes and earthquakes are generated. When plates move apart, ocean trenches can form.

Moving tectonic plates

Earth's crust is made up of seven major **tectonic plates** that float and move on the semi-molten rocks in the **mantle**. As the plates move apart (**diverging margins**) or collide (**converging margins**), landforms such as mountains, ridges and ocean trenches are created.

Most of the world's volcanoes are located along converging margins or subduction zones. **Subduction** occurs when a continental plate, such as the North American Plate, converges with an oceanic plate, such as the Caribbean Plate. The heavier oceanic crust bends and moves down toward the upper mantle (see page 26). The edge of the plate melts into magma that feeds volcanoes along the converging margin, such as Mount Pinatubo in the Philippines.

Most earthquakes occur near the edges of converging plates or along **transform margins**, which is where the plates slide past one another. Here the plates can become stuck and then break free with a sudden movement that triggers an earthquake. California's San Andreas Fault is a transform margin between the North American and Pacific Plates that suffers many earthquakes.

The world's tectonic plates

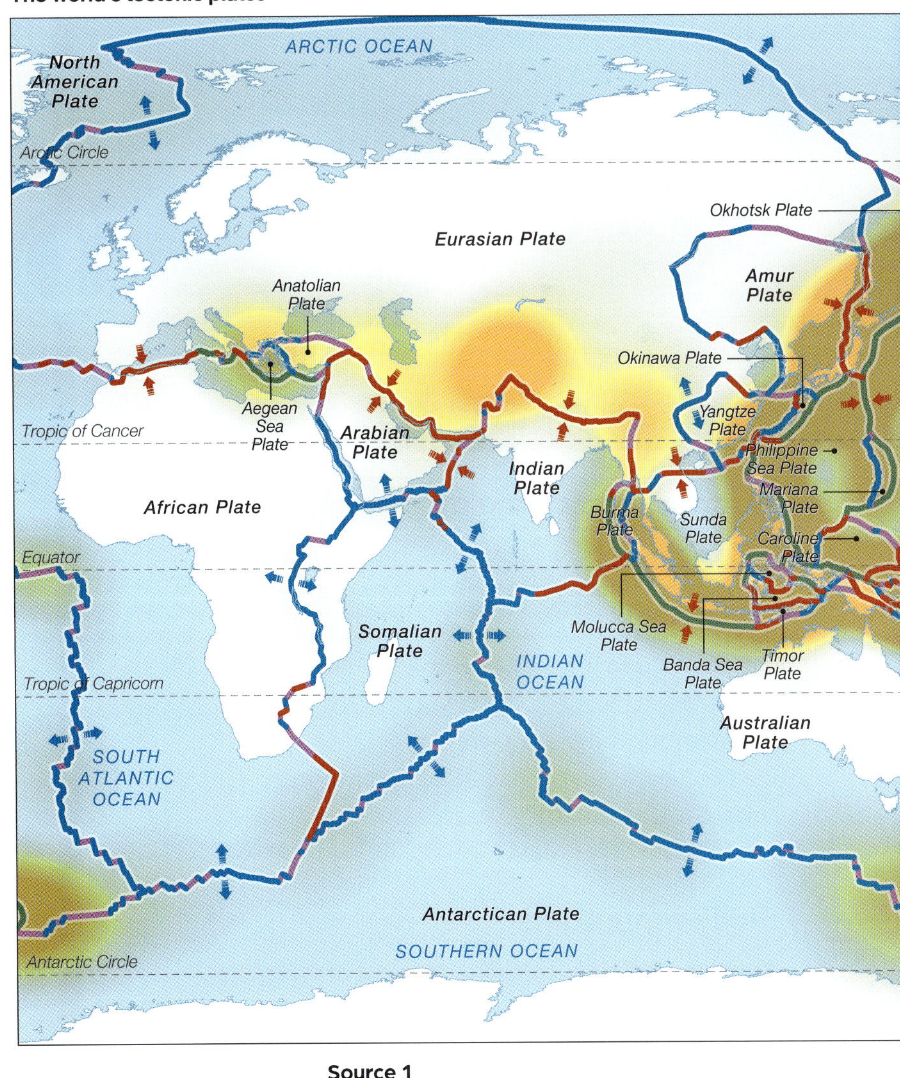

Source 1

The world's tectonic plates – seven major plates and some of the minor plates

Source 2

Tectonic plate movement

Source: Matilda Education Australia, Bird/UCLA

Learning ladder G1.4

Show what you know

1 Briefly explain one key difference difference between the formation of mountains and volcanoes.

2 What landform can be created along diverging plate boundaries?

Interconnections

Step 1: I can identify and describe interconnections

3 Source 1: Use PQE to help with the following.
   a Along which tectonic plate boundaries do earthquakes and volcanoes occur?
   b Quantify this pattern with a count or estimated percentage, or use of the legend.
   c Give an example of an exception to the pattern.

Step 2: I can explain interconnections

4 Draw a block diagram of Earth's crust. Annotate your diagram to explain the process of tectonic plate movement that forms a mountain or volcano.

Step 3: I can identify and explain the implications of interconnections

5 Using SHEEPT, explain the interconnection between the location of tectonic plates and the presence of earthquakes, mountain ranges and volcanoes.

Step 4: I can evaluate the implications of significant interconnections

6 The phreatic volcanic eruption (see page 29) on New Zealand's White Island killed 21 tourists in December 2019. What interconnection made this eruption particularly dangerous?

Block diagrams, page 144
PQE, page 138
SHEEPT, page 140

How did the Himalayas form?

Central Asia is dominated by the highest land on Earth – the Plateau of Tibet and the Himalayan mountains. This area is home to 90 of the world's 100 highest peaks, including Mount Everest, the world's highest mountain at 8848 metres above sea level.

The Himalayan Mountains were formed by the collision of the Indo-Australian and Eurasian **tectonic plates** (see pages 20–21). As the Indo-Australian plate moved north, it bumped into the southern edge of the Eurasian plate. The land crumpled and was pushed upwards. In just 50 million years, the Himalayan Mountains have risen from sea level to nearly nine kilometres high.

The Himalayas now stop the movement of the **monsoonal** rains from the south, making the Plateau of Tibet one of the driest regions in the world.

Source 1

The southern edge of the Eurasian tectonic plate was crumpled and buckled up above the Indo-Australian plate. Neither of these two continental plates were heavy enough to be subducted. This caused the crust to thicken because of folding and faulting (page 18). Earth's crust here is about 75 kilometres thick – twice its average thickness. This process marked the end of volcanic activity in the region.

Source 2

The collision of the Indo-Australian tectonic plate with the Eurasian plate formed the Himalayan Mountains. The Indo-Australian plate still travels northward at 3.7 centimetres per year and the Himalayas are still growing at 1 centimetre per year.

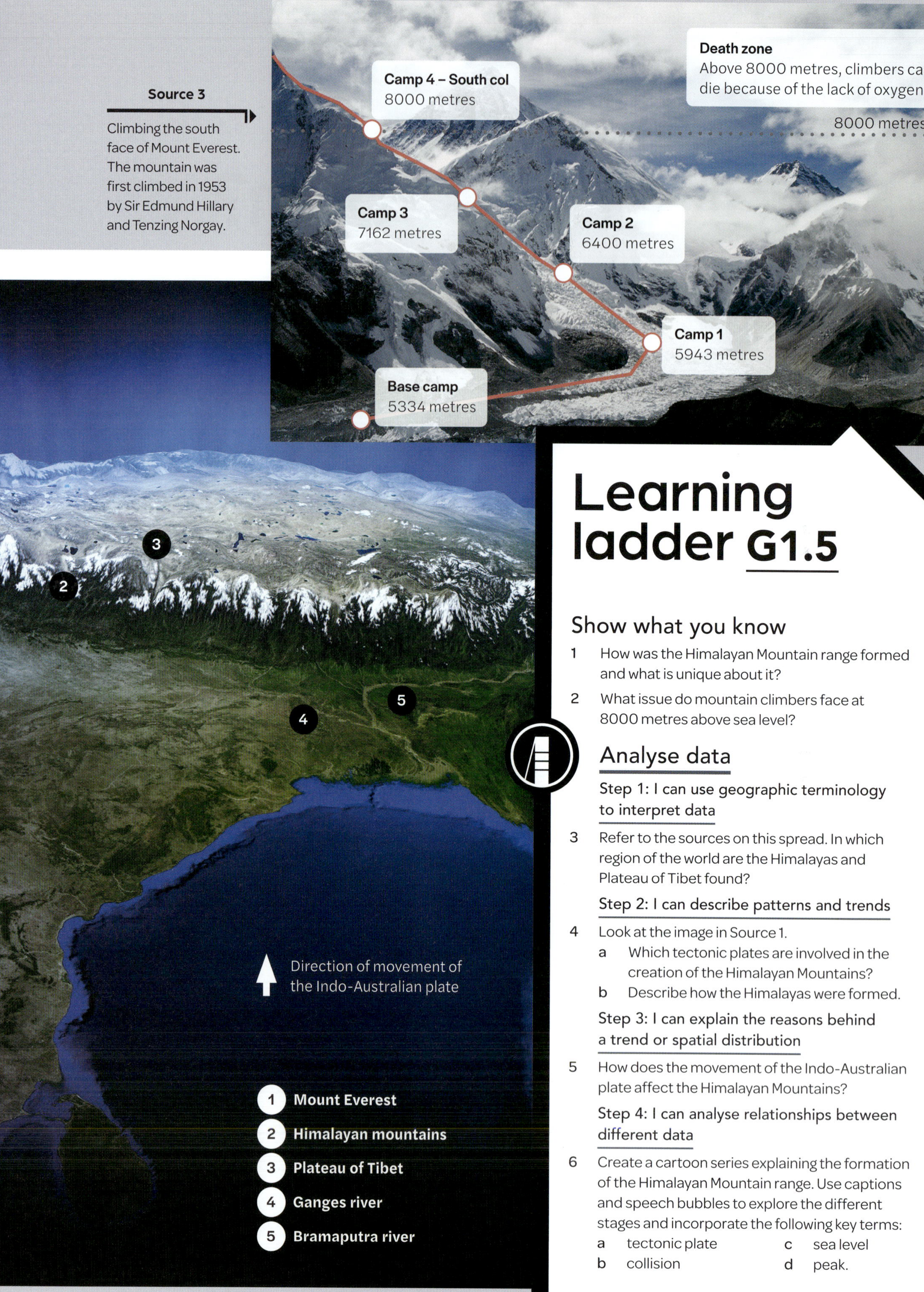

Source 3

Climbing the south face of Mount Everest. The mountain was first climbed in 1953 by Sir Edmund Hillary and Tenzing Norgay.

Learning ladder G1.5

Show what you know

1 How was the Himalayan Mountain range formed and what is unique about it?

2 What issue do mountain climbers face at 8000 metres above sea level?

Analyse data

Step 1: I can use geographic terminology to interpret data

3 Refer to the sources on this spread. In which region of the world are the Himalayas and Plateau of Tibet found?

Step 2: I can describe patterns and trends

4 Look at the image in Source 1.
- a Which tectonic plates are involved in the creation of the Himalayan Mountains?
- b Describe how the Himalayas were formed.

Step 3: I can explain the reasons behind a trend or spatial distribution

5 How does the movement of the Indo-Australian plate affect the Himalayan Mountains?

Step 4: I can analyse relationships between different data

6 Create a cartoon series explaining the formation of the Himalayan Mountain range. Use captions and speech bubbles to explore the different stages and incorporate the following key terms:
- a tectonic plate
- b collision
- c sea level
- d peak.

How do volcanoes form?

Volcanoes form when magma from beneath Earth's surface pushes through Earth's crust along tectonic boundaries and at hot spots in Earth's mantle.

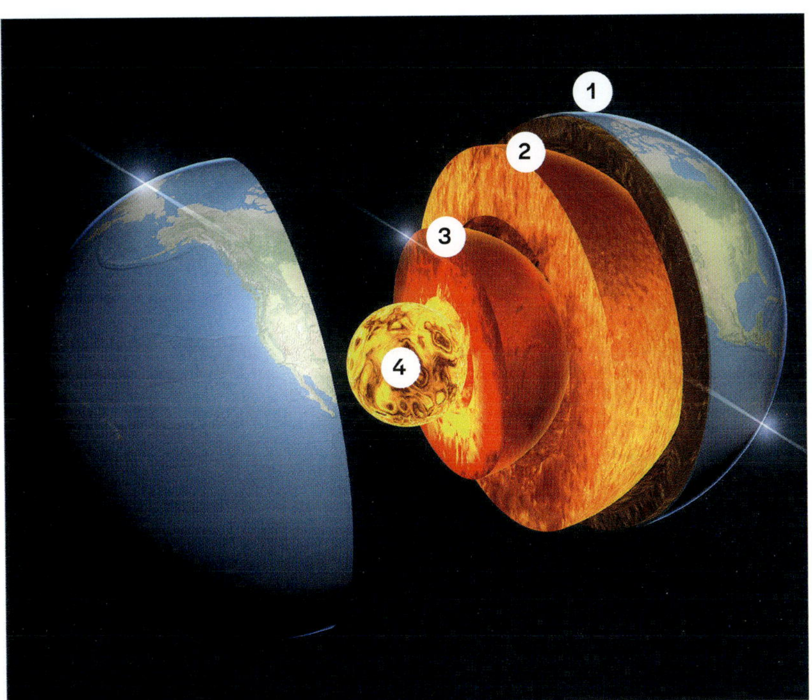

Source 1

Structure of Earth

1. The crust is broken into tectonic plates that form mountains and cause earthquakes.
2. Parts of the solid mantle can melt into magma and move to the surface through volcanoes.
3. With temperatures up to 6000°C, the outer core is the only completely liquid layer.
4. Earth's inner core is solid iron and nickel with temperatures up to 10 000°C.

Source 2

A stratovolcano is a cone-shaped volcano built up by many layers of hardened lava, tephra, pumice and ash. Active stratovolcanoes are steep and can have explosive eruptions. The lava flowing from stratovolcanoes is usually thick and sticky, so it cools and hardens around the cone. Stratovolcanoes are also known as composite volcanoes because they are built up in layers from each eruption.

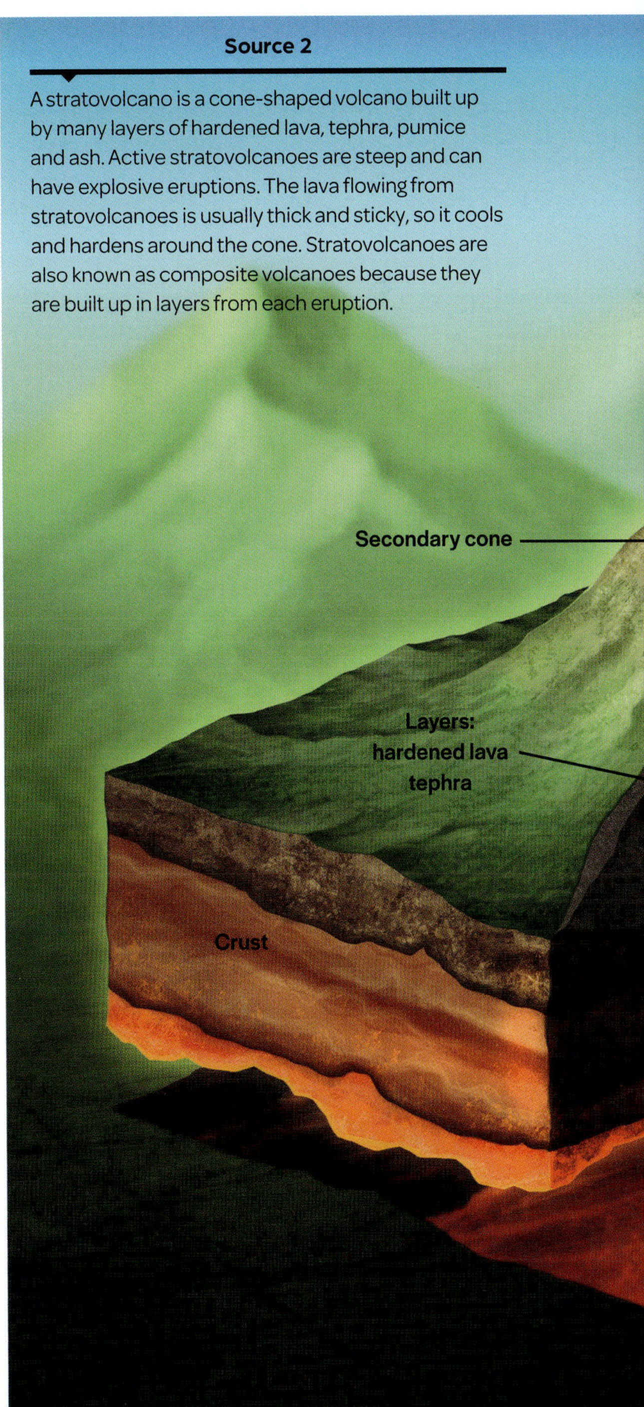

Volcanoes

Volcanic mountains form when **magma** – melted rock from below Earth's surface – is pushed through an opening in Earth's crust. This material can build up to form a cone-shaped mountain known as a **volcano**.

Volcanoes are mainly found along **tectonic plate** boundaries where plates collide and one plate is forced downwards into the Earth's **mantle** in a process called **subduction** (page 20). The edge of the sinking plate melts into magma to feed the volcanoes along the plate boundary.

Ash, steam and smoke
Crater
Primary cone
Lava flow
Main vent
Sill
Magma chamber

Hot spot volcanoes are found far from plate boundaries. They occur when particularly hot areas of the mantle burn through Earth's crust, such as Mount Elephant in Victoria (discussed on page 26).

The material that spews from a volcano builds up Earth's surface, creating new landforms. Each volcanic eruption brings new material to the surface as **tephra** (rocks and ash) or **lava** (magma that reaches Earth's surface). Across Earth's surface, lava flows cover material from previous eruptions to build the height of the land in layers.

Volcanoes are classified as **active** (erupting or likely to erupt), **dormant** (inactive but may erupt in the future) and **extinct** (unlikely to erupt again).

Learning ladder G1.6

Show what you know

1 Source 1: Which of Earth's layers is completely liquid?

2 Source 2: What is the name of the melted rock from the mantle layer that feeds volcanoes? What is it called when it reaches the surface?

Spatial characteristics

Step 1: I can identify and describe spatial characteristics

3 Explain why volcanoes can only be present in certain *places* on Earth.

Step 2: I can explain spatial characteristics

4 Create a photo essay of different volcanoes around the world. For each photo state the location and type of the volcano in that place, as well as whether it is active, dormant or extinct.

Step 3: I can explain processes influencing places

5 Explain why stratovolcanoes form in layers.

Step 4: I can predict changes in the characteristics of places

6 Research how volcanologists predict volcanic eruptions. Is it an exact science?

HOW TO

Photo essays, page 145

Does Victoria have any volcanoes?

The Western Victorian Volcanic Plains stretch more than 400 km from Melbourne in the east to Mount Gambier in the west. They comprise the third-largest area of volcanic plains in the world, with more than 400 volcanoes and extensive lava flows.

Western Victorian Volcanic Plains

The Western Victorian Volcanic Plains cover more than 10 per cent of the state of Victoria. These **basalt** plains were formed by volcanic eruptions over the last six million years; the most recent eruption was 5000 years ago at Mount Gambier.

Volcanoes such as Mount Elephant and Tower Hill erupted explosively with a huge ash cloud and fire fountains creating volcanic cones. Tower Hill is actually a group of **scoria cones** in a large **maar crater**. Maar craters are formed by the explosive interaction between magma and groundwater; they are usually shallow, with steep sides and often fill with water.

Mount Elephant

Mount Elephant at Derrinallum in Victoria is a 240-metre scoria cone formed by a **dormant** volcano. Recent scientific testing estimated that it last erupted 550 000 years ago. Scoria cones are small volcanoes with relatively steep sides, usually formed as the result of a single volcanic eruption.

Source 1

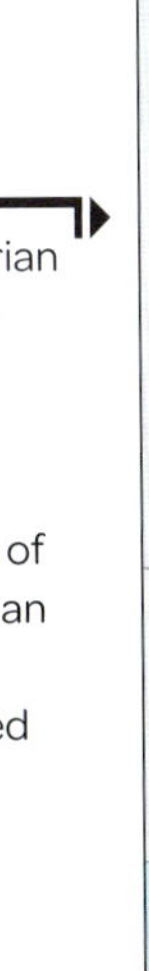

The Western Victorian Volcanic Plains are the third-largest volcanic plains in the world, with an average lava depth of 60 metres. More than 400 volcanic sites have been identified in this region.

Western Victorian Volcanic Plains

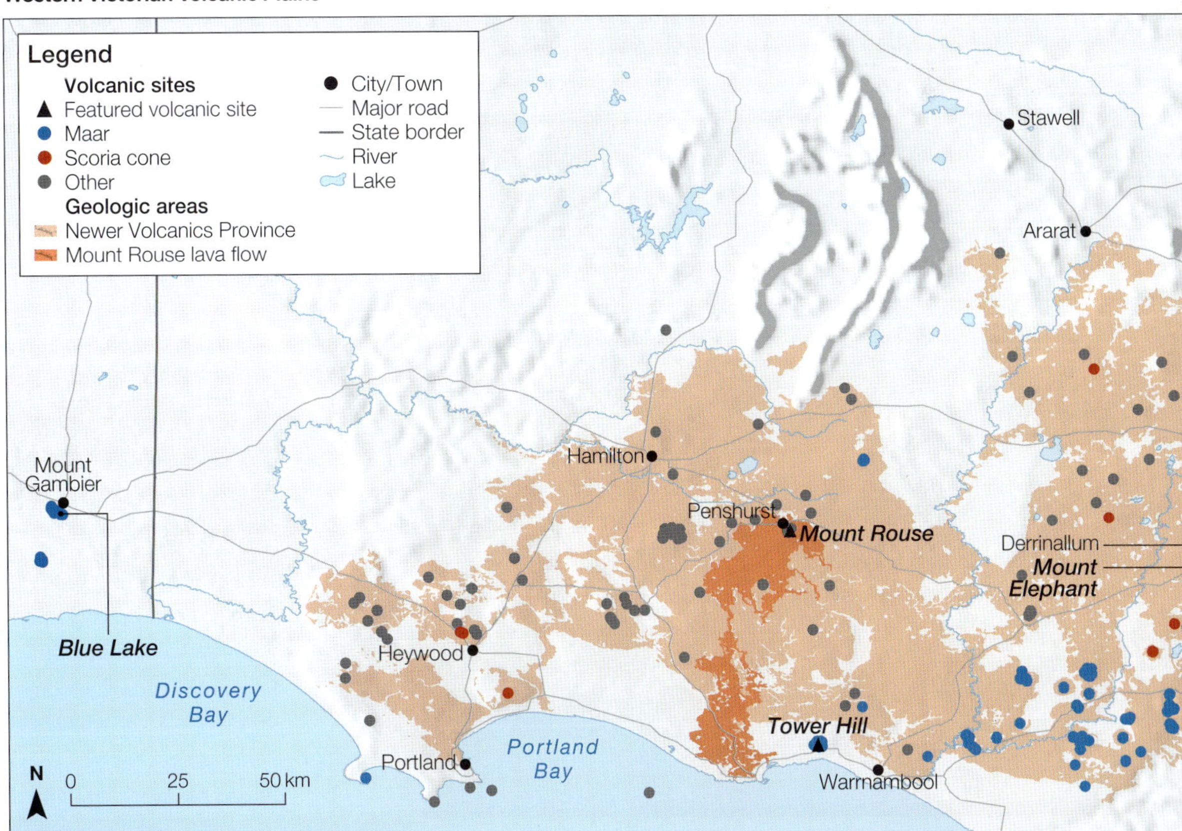

Source: Matilda Education Australia, Victorian State Government, Julie Boyce

Source 3

The eruption of Mount Elephant breached (broke) the crater at the top of the mountain. The quarry in the foreground was a scoria mine, used to make many of the roads and buildings in and around the town of Derrinallum.

Source 2

Lake Bullen Merri near Camperdown is a shallow maar volcanic crater formed by a volcanic eruption with little lava.

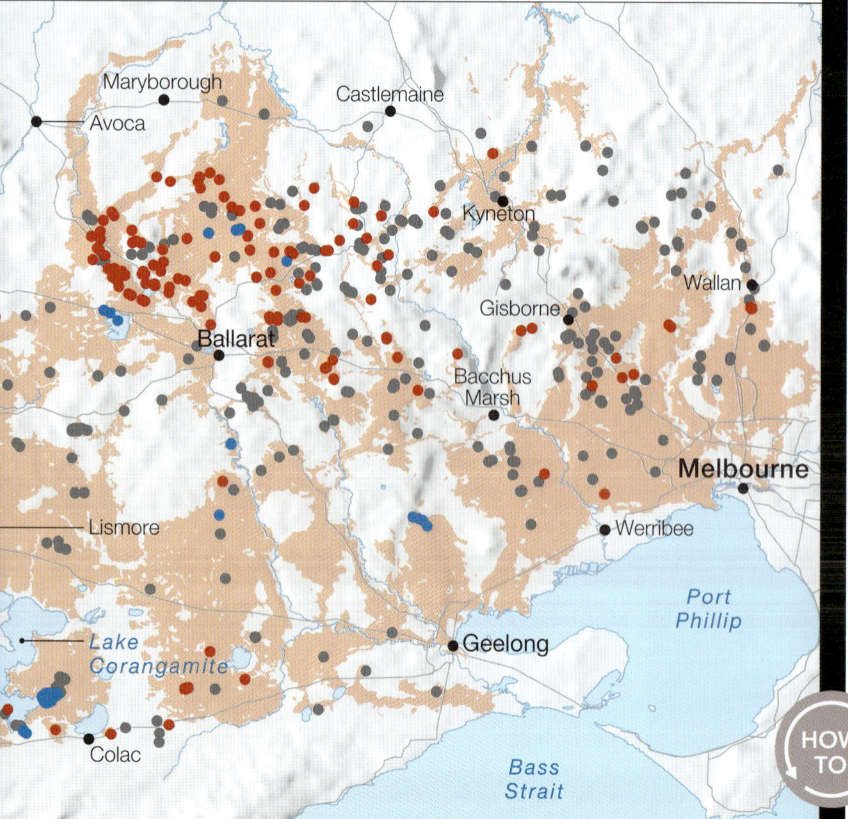

Learning ladder G1.7

Show what you know

1. When did Mount Elephant last erupt?
2. Is Mount Elephant extinct, dormant or active? Define all three of these terms.

Collect, record and display data

Step 1: I can collect, record and display data in simple forms

3. Suggest how the oblique aerial image in Source 2 was collected.

Step 2: I can recognise and use different types of data

4. The map in Source 1 is an example of a secondary source of data. Use the following questions to help demonstrate your understanding of this map.
   a. Where are the volcanic plains located in Victoria?
   b. Describe the location of the greatest concentration of scoria cones in relation to Melbourne.
   c. Where is the greatest concentration of maar craters? What other volcanic feature is also found in this location?

Step 3: I can choose, collect and display appropriate data

5. Make a sketch map of Mount Elephant from Source 3 and include these labels:
   a. scoria
   b. cone
   c. breach
   d. crater
   e. quarry.

Step 4: I can use data to support claims

6. Consider the map in Source 1. How helpful do you think this map would be when investigating the presence of volcanoes at Tower Hill? Justify your response and suggest alternatives.

HOW TO

Sketching, page 142

What happens when a volcano erupts?

Volcanic eruptions are one of the most dangerous natural hazards on Earth. Explosive eruptions can completely destroy the surrounding area and ash clouds can travel for thousands of kilometres. Other eruptions may just produce small ash clouds or slow-flowing streams of lava.

The main categories of **volcanic eruptions** are:

1. **effusive eruptions** where magma flows out of the volcano as a lava, also known as Hawaiian eruptions
2. **explosive eruptions** are either where magma full of bubbling gas explodes as it reaches the surface, also called a Plinian eruption, or where magma reacts with water to create a steam-blast, also called a phreatic eruption.

The **magma** that feeds **volcanoes** along plate boundaries generally contains a lot of gas. If magma moves up through the volcano's vent slowly, the gases are released slowly and escape safely. However, if magma moves to the surface quickly, most gases are retained and will explode if there is enough pressure. Explosive volcanic eruptions push huge amounts of ash and gas high into the air. Powerful eruptions can launch large rocks, known as **volcanic bombs**, into the air.

Mount Pinatubo erupts

When Mount Pinatubo erupted in the Philippines in 1991, it ejected more than 5 cubic kilometres of material from its vent and the ash cloud rose 35 kilometres into the air. An ash and sulfur-dioxide gas cloud travelled several times around the globe, blocking the sun and lowering world temperatures by 0.5°C for a year.

Source 1 Different types of volcanic eruptions

Source 2

The Mount Pinatubo eruption in 1991 was the second-largest volcanic eruption of the 20th century.

A **pyroclastic flow** raced down Mount Pinatubo, wiping out villages and everything else in its way. A pyroclastic flow is a huge avalanche of hot gas and rock particles, known as **tephra**, that can move at speeds of up to 700 kilometres per hour and heat to temperatures of 1000°C, destroying everything is its path. Near Mount Pinatubo, valleys were filled with ash and rock up to 200 metres thick. Heavy rains washed this material onto lowlands in fast-moving mudflows called **lahars**, creating even more damage than the eruption. More than 700 lives and 100 000 homes were lost in the Mount Pinatubo eruption.

Steam-blast (or phreatic) eruption
Steam-blast eruptions result from water or cold ground coming into contact with magma. White Island volcano in New Zealand erupted in this way in 2019.

Learning ladder G1.8

Show what you know

1 How does the speed of moving magma influence volcanic eruptions?

2 What is a pyroclastic flow and why is it so dangerous?

Analyse data

Step 1: I can use geographic terminology to interpret data

3 Source 2: Was the eruption of Mount Pinatubo in 1991 an effusive or explosive eruption? Explain.

Step 2: I can describe patterns and trends

4 What impact did Mount Pinatubo's eruption have:
  a locally?
  b globally?

Step 3: I can explain the reasons behind a trend or spatial distribution

5 Refer to the map of Hawaii's lava danger zones on page 30. Explain why different areas have different hazard-level ratings. Are you surprised by the locations of any of the zones ranked as a low–medium hazard? Why or why not?

Step 4: I can analyse relationships between different data

6 Outside the classroom, use sand, papier-mâché or a cylindrical container to build a mini volcano-shaped mountain. Add two spoonfuls of baking soda and about a spoonful of dish soap to your volcano. When you are ready, add about 30 mL of vinegar and observe your volcano 'erupt'. As a class, compare this basic experiment to the real process of eruption.

What was the impact of the Kīlauea eruption?

Kīlauea Erupts

Tuesday 8 May 2018, Hawaii, USA

Hawaii's Kīlauea volcano is causing a state of emergency on the Big Island, spewing lava and dangerous fumes and destroying everything in its path.

Lava destroys homes

A dangerous lava flow has already destroyed 26 homes and forced 1700 people to evacuate. The 1247-metre Kīlauea volcano began erupting on Thursday afternoon, following a magnitude 5 earthquake. Another earthquake on Friday was measured at magnitude 6.9, the most powerful earthquake to hit Hawaii since 1975. Kīlauea has been erupting continuously since 1983 and is one of five currently active volcanoes on Hawaii's Big Island.

Twelve volcanic vents have opened, sending large flows of lava into residential areas near Pahoa. People in the affected area have been ordered to evacuate, but not all people have left.

Sixty-five-year-old Sam Knox lives less than 100 metres from a volcanic fissure. He has decided not to leave, even as lava flows across his neighbour's property. 'It was roaring sky high. It was incredible ... rocks were flying out of the ground ... I decided to stay because I wanted to experience this in my life,' he said.

People are also at risk from deadly sulfur-dioxide gases produced by Kīlauea. Winds from the east have spread the toxic gas as far as 100 kilometres across the southern part of the island.

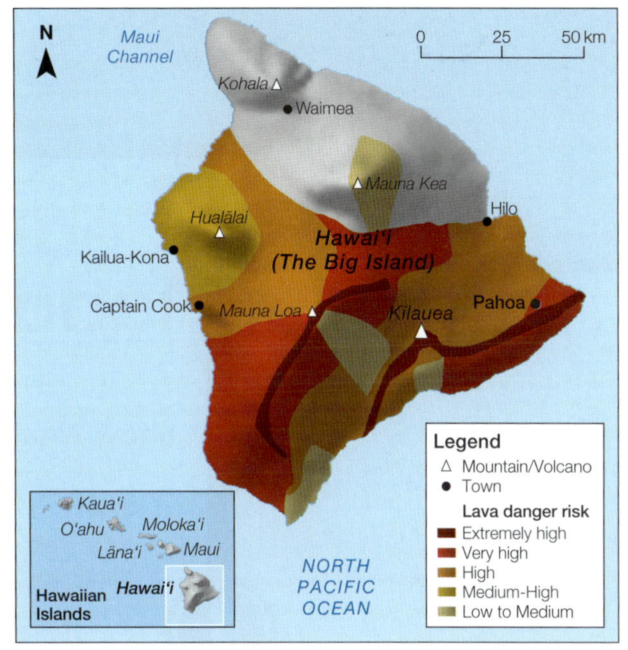

Source 1 Lava danger zones on Island of Hawaii

Source: Matilda Education Australia, ESRI, USGS, Hawaii State GIS Program

No end in sight

Though the rate of movement is slow, there is no indication when the lava might stop or how far it might spread. 'There's more magma in the system to be erupted. As long as that supply is there, the eruption will continue,' volcanologist Wendy Stovall said.

Aerial footage showed orange streams of lava bursting through openings in the ground and moving through residential areas, covering roads and starting small fires.

Offerings to a goddess

Hawaiians believe the goddess Pele lives in Kīlauea's crater. She is known to be unpredictable, so residents in the disaster zone are leaving gifts for her outside their homes to keep her happy.

Source: Author article based on multiple press reports.

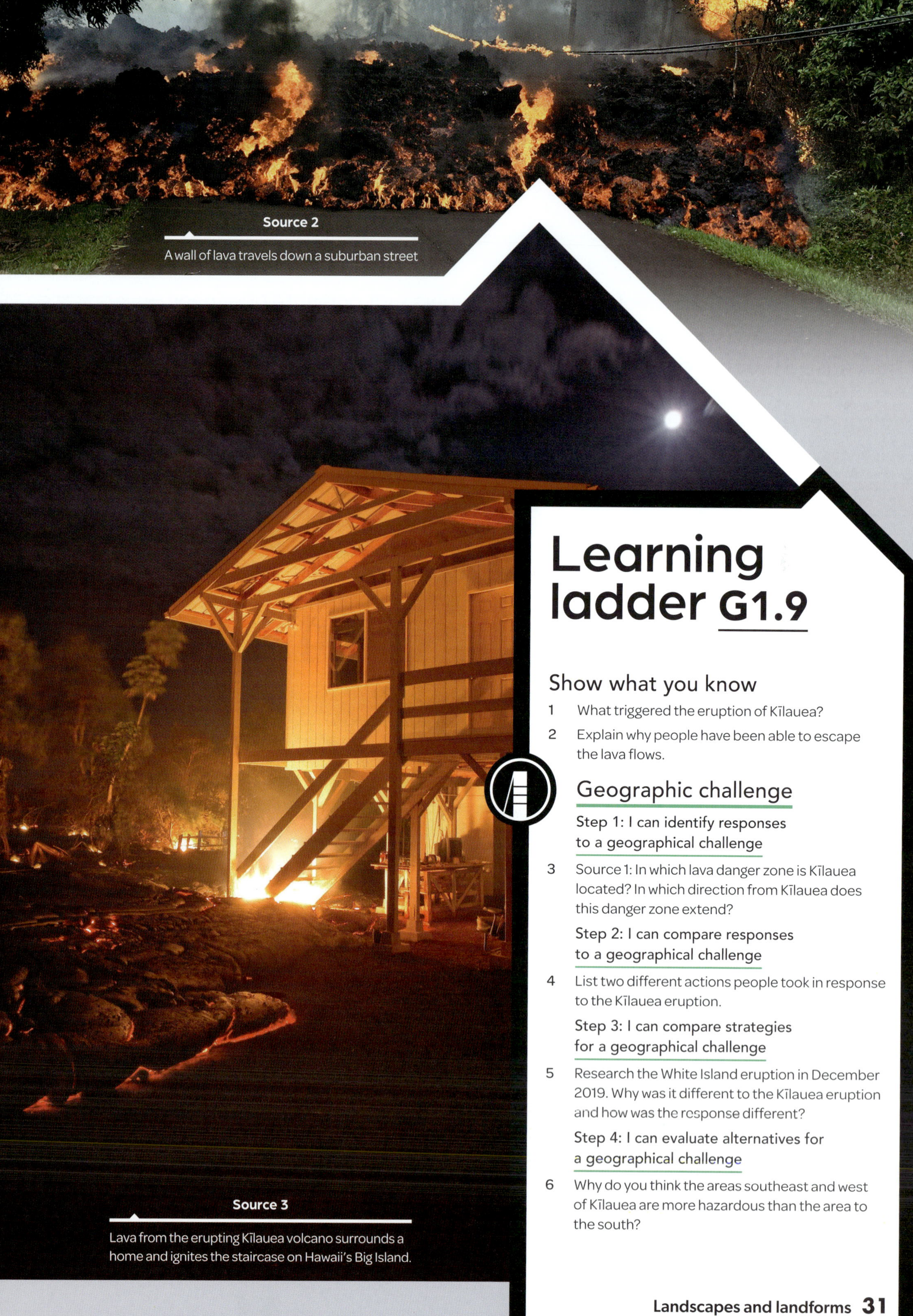

Source 2

A wall of lava travels down a suburban street

Source 3

Lava from the erupting Kīlauea volcano surrounds a home and ignites the staircase on Hawaii's Big Island.

Learning ladder G1.9

Show what you know

1 What triggered the eruption of Kīlauea?

2 Explain why people have been able to escape the lava flows.

Geographic challenge

Step 1: I can identify responses to a geographical challenge

3 Source 1: In which lava danger zone is Kīlauea located? In which direction from Kīlauea does this danger zone extend?

Step 2: I can compare responses to a geographical challenge

4 List two different actions people took in response to the Kīlauea eruption.

Step 3: I can compare strategies for a geographical challenge

5 Research the White Island eruption in December 2019. Why was it different to the Kīlauea eruption and how was the response different?

Step 4: I can evaluate alternatives for a geographical challenge

6 Why do you think the areas southeast and west of Kīlauea are more hazardous than the area to the south?

What are geohazards?

Geomorphic hazards, or geohazards, are dangers that originate at or near the Earth's surface, such as volcanic eruptions, earthquakes and landslides. They pose a great threat to people and environments.

Volcanoes

Volcanoes pose a huge threat to nearby communities. Dangerous **pyroclastic** flows, **volcanic bombs**, **lava flows** and volcanic gases are an immediate threat to people. Other **geomorphic hazards** can follow a volcanic eruption, including **lahars** (mudflows), **landslides** and even **tsunamis** caused by ocean-floor volcanic eruptions.

Earthquakes

Earthquakes are one of the deadliest geomorphic hazards. An earthquake is caused when **seismic waves** shake the surface of the Earth. These waves are generated from the sudden release of energy in or beneath Earth's crust. The shaking can cause massive damage to buildings and large death tolls. Earthquakes can also trigger landslides and tsunamis.

Source 1

Ash-covered Aeta people were forced from their villages on the slopes of Mount Pinatubo (see pages 28–29) following a volcanic eruption in 1991. More than 200 000 Aeta lost their homes and were forced into resettlement camps or moved to urban areas.

Source 2

A devastated street in Port-au-Prince, Haiti, a week after the powerful 2010 earthquake.

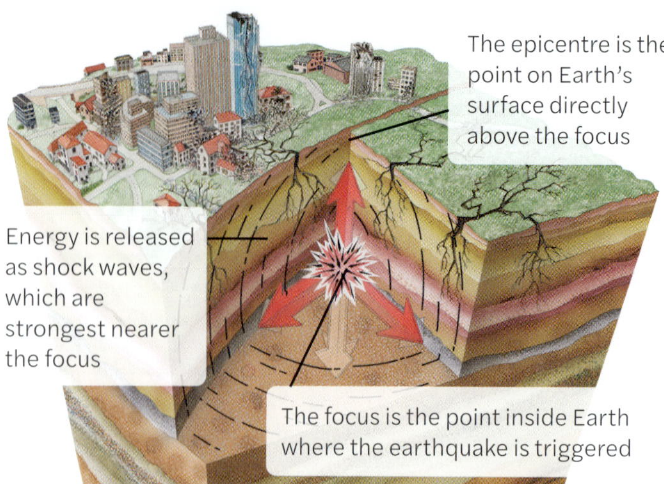

Source 3

An earthquake in action

A major earthquake struck Haiti's capital, Port-au-Prince in January 2010. Up to 85 000 people died and another million were made homeless. Haiti is a **less economically developed country (LEDC)** and most of its population live in poverty. Those left homeless by the earthquake could not afford to move elsewhere. In the devastated urban areas, displaced Haitians were forced into crowded and unsanitary camps in city parks, and they were living in self-made structures or donated tents.

Humanitarian aid was supplied by organisations such as the United Nations and the Red Cross. Countries sent doctors, relief workers and supplies in the wake of the disaster in Haiti. Aid agencies struggled to supply the most basic needs and in October contaminated food and water caused an outbreak of cholera in the camps, claiming hundreds of lives. Two years after the earthquake, more than half a million people were still living in refugee camps.

Landslides

A landslide is the mass movement of soil, rock and debris down a slope. In most cases, landslides are caused by extended rainfall or flooding, where the soil becomes saturated and heavy. Landslides can also be triggered by earthquakes, volcanic eruptions or cutting into slopes for buildings and roads. In 1997, a landslide at Thredbo in New South Wales destroyed two buildings and killed 18 people. Miraculously, rescue workers pulled the sole survivor Stuart Diver from the rubble after three days in sub-zero temperatures.

Learning ladder G1.10

Show what you know

1 List four geomorphic hazards.
2 Source 3: How do earthquakes work and why do they cause so much damage?
3 List one other geomorphic hazard that earthquakes can trigger.
4 Summarise why mountain landscapes are prone to landslides.
5 Connect to the following site: http://mea.digital/W9t2. Explore the locations of historical and recent geohazards and discuss as a class any clear patterns.

Geographic challenge

Step 1: I can identify responses to a geographical challenge

6 List the threats from a volcanic eruption.

Step 2: I can compare responses to a geographical challenge

7 What made the earthquake disaster in Haiti more difficult to deal with? What more could have been done?

Step 3: I can compare strategies for a geographical challenge

8 Compare the response to the Mount Pinatubo and Haiti disasters. What similarities are there?

Step 4: I can evaluate alternatives for a geographical challenge

9 Source 2: What rules need to be made for buildings in earthquake-prone areas? Why might this be difficult to enforce in countries like Haiti?

How does weathering affect landforms?

When rock is exposed on Earth's surface it can be broken down by water, temperature and the roots of plants. Moving ice, water and wind can further wear down rock and carry it away.

Weathering and erosion

Landforms change because of **weathering** and **erosion**. Weathering is a process where rock is worn away, broken down or dissolved into smaller and smaller pieces by water, heat and cold.

Weathering involves two processes that work to break down rocks on the surface of Earth. **Chemical weathering** is a chemical change in a rock, while **mechanical weathering** involves rocks being broken into pieces.

A weathered rock particle is loosened but stays in its place. When the particle begins moving, we name this erosion, or **mass wasting**. Mass wasting is simply the movement of rocks downhill due to gravity, such as rock falls and landslides. Erosion is where rock particles are moved by flowing air (wind), water (rivers, sea and rain) or ice (glaciers). Erosion acts more quickly on softer rocks. Harder rocks remain as landforms such as mountains or **headlands**.

Mechanical weathering

Mechanical weathering is where large rocks on Earth's surface are broken down into smaller ones. Mechanical weathering is caused by:

- *temperature* – cold nights and hot days cause rocks to expand and contract, which makes them crack and break apart.
- *ice* – water seeps into cracks in rocks. When the water freezes it expands by 9 per cent, splitting the rocks.
- *roots and plants* – vegetation pushes into rocks and acts as a wedge to push the rocks apart.

Source 1

The process of erosion

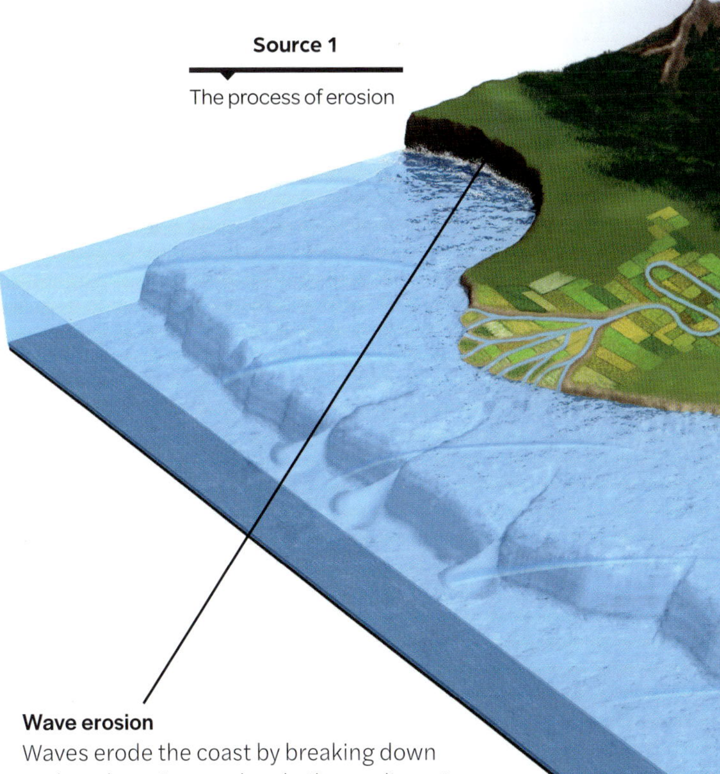

Source 2

Mechanical weathering is the key agent that shapes the amazing rock features of Monument Valley in the USA. The hot days and cold nights enable the rocks to expand and contract, which causes cracks that eventually split sections of rock. Wind erosion picks up the small dry weathered rock particles and blasts them against larger rocks, which further shapes them.

Source 3

Frost wedging is a form of mechanical weathering that occurs when water freezes and expands in a crack in a rock.

Ice

Water erosion

Water erosion occurs from the force of the flow of water and the eroded rock and other material carried in the water. Rocks or debris carried by a river erode its banks and bed as they crash into it.

Wind erosion

Wind erosion is particularly active in dry desert areas. Light objects such as sand and pebbles are swept up and carried by the wind. The small rock particles are thrown against landforms, eroding materials off them. The newly eroded material is also swept away by the wind to sandblast the desert landscape into amazing shapes.

Source 4

The pressure of expanding plant roots causes rocks to crack and break away.

Antelope Canyon

Mechanical weathering processes such as temperature change and frost wedging opened wide cracks in the soft sandstone rock of Antelope Canyon. Flash floods carrying sand, rocks, logs and other debris eroded large amounts of rock to further widen the cracks. Over time, erosion has cut and smoothed a skinny deep slot canyon. The amazing shapes of Antelope Canyon are very popular with tourists.

Chemical weathering

Chemical weathering decomposes rocks and minerals. Chemicals in water or the reaction of water with minerals in rocks either form or destroy minerals in the rock, breaking down the bonds holding rocks together. This weathering is common in locations where there is a lot of water to help the chemical reactions take place.

The most common types of chemical weathering are:

- *hydrolysis* – hydrogen in water reacts with minerals in the rocks to form new minerals. For example, hard granite rock changes to softer clay through hydrolysis. Clay is more prone to mechanical weathering and erosion than granite.
- *carbonation* – rainwater contains carbonic acid because carbon dioxide in the atmosphere dissolves in it. This carbonic acid reacts with minerals in some rocks, causing the rocks to weather.
- *oxidation* – when oxygen combines with other elements in rocks, new types of rock are formed. As with hydrolysis, these new rocks are generally softer and easier for mechanical weathering to break apart.

Source 5

Antelope Canyon has been formed by a combination of weathering and erosion in the semi-desert region of Arizona, USA.

Karlu Karlu

Karlu Karlu (the Devil's Marbles) in the Northern Territory is a local example of chemical weathering. The granite rock has been shaped into round boulders by a chemical weathering process known as **spheroidal weathering**. Water seeps into cracks and chemical reactions cause the rocks to disintegrate along the edges. The cracks are opened wider, allowing even more water to reach the surfaces. Corners are attacked by water from more than one direction and wear away more rapidly. The sharp corners are rounded, forming spherical rocks.

Hard rock with regolith (softer unconsolidated rock)

Hard rock

Regolith

Rain water carrying carbonic acid reacts with regolith and breaks it down

Weathered regolith leaves rocky bits standing alone

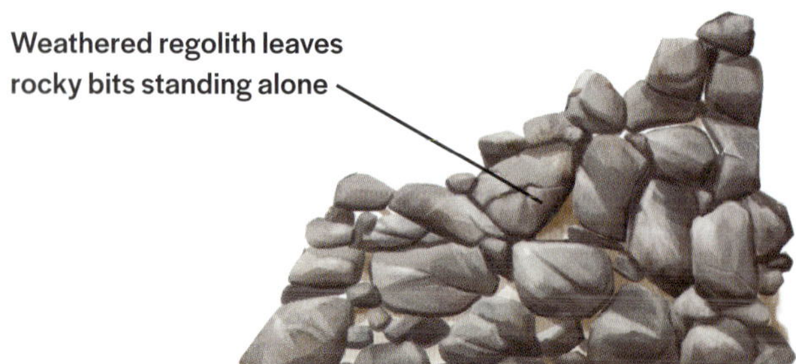

Source 6

Chemical weathering causes a chemical change in the rock that enables softer rock to be worn away, leaving harder rock behind.

Source 7

Karlu Karlu (the Devil's Marbles) has been shaped by a chemical weathering process known as spheroidal weathering, leaving these amazing balancing boulders behind.

Learning ladder G1.11

Show what you know

1 Define the term 'weathering'.

2 What are the two types of weathering and how are they different?

Spatial characteristics

Step 1: I can identify and describe spatial characteristics

3 Source 4: Draw a two-stage flow diagram to show how plants can cause mechanical weathering.

Step 2: I can explain spatial characteristics

4 What is the interconnection between weathering and erosion?

Step 3: I can explain processes influencing places

5 Weathering is due to a range of *interconnecting* factors, which ultimately leads to a *change* in landforms over time. Discuss.

Step 4: I can predict changes in the characteristics of places

6 Source 5: Describe the change over time in this location and describe the processes that have caused the changes.

Flow diagrams, page 144

What are deserts?

Deserts are dry environments. They can be hot or cold and be rocky, sandy or even covered in ice. Rocky deserts can be weathered and then eroded by water and wind into amazing shapes.

Different desert landscapes

Deserts cover about thirty per cent of Earth's surface and receive less than 250 millimetres of rain per year. Most are hot deserts that lie between the latitudes of 15 and 30 degrees north and south of the equator. Cold deserts are found in temperate areas or high **plateaus**, while polar deserts are localised at the North and South poles.

Desert **landscapes** can vary greatly. Some deserts are seas of sand; others are vast, stony plains; others have spectacular rock formations shaped by **weathering** and **erosion**. Plateaus of rock weather and erode into isolated **mesas**, **buttes** and chimney rocks. Natural arches form from weathering along joints that leave narrow walls of rock. When the lower wall erodes, an arch results. **Inselbergs** such as Uluru (see pages 40–41) are islands of rock that remain after a long period of erosion of the surrounding soft rock.

The world's deserts are expanding through a process known as **desertification** (see pages 44–45). This occurs when land along the edge of a desert becomes damaged by drought and through overuse by humans.

Source 1

Desert landforms

❹ *Arch* – a natural bridge weathered and eroded through a solid rock cliff over time.

❶ *Alluvial fan* – semicircular build-up of eroded material deposited by water and wind at the end of **wadis** (desert valleys).

❷ *Butte* – isolated hill with vertical sides and a small flat top. Buttes are smaller than mesas.

❸ *Inselberg* – isolated steep-sided mountain, uncovered as the softer land around it is eroded away.

World deserts

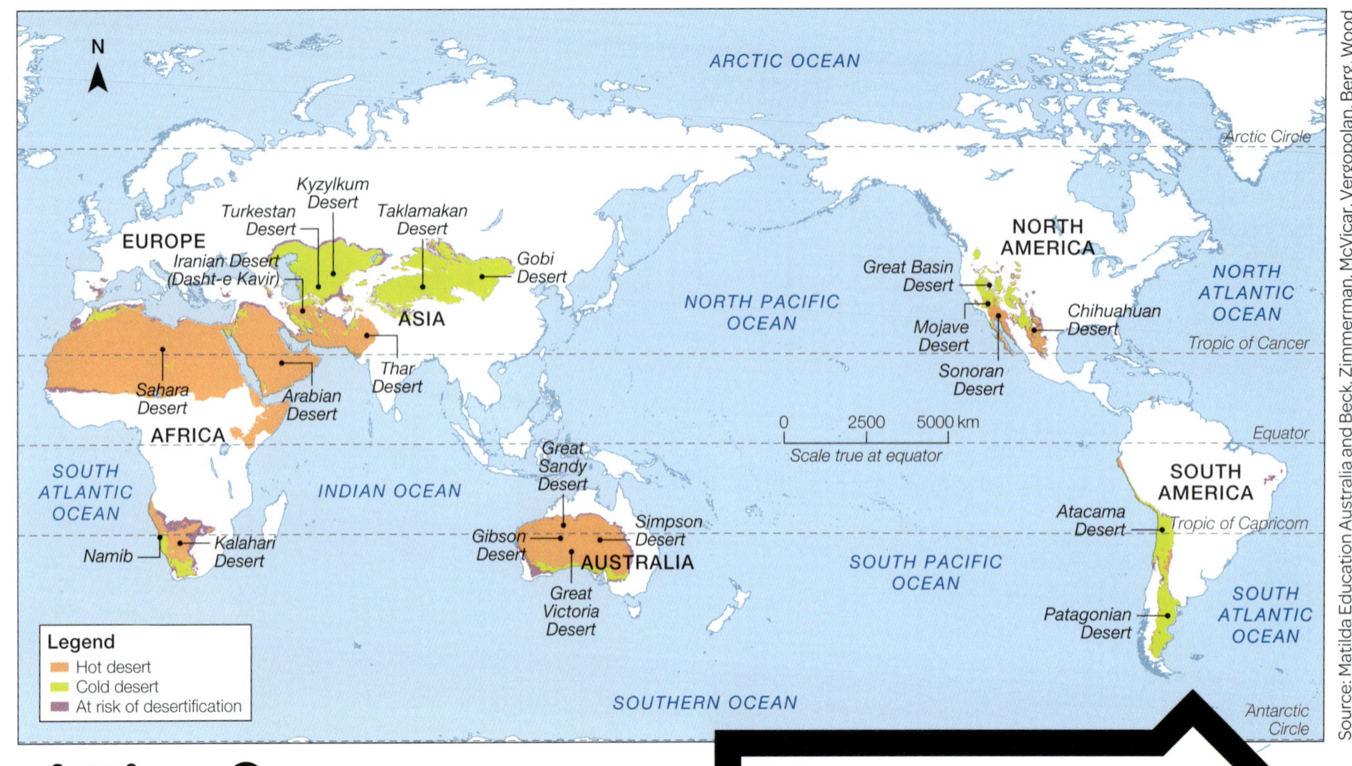

Source: Matilda Education Australia and Beck, Zimmerman, McVicar, Vergopolan, Berg, Wood

Source 2

World deserts

5 *Mesa* – broad, isolated flat-topped landform found in desert areas. They are the eroded remains of a wide plateau.

6 *Chimney rock* – thin pillars of rock that are the remains of buttes. Evaporating water creates a hard crust on top to protect it from chemical weathering as the column wears away.

Learning ladder G1.12

Show what you know

1 Explain how deserts can be classified as both hot and cold.

2 Outline how desert landscapes can vary.

Analyse data

Step 1: I can use geographic terminology to interpret data

3 Describe the geographical characteristics of a desert.

Step 2: I can describe patterns and trends

4 Describe, using PQE, the distribution of deserts on a global scale.

Step 3: I can explain the reasons behind a trend or spatial distribution

5 Using SHEEPT, explain the distribution of hot and cold deserts on a global scale.

Step 4: I can analyse relationships between different data

6 Source 1: Look at the photographs of the mesa, butte and chimney rocks. Analyse the relationship between these landforms.

HOW TO

PQE, page 138
SHEEPT, page 140

How did Uluru form?

Uluru and nearby Kata Tjuta are inselbergs that were uncovered as the land around them eroded away.

About 500 million years ago, colliding **tectonic plates** caused the land in Central Australia to **fold** and be pushed upwards (pages 18–19). The layers of sedimentary rock were gradually worn away by **weathering** and **erosion** (pages 34–37), leaving only the hardest rocks showing as the **inselbergs** of Uluru and Kata Tjuta on the surface. Uluru is the largest surface rock in the world. It rises 348 metres above the surrounding land and covers an area of 3.3 square kilometres.

Uluru and Kata Tjuta are sacred to the Pitjantjatjara Anangu, who have their own version of how they formed. The information has been passed down over thousands of years through stories, songs, ceremonies, dances and art. For the Anangu people, Uluru isn't just a rock, it's a living place – with the marks of the creation beings shown everywhere.

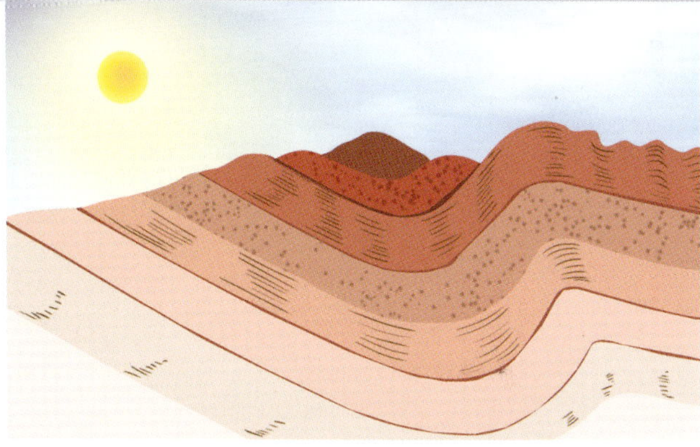

1 Fold mountains (see page 18) are pushed upwards due to pressure in Earth's crust. Layers of sedimentary rock are bent and pushed upwards to create fold mountains.

2 Over time, the processes of weathering and erosion wear away the softer rock, leaving only the harder rocks showing on the surface of the land.

Source 1

The formation of Uluru and Kata Tjuta

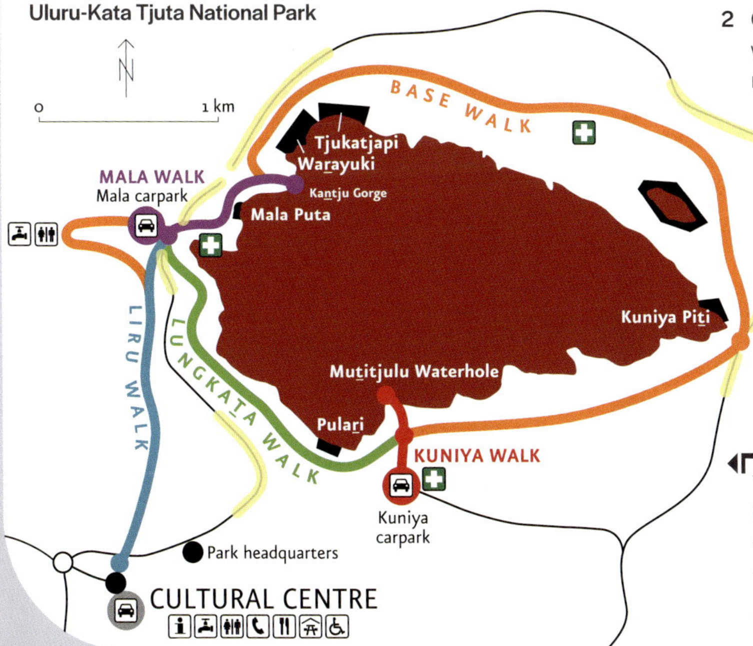

Source 2

This image of Uluru is in the *plan view* – directly above. Maps are drawn from the plan view because the scale is constant over the area shown. There is no foreground or background in plan-view images and maps.

Source: Matilda Education Australia, Parks Australia

Source 3

This is an *oblique aerial view* of Uluru. These angled views are good at showing the shape of landforms, as they show the foreground and the background. Uluru is featured in the foreground. Kata Tjuta appears much smaller because it is in the background.

Learning ladder G1.13

Show what you know

1 What record does Uluru hold?

2 What spiritual connection do the Anangu people have with Uluru?

Collect, record and display data

Step 1: I can collect, record and display data in simple forms

3 Create a statistics bank that contains five pieces of quantitative data about Uluru and its formation.

Step 2: I can recognise and use different types of data

4 Outline the advantages and disadvantages of using oblique aerial views to study a landform.

5 What is the key advantage of using maps and images in plan view?

Step 3: I can choose, collect and display appropriate data

6 Go to the contour map of Uluru, Source 26 on page 151. Draw a cross-section of Uluru from:

a Point A to B

b Point C to D.

Step 4: I can use data to support claims

7 What role did the following have on the formation of Uluru?

a Movement of tectonic plates

b Weathering and erosion

Cross-sections, page 151

How do landforms attract tourists?

Images of Australian landforms and iconic outback landscapes are marketed to tourists from all over the world. Tourism Australia uses our vast outback and unique landforms such as Uluru and the Great Barrier Reef to attract tourists to Australia. Tourists are more likely to come to Australia when the value of the Australian dollar is low compared to other currencies.

Marketing Australia

Tourism Australia promotes Australia's **landforms** and **landscapes** and the Australian lifestyle to the world. Like other businesses, Tourism Australia needs to find something distinctive and special about its product to compete with other countries for tourist dollars. In 1986, actor and comedian Paul Hogan developed the character of the popular bush larrikin Crocodile Dundee, with successful films that promoted interest in travel to Australia.

The Crocodile Dundee films became tourism advertisements for Australia. Outback landscapes and landforms were filmed in Kakadu National Park, in the Northern Territory. The films created great interest in Australia from the USA. In the 1980s, airfares reduced and the value of the Australian dollar against the American dollar was low, so American tourists could travel and stay relatively cheaply in Australia.

In 2018, Tourism Australia launched a reboot of the Crocodile Dundee campaign with a $36 million advertising campaign featuring well-known Australian actors such as Chris Hemsworth, Margot Robbie and Hugh Jackman. The campaign aimed to increase the current 780 000 American visitors to Australia.

Source 1 Crocodile Dundee rebooted by Tourism Australia

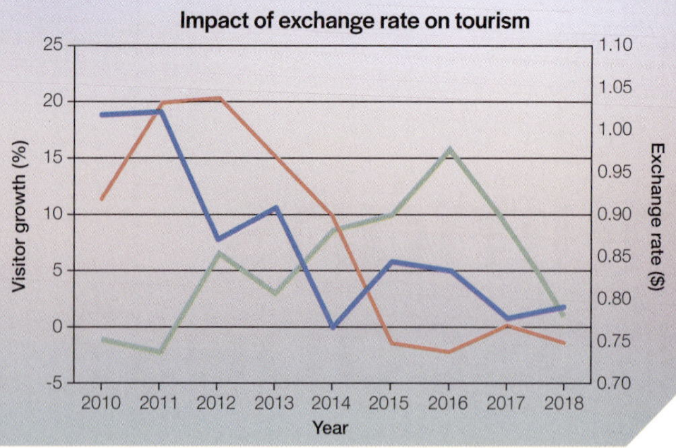

Source 2 The relationship between the exchange rate and tourism

Exchange rates

An **exchange rate** is how much one **currency** is valued compared to another currency. Exchange rates continually respond to changes in a country's **economy**. Higher-valued currencies are generally associated with confidence in a government and its economy; for example, currency investors are confident in a country's low rates of **inflation**.

The value of a currency is determined by the movement of money in and out of a country. When a currency is in high demand, this leads to an increase in its value.

The exchange rate defines the amount of a foreign currency you can buy with one Australian dollar. In 2012, when the Australian dollar was in demand, you would receive $1.04 American dollars for every Australian dollar. The average exchange rate with the American dollar in 2018 was just 75 cents. This means that you receive 75 American cents for every Australian dollar.

Exchange rates and tourism

When travelling, exchange rates greatly influence the prices you will pay in a foreign country. When the Australian dollar is high compared to other countries, Australians are encouraged to travel overseas because they can buy more with their money, but international tourists to Australia are discouraged because their currency is not worth as much when converted to Australian currency.

Consumers such as tourists respond to changing prices in the market. Australian tourism benefits when the Australian dollar is low against other currencies. In 2016, when the Australian dollar exchange rate was just 74 cents against the American dollar, USA tourist numbers to Australia grew by nearly 16 per cent. In 2011 when the Australian dollar exchange rate was $1.03 against the American dollar, USA tourist numbers fell by more than 2 per cent.

Learning ladder G1.14

Show what you know

1 How are landforms and landscapes important in promoting tourism to Australia?

Economics and business

Step 1: I can recognise economic information

2 Define these terms from the information on these pages: currency, exchange rate, inflation.

Step 2: I can describe economic issues

3 Source 1: Why do you think Tourism Australia turned to famous Australian actors to help them in marketing Australia?

Step 3: I can explain issues in economics

4 What impact do exchange rates have on the Australian tourism industry?

Step 4: I can integrate different economics topics

5 Look carefully at the line graph in Source 2.
   a When was the best time for Australians to travel overseas and why?
   b When was the best time for American tourists to travel to Australia and why?
   c Compare the USA and Australian tourist growth trends shown on the graph. Why are they different?

Step 5: I can evaluate alternatives

6 Using Source 2 as a reference, why do you think Tourism Australia chose 2018 to make a large investment in attracting American tourists to Australia? Could there have been a better time?

How do humans degrade landscapes?

As humans change landscapes, the land can be degraded through processes such as deforestation and desertification. Natural vegetation is cut down for fuel and timber and replaced with farms, mines and towns.

Soil erosion

Deforestation and overgrazing account for two-thirds of Australia's land degradation. Vegetation binds the soil together with its roots, so when trees are cut or plants are trampled by cattle the soil is opened to **erosion** (see pages 34–37).

Water can remove unprotected soil in sheets or cut gullies into the earth. Wind can pick up dry soil and deposit it great distances away. The eroded soil can clog rivers and the deforested land can no longer hold onto water, causing more erosion and flooding. Worldwide, half of Earth's topsoil has been lost in the last 150 years.

Desertification

Desertification occurs on the edges of deserts, when fragile land is over-farmed and overgrazed, causing the desert to spread. Desertification can also be caused by drought.

Source 1

a Forested hillside binds the soil together and takes up water through its roots.

b Deforested hillside leaves the soil unprotected and easily eroded by water and wind.

Land degradation in Australia

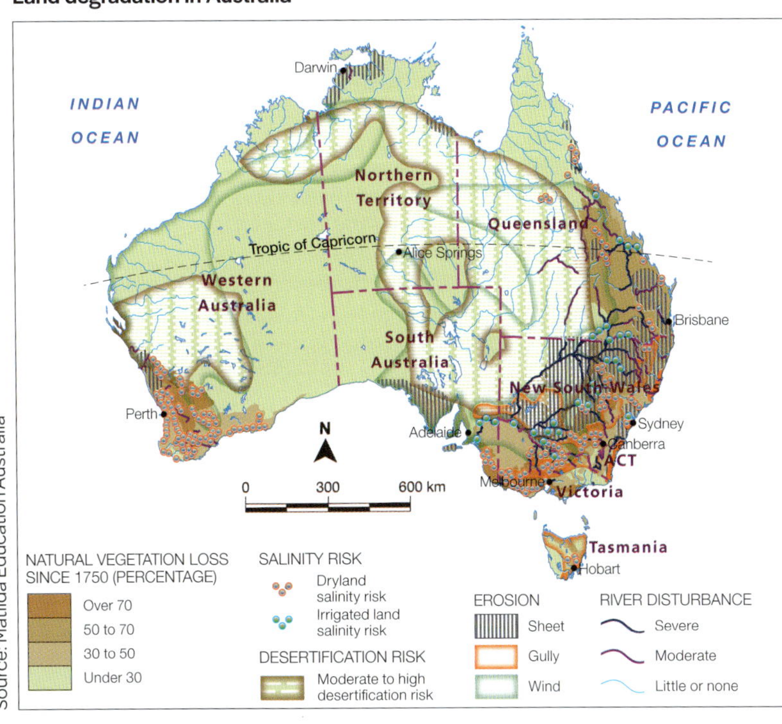

Source 2

Land degradation in Australia

Source 3

The bare hills surrounding Queenstown in Tasmania resemble a lunar landscape. The rainforest was cut down to fuel copper smelters. Toxic sulfur fumes from the smelters further destroyed the vegetation. The heavy west coast rain eroded the hills of their soil and a man-made desert was left.

Source 4

Desertification

Source 5

This aerial view shows the sand-covered streets of Nouakchott, the capital city of Mauritania in Africa. In 1960, Nouakchott was many days' walk from the Sahara Desert, but now the advancing sand dunes have covered outlying towns and forced people and animals to come to Nouakchott, putting further pressure on the land.

Learning ladder G1.15

Show what you know

1 What is deforestation and why does it lead to increased erosion?

2 What is desertification and how is it caused?

3 List three similarities between desertification and deforestation.

Geographical challenge

Step 1: I can identify responses to a geographical challenge

4 Source 5: Why have people been forced to relocate to Nouakchott?

Step 2: I can compare responses to a geographical challenge

5 Source 4: What are the key causes of desertification and how are humans affected? What plans can be put in place to reduce desertification and repair the land?

Step 3: I can compare strategies for a geographical challenge

6 Create an infographic to raise awareness of how humans are acting to degrade landforms and landscapes. Include the following key ideas in your graphic:

a causes of degradation

b examples of locations undergoing severe degradation

c ways we can reduce our impacts on the landscape.

Step 4: I can evaluate alternatives for a geographical challenge

7 Source 2: Identify the areas of Australia affected by salinity. Draw a block diagram to show how salinity occurs and present some alternatives for reducing salinity.

HOW TO

Block diagrams, page 144
Infographics, page 150

G1.16

civics+citizenship

How can we take action to save landscapes?

To help improve a landscape or stop its destruction, people can let their voices be heard by protesting or lobbying members of parliament through letters and petitions.

Participating in democracy

Democracy is government by the people. In order for people to participate in democratic government and influence issues such as protecting natural landscapes, they can take actions to enable their voices to be heard. These actions include:

- voting for government representatives that support their views
- contacting their local member of parliament to let them know their views
- taking **direct action** such as protesting, strikes or signing a petition
- contacting or joining an interest group that will lobby elected representatives to make changes in the laws to support their views.

Direct action

People raise awareness about environmental and other issues through direct action. You can take direct action to have your voice heard even if you have not reached the voting age of 18 years. Direct action is an effective way of raising awareness about issues. Today, people use social media to quickly spread messages about direct action events and bring issues to people's attention.

Protests are an effective form of direct action where protesters march or hold a **demonstration** at the site of an issue such as a deforested area or the steps of parliament.

Protests can attract large numbers of people with placards, t-shirts and badges showing slogans to get their message across. Protest organisers

Source 1

Protesters gather outside Queensland's parliament to protest against deforestation. Lobby group, the Queensland Conservation Council, stated that '... Our native woodlands have been exposed to unnecessary and unrestrained land clearing'.

Source 2

Australian land damaged by deforestation for cattle grazing

invite the media to their demonstrations to record speeches and beam the protest to a wider audience in Australia and around the world.

In May 2018, protesters demonstrated against deforestation outside Queensland's parliament. Protesters erected a large billboard to show how deforestation has a great impact on native animals. Farmers who held an opposing view also gathered to demonstrate against proposed changes to land-clearing laws. Both groups were trying to influence the members of the Queensland parliament who were about to vote on the issue.

Lobby groups

Lobbying is any attempt by individuals or groups to influence the decisions of government. Lobby groups work on behalf of a particular cause to influence political decisions. In Australia, there are many conservation organisations that work to influence public and government opinions on issues such as deforestation, pollution and the impact of introduced species.

The World Wildlife Fund (WWF) is a lobby group that works around the world to protect endangered species and habitats. It tries to influence political decisions about the environment through media campaigns or speaking directly with federal or state members of parliament. The WWF website gives interested people the chance to email a message to members of parliament on issues such as protecting koalas from **deforestation**.

The WWF also organises online **petitions** to gather as many signatures as possible to lobby national governments and the United Nations to support causes such as stopping plastics from polluting our oceans and endangering wildlife. Petitions are used to persuade politicians that there is large support for a cause. Anyone can start a petition, so it is a popular form of direct action to help show rising support for an issue.

Learning ladder G1.16

Show what you know

1 How can people let others know what they think about an issue?

2 What is a lobby group and how do they operate?

Civics and citizenship

Step 1: I can identify topics about society

3 What is direct action and why is it a democratic right of citizens?

Step 2: I can describe societal issues

4 Explain why deforestation is an issue and the problems that are caused by deforestation.

Step 3: I can explain issues in society

5 Look at the photograph in Source 1.
- a What form of direct action is shown here?
- b What are they doing to get their message across?
- c Who were they trying to influence and why?

Step 4: I can explain different points of view

6 An opposing group held a protest at the same time as the protesters in Source 1. Research the web to find out who the opposing group represented. What views might they have about land clearing?

Step 5: I can analyse issues in society

7 Analyse how a lobby group operates by visiting http://mea.digital/pdPo.
- a What do you think this lobby group is trying to achieve?
- b What petitions or letters to members of parliament is the lobby group currently organising?

Masterclass

Learning Ladder

Work at the level that is right for you or level-up for a learning challenge!

Global distribution of volcanoes and earthquakes

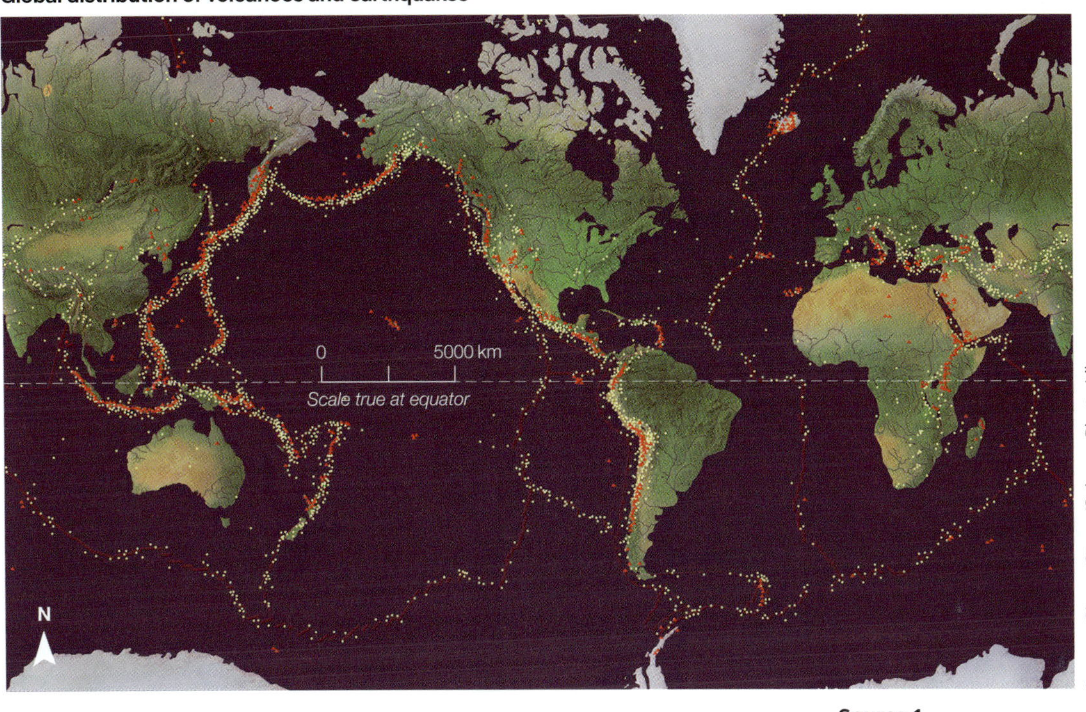

Source 1

Global distribution of volcanoes (red dots) and earthquakes (yellow dots)

Hawaii contour map

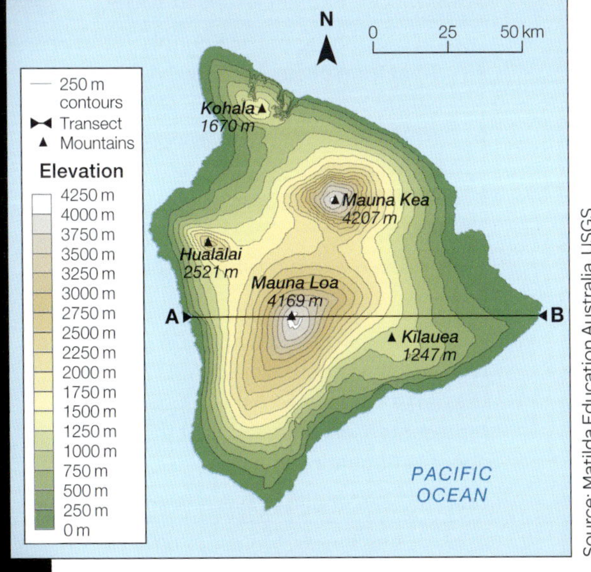

Source 2

A contour map of Hawaii featuring Mauna Loa volcano.

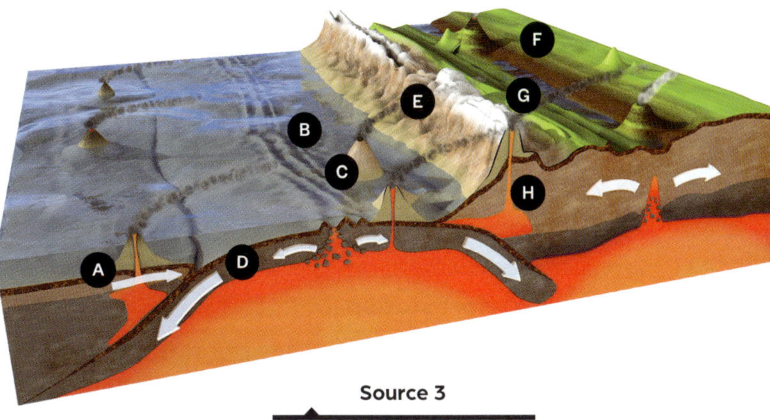

Source 3

Block diagram of a landscape

Source 4

Rescue teams work carefully through rubble in the search for missing people following the 6.3 magnitude earthquake that hit Christchurch, New Zealand in 2011.

Step 1

a I can identify and describe spatial characteristics

Describe the *place* photographed in Source 4.

b I can identify and describe interconnections

With reference to Source 1, identify one key pattern of earthquake risk on a global scale.

c I can identify responses to a geographical challenge

Source 4: What response is shown to the Christchurch earthquake disaster in 2011?

d I can collect, record and display data in simple forms

Refer to Source 1. Identify what the triangles and circles represent on the map.

e I can use geographic terminology to interpret data

Source 2: Where is the land under 250 metres located on this island?

Step 2

a I can explain spatial characteristics

Source 2: What are the key landforms on this Hawaiian island?

b I can explain interconnections

Outline two strong interconnections with the subduction zone shown in Source 1.

c I can compare responses to a geographical challenge

Source 4: Compare this image with the case study of Haiti on page 33. What additional issues did the Haitian earthquake provide?

d I can recognise and use different types of data

Refer to Source 1. List the continents that currently have hot spots.

e I can describe patterns and trends

Look carefully at Source 3. Match the letters A–H with the following features: magma, fold-mountain, volcanic crater, subducting plate, converging plate margin, diverging plate margin, hot-spot volcano, mid-ocean ridge.

Step 3

a I can explain processes influencing places

Refer to Source 2. Explain the processes that have helped to form this landscape.

b I can identify and explain the implications of interconnections

Source 1. What are the implications for New Zealand given its location on a tectonic plate boundary?

c I can compare strategies for a geographical challenge

What action did authorities and individuals take in response to the 2018 Kīlauea volcanic eruption (pages 30–31)?

d I can choose, collect and display appropriate data

Using Source 2, create a cross-section for transect A–B.

e I can explain the reasons behind a trend or spatial distribution

Use SHEEPT to explain the pattern shown in Source 2 on page 43.

Masterclass

Step 4

a I can predict changes in the characteristics of places

Predict what will happen to Nouakchott if desertification remains unchecked (see page 45).

b I can evaluate the implications of significant interconnections

There is an interconnection between converging plates with both fold mountains and volcanoes. Why are different landforms created from the same process?

c I can evaluate alternatives for a geographical challenge

What are some key challenges for aid agencies when natural disasters occur in LEDCs? What alternatives might be considered to help (see page 33)?

d I can use data to support claims

The photograph in Source 4 was taken after the 2011 earthquake in Christchurch, New Zealand. The disaster had both economic and social impacts. Discuss these with reference to evidence in the photo.

e I can analyse relationships between different data

Compare the map in Source 1 with the block diagram in Source 3. Explain how they represent the process of subduction in different ways.

Step 5

a I can analyse the impact of change on places

Source 4. Analyse the short- and long-term impact on Christchurch following the 2011 earthquake.

b I can explore spatial association and interconnections

Explain why earthquakes follow a clustered, linear distribution on a global scale.

c I can plan action to tackle a geographical challenge

List the methods that can be used to tackle the issue of deforestation and select the most tactical. Justify your decision (see pages 46–47).

d I can evaluate data

Justify whether you think Source 1 would be reliable in predicting and planning for an earthquake event. What additional information would be required?

e I can draw conclusions by analysing collected data

Source 4 depicts the outcome of the 2011 Christchurch earthquake in New Zealand. Based on your knowledge of geohazards and tectonic plates, explain whether you think that the local community was adequately prepared for this event. If not, what could have been done to reduce the negative outcomes of this earthquake?

Capstone

How can I understand landscapes and landforms?

In this chapter, you have learnt a lot about landscapes and landforms. Now you can put your new knowledge and understanding together for the capstone project to show what you know and what you think.

In the world of building, a capstone is an element that finishes off an arch or tops off a building or wall. That is what the capstone project will offer you, too: a chance to top off and bring together your learning in interesting, critical and creative ways. You can complete this project yourself, or your teacher can make it a class task or a homework task.

mea.digital/GHV8_G1

Scan this QR code to find the capstone project online.

G2

Coastal landscapes

HOW DO WAVES SHAPE LANDFORMS?

page 56

spatial distributions and patterns

page 68

HOW ARE COASTS USED?

thinking locally

page 70

WHAT ARE THE EFFECTS OF EROSION?

economics + business

page 72

HOW DO TOURISTS VALUE THE COAST?

How can I learn about coastal landscapes?

Coastal landscapes are continually shaped and reshaped by both natural and human forces. Shifting sand, waves and other erosional systems such as wind, ice and rain all have a huge impact on coastal landscapes over time. Humans are now beginning to shape these landscapes to prevent further erosion, as well as to meet the rising demands for housing, trade and tourism.

learning ladder

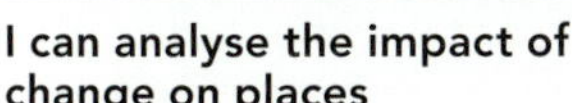

Step	Spatial characteristics	Interconnections	Geographical challenge
step 5	**I can analyse the impact of change on places** I can analyse and evaluate the implications of changing coastal land use over time and at different scales and calculate its impact on people and environments.	**I can explore spatial association and interconnections** I can compare distribution patterns and the interconnections between them; e.g. the erosional landforms along limestone coastal areas.	**I can plan action to tackle a geographical challenge** I can frame questions, evaluate findings, plan actions and predict outcomes to tackle a coastal-based geographical challenge.
step 4	**I can predict changes in the characteristics of places** I can predict changes in the characteristics of places over time due to coastal erosion and geohazards such as tsunamis.	**I can evaluate the implications of significant interconnections** I can identify, analyse and explain key coastal-based interconnections within and between places, and evaluate their implications over time and at different scales.	**I can evaluate alternatives for a geographical challenge** I can weigh up alternative views and strategies on a coastal-based geographical challenge using environmental, social and economic criteria.
step 3	**I can explain processes influencing places** I can explain the series of actions leading to change in a place, such wave action.	**I can identify and explain the implications of interconnections** I can identify, analyse and explain coastal-based interconnections and explain their implications.	**I can compare strategies for a geographical challenge** I can compare strategies for a geographical challenge, taking into account a range of factors and predicting the likely outcomes.
step 2	**I can explain spatial characteristics** I can identify concepts of Space, Place, Interconnection, Change, Environment, Scale and Sustainability (SPICESS) when I read about coastal landscapes.	**I can explain interconnections** I can describe and explain interconnections and their effects, such as wave action and soft coastal rock.	**I can compare responses to a geographical challenge** I can identify and compare responses to a geographical challenge and describe its impact on different groups.
step 1	**I can identify and describe spatial characteristics** I can talk about spatial characteristics at a range of scales; e.g. headlands and bays.	**I can identify and describe interconnections** I can identify and explain simple interconnections involved in phenomena such as wave direction and sand build-up.	**I can identify responses to a geographical challenge** I can find responses to a geographical challenge such as coastal erosion and understand the expected effects.

Source 1

Shodo Island in Japan is a tombolo formed by longshore drift depositing a sand spit to link the island to the mainland.

Warm up

Spatial characteristics

1 Looking at Source 1, what geographic characteristics can you identify? Describe how they may have occurred using the SPICESS terms of space and place.

Interconnections

2 Source 1: How does the spit provide an interconnection between the mainland and the island?

Geographical challenge

3 What geohazards threaten coastlines?

Collect, record and display data

4 With a partner, discuss what you think might be some of the most devastating types of natural forces that affect coastal landscapes. Pick one, and research some examples of it.

Analyse data

5 Looking at the plan aerial views of the Maroochy River in Source 1 on pages 66–67, identify at least one key change that occurred between 1973 and 2003.

I can evaluate data
I can determine whether data presented about coastal landscapes is reliable and assess whether the methods I used in the field or classroom were helpful in answering a coastal-based research question.

I can use data to support claims
I can select or collect the most appropriate data and create specialist maps and information using ICT to support investigations into coastal landscapes.

I can choose, collect and display appropriate data
I can select useful sources of coastal data and represent them to conform with geographic conventions.

I can recognise and use different types of data
I can define the terms primary, secondary, qualitative and quantitative data and represent data in more complex forms.

I can collect, record and display data in simple forms
I can identify that maps and graphs use symbols, colours and other graphics to represent data.

Collect, record and display data

I can draw conclusions by analysing collected data
I can summarise findings and use collected data to support key patterns and trends I have identified for coastal-based research.

I can analyse relationships between different data
I can use multiple data sources, overlays and GIS to find relationships that exist in patterns of coastal land use.

I can explain the reasons behind a trend or spatial distribution
I can identify Social, Historical, Economic, Environmental, Political and Technological (SHEEPT) factors to help me explain patterns in data.

I can describe patterns and trends
I can identify Patterns, Quantify them and point out Exceptions (PQE) to describe the patterns I see.

I can use geographic terminology to interpret data
I can identify increases, decreases or other key trends on a map, graph or chart about coastal landscapes.

Analyse data

How are coastal landscapes formed?

Coastal **landscapes** are formed through a range of natural processes over thousands of years. As humans began developing coastlines for residential, trade and recreational purposes, we also began shaping this fragile landscape. Coastlines are excellent examples of how both natural and human forces can alter landforms and landscapes over time.

Source 1

These **training walls** (see pages 74–77) built to protect the river mouth at Narooma, NSW are one example of how humans have altered the coastline.

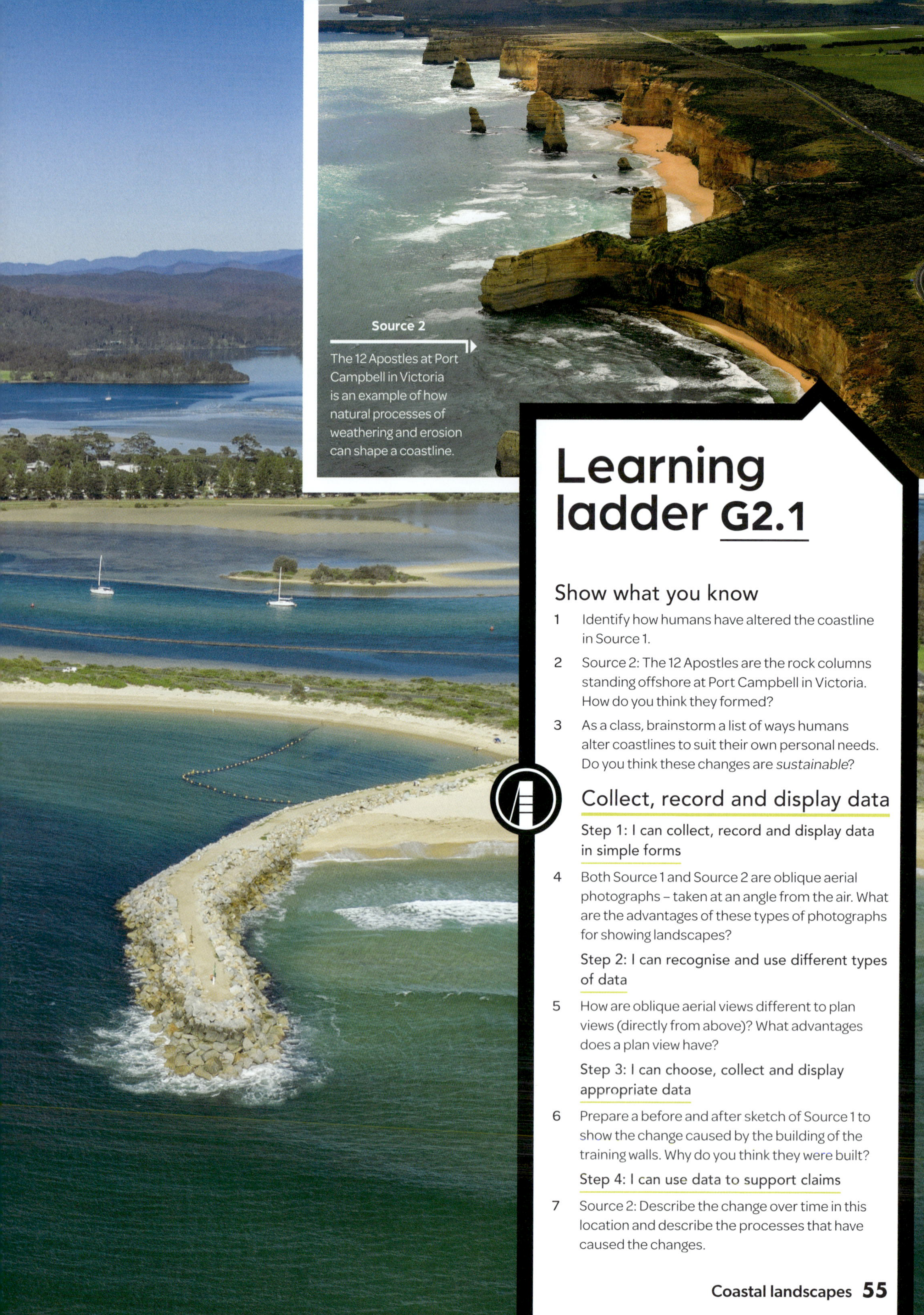

Source 2

The 12 Apostles at Port Campbell in Victoria is an example of how natural processes of weathering and erosion can shape a coastline.

Learning ladder G2.1

Show what you know

1 Identify how humans have altered the coastline in Source 1.

2 Source 2: The 12 Apostles are the rock columns standing offshore at Port Campbell in Victoria. How do you think they formed?

3 As a class, brainstorm a list of ways humans alter coastlines to suit their own personal needs. Do you think these changes are *sustainable*?

Collect, record and display data

Step 1: I can collect, record and display data in simple forms

4 Both Source 1 and Source 2 are oblique aerial photographs – taken at an angle from the air. What are the advantages of these types of photographs for showing landscapes?

Step 2: I can recognise and use different types of data

5 How are oblique aerial views different to plan views (directly from above)? What advantages does a plan view have?

Step 3: I can choose, collect and display appropriate data

6 Prepare a before and after sketch of Source 1 to show the change caused by the building of the training walls. Why do you think they were built?

Step 4: I can use data to support claims

7 Source 2: Describe the change over time in this location and describe the processes that have caused the changes.

How do waves shape landforms?

Waves are one of the strongest forces shaping coastal landforms. Waves can be generated by wind or movements in Earth's surface. They can travel for thousands of kilometres through water, before breaking in shallower water near coasts, building or scouring away the coastline in the process.

How do waves begin?

Waves travel through water. The water moves up and down but does not actually travel with the wave. It is the energy that travels with the wave. The energy required to create waves most commonly comes from wind, but may also come from **earthquakes** or volcanic eruptions.

When wind blows across water it creates ripples on the surface. The wind forces the water to rise and gravity pulls it back down again. Water rising and falling creates a circular movement of water particles beneath the surface. On the surface, we see the top of this orbit, known as the **crest**, and the zone where it falls, called the **trough**.

The orbiting waves can travel for long distances as **swells** through the ocean. A swell is a collection of waves formed by winds. The size of the swell is measured from the crest to the trough. The largest swell ever measured in the southern hemisphere was a 23.8 metre swell off New Zealand's Campbell Island in May 2018.

Why do waves break?

As waves move into increasingly shallower water near the coastline or offshore **reefs**, the bottom of the wave orbit comes into contact with the sea floor and slows down. The wavelength is shortened and the crest of the wave overtakes the bottom of the wave. It spills forward as a **breaking wave**. Waves will generally start to break when they reach a water depth of 1.3 times the wave height.

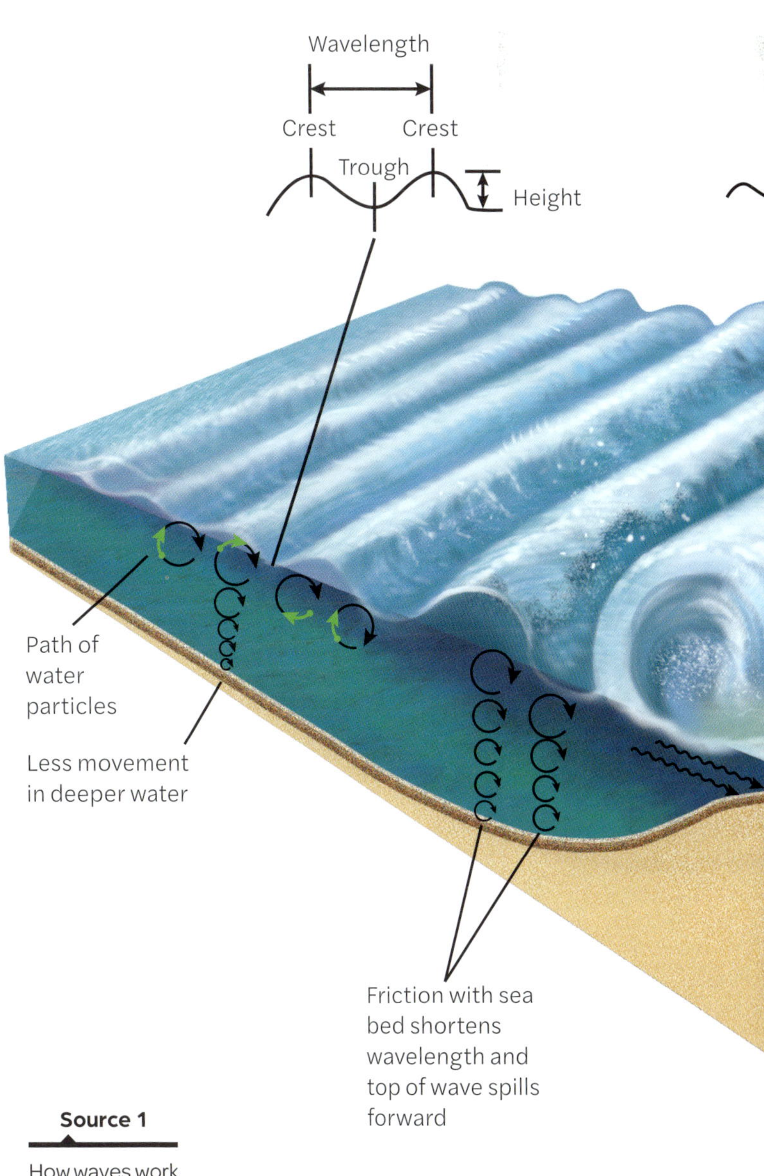

Source 1
How waves work

Source 2

A surfer uses the energy of the breaking wave to be propelled down the face of the wave and into the shallow water near the beach.

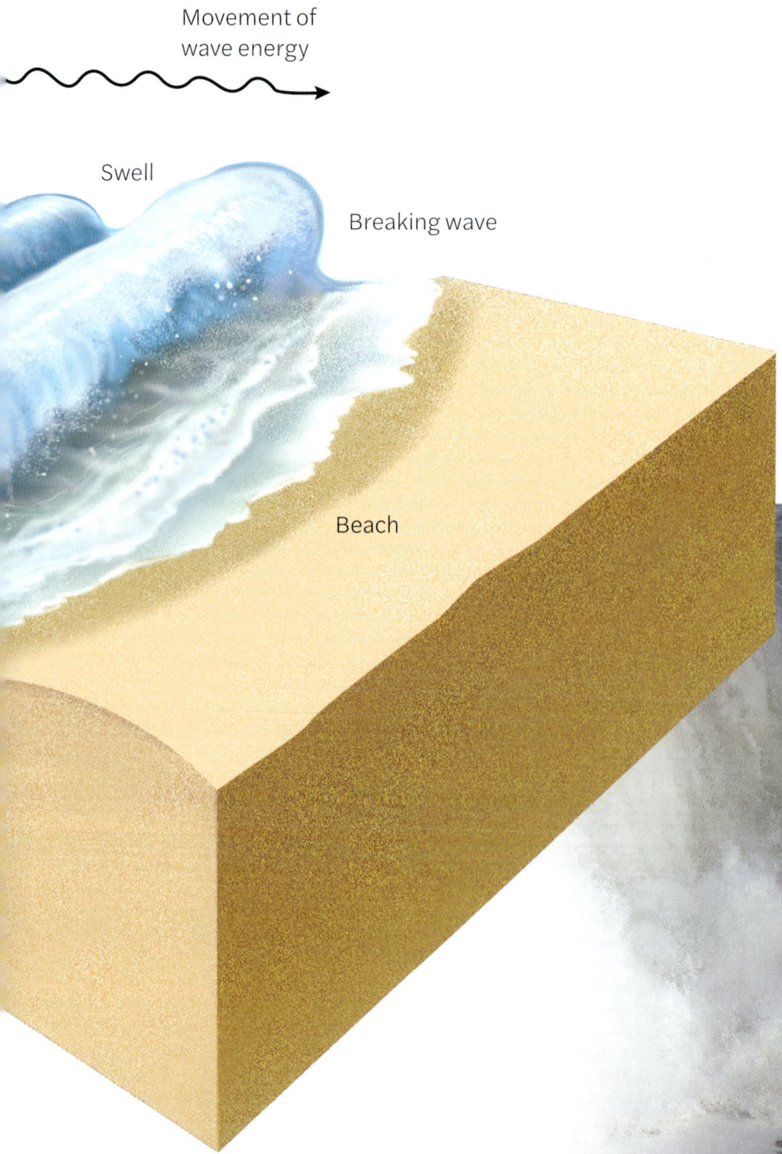

What conditions do surfers look for?

Groundswell waves are steeper and faster. They are the result of strong wind storms in the open ocean that create longer wavelength waves that move quickly into shallow water before breaking. Wind swell waves are the smaller, less powerful waves formed by local winds.

Offshore winds create the best waves as wind blows against the top of the wave to delay it from breaking. Onshore winds push the top of the wave forward, causing the wave to break before the usual breaking depth is reached.

Source 3

Australian surfer Owen Wright surfs groundswell waves at Bells Beach, Victoria

How do different waves shape landforms?

Coastal landforms are both created and worn away by wave action. **Destructive waves** destroy and erode the coastline. Destructive waves are large and powerful; they constantly crash onto the coastline, wearing away cliffs and digging out areas of beach.

Destructive waves travel across open ocean and build a significant amount of energy that hits coastal cliffs and beaches with great force. Cliffs can be carved into amazing erosional landforms such as headlands, caves, arches and stacks.

Constructive waves are smaller and less frequent. They deposit sand on the beach rather than scouring it away. Constructive waves are responsible for creating depositional landforms such as beaches, sand dunes, sand bars, spits and tombolos.

The amount of eroded material taken away or deposited on the coastline depends on the water flowing onto the beach – **swash** – and returning to the sea – **backwash**. Destructive waves have weak swash, but strong backwash that removes material. Constructive waves have strong swash and weak backwash, helping it to build up sand.

Erosional landforms

Destructive waves create erosional landforms including:

- *cliffs* – the almost vertical rock face along many coastlines
- **headlands** – narrow, high land jutting out into the sea, formed of harder rock that resists erosion
- *caves* – formed as waves bend around headlands and attack softer rock
- *arches* – formed when caves on either side of a headland join, leaving a gap
- **stacks** – pillars of rock separated from cliffs or a collapsed arch
- **wave-cut platforms** – flat eroded areas of rock at the base of coastal cliffs.

Source 4

Coastal landforms

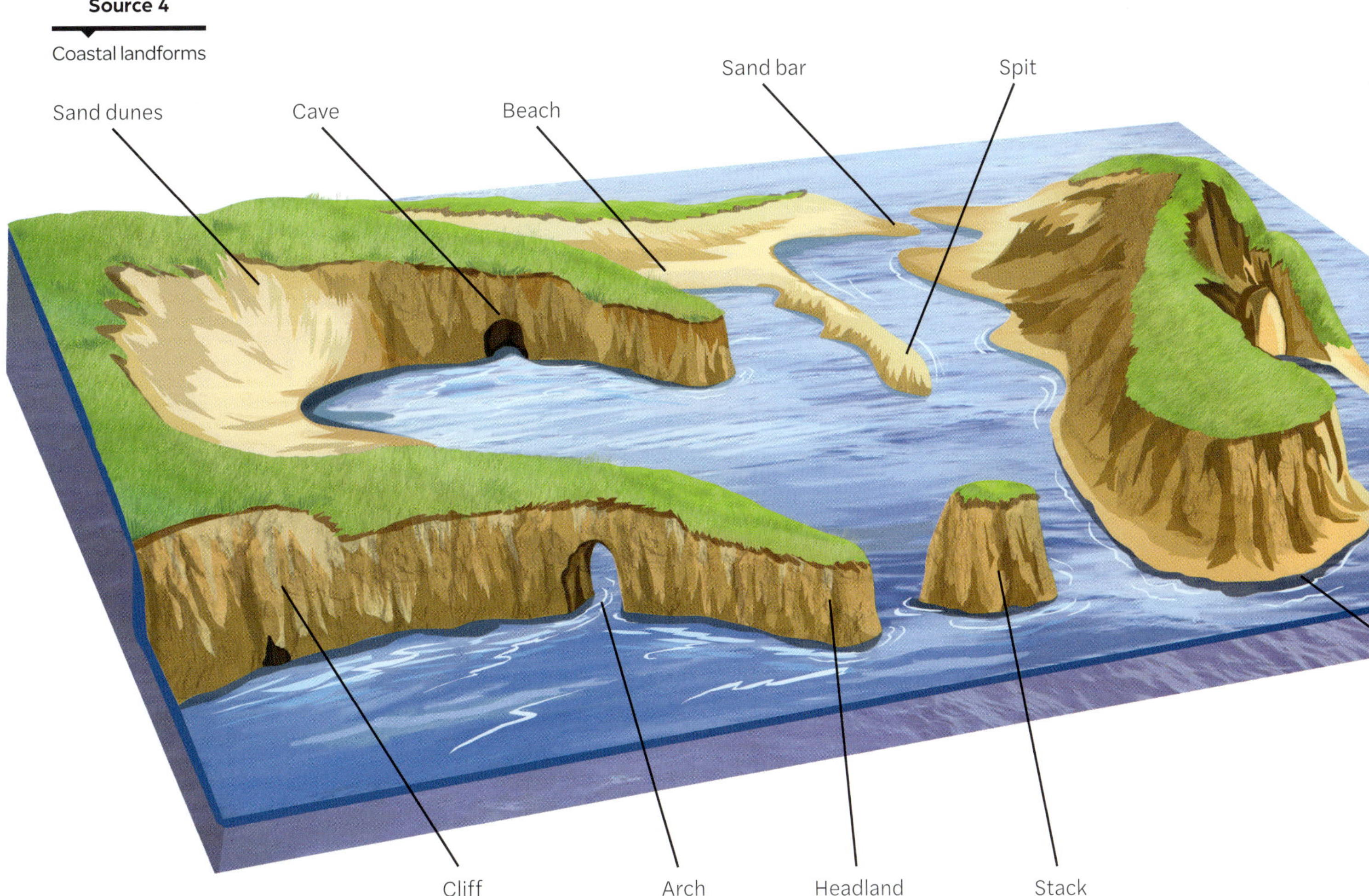

Source 5

The action of waves and other erosional forces such as wind, ice and rain are continually shaping and reshaping our coastlines. The famous Elephant Rock sits on the Tongaporutu coastline in New Zealand. It was a double arch in the shape of an elephant until continued erosion and an earthquake removed the elephant's trunk in 2016. Elephant Rock sits alongside the Three Sisters stacks in the same location. One of the Three Sisters disappeared following a storm in 2003, but it was replaced with another sister that became separated from the Tongaporutu cliff face in 2013.

Depositional landforms

Constructive waves create depositional landforms, including:

- *beaches* – made up of loose particles such as sand and pebbles deposited by waves
- *sand dunes* – hills of sand that form at the back of beaches from wind-blown sand
- *sand bars* – ridges of deposited sand in the ocean that are exposed at low tide
- **spits** – narrow bars of deposited sand linked to the coast
- **tombolos** – form when waves deposit a bar of sand to link an island to the coast.

Tombolo

Wave-cut platform

Learning ladder G2.2

Show what you know

1 Identify two ways that different wave types can *change* coastlines.

Spatial characteristics

Step 1: I can identify and describe spatial characteristics

2 Sketch the scene in Source 2 on page 55. Annotate the sketch with at least three coastal features highlighted on this spread.

Step 2: I can explain spatial characteristics

3 Outline how the *place* Elephant Rock has *changed* over time.

Step 3: I can explain processes influencing places

4 Construct a flow diagram showing the process from wave formation to wave break.

Step 4: I can predict changes in the characteristics of places

5 Create a Venn diagram comparing the similarities and differences between destructive and constructive waves.

Flow diagrams, page 144

How do bays form?

Headlands and bays form along coastlines that have alternating bands of hard and soft rock. The hard rock is more resistant to erosion and wears away more slowly, leaving a headland that extends out into the sea. Softer rock is less resistant to erosion and bays are carved out between headlands.

Different rates of erosion

Headlands take the full force of waves, because they jut out from the coastline. As waves approach land they bend to the shape of the coast in a process known as **refraction**. Wave energy is concentrated on the headlands, which erodes into landforms such as caves, arches and **stacks** (see pages 58–59).

Waves in bays are more spread out and the wave energy is reduced, allowing for sand to be deposited. Beaches are usually found in shallow bays where waves have less energy.

How bays and headlands form over time

Refer to Source 3 at right.

1. Waves attack the alternating bands of hard and soft rock along a coastline.
2. Destructive waves erode the cliff causing the cliff to retreat.
3. Softer rock erodes more quickly to create bays. The resistant hard rock sticks out to sea as headlands.
4. The protected bays allow deposition of sand to form beaches.

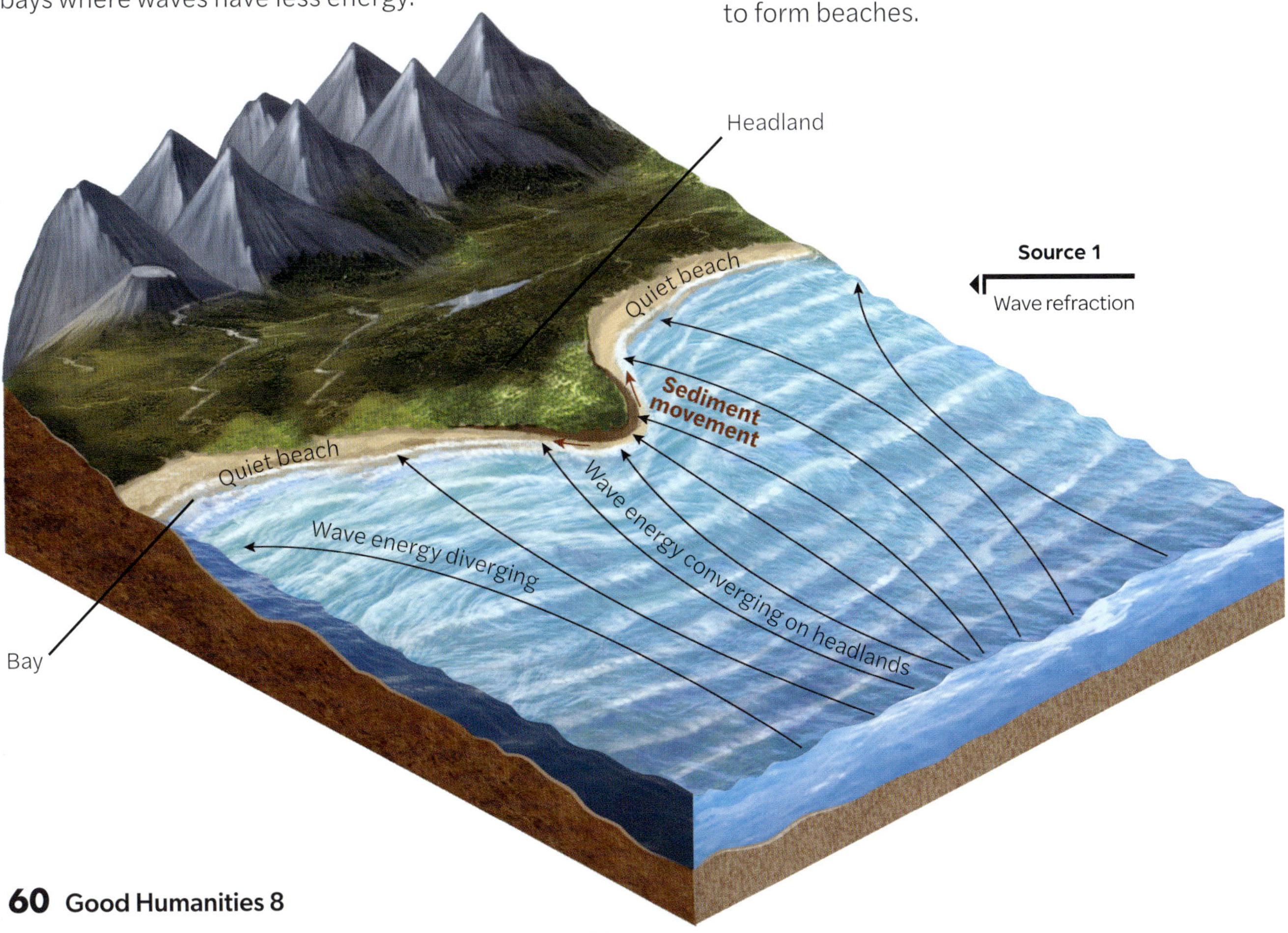

Source 1 Wave refraction

Source 2

Wineglass Bay on the east coast of Tasmania is a calm body of water protected by two granite headlands. Granite is a very hard rock that is resistant to erosion. The white quartz sand is eroded material from the granite rocks, which are rich in quartz.

Source 3

The formation of bays and headlands over time

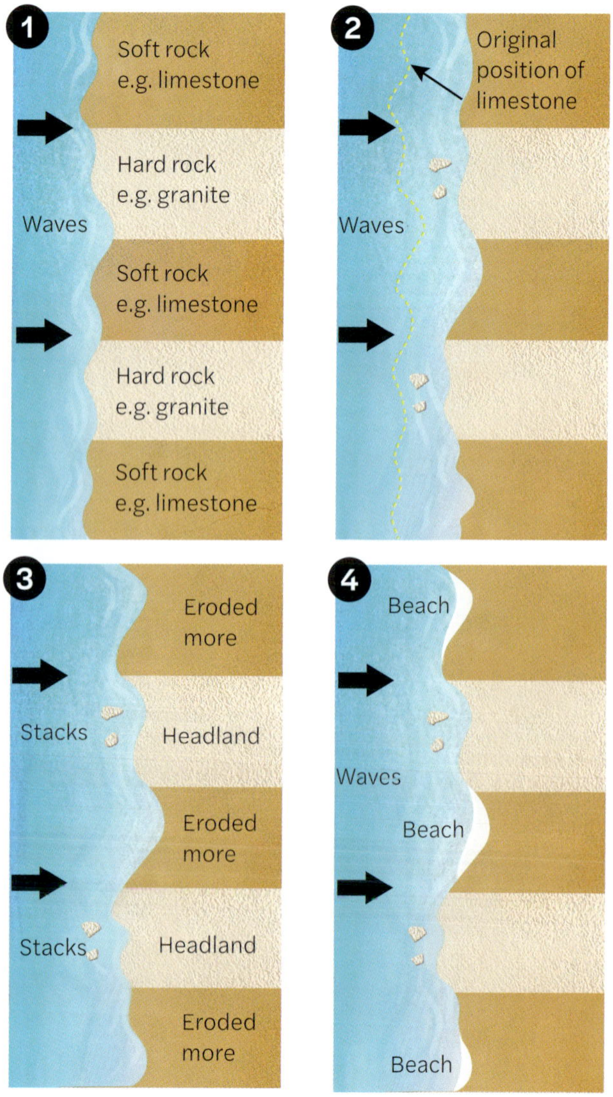

1

Soft rock e.g. limestone

Hard rock e.g. granite

Waves

Soft rock e.g. limestone

Hard rock e.g. granite

Soft rock e.g. limestone

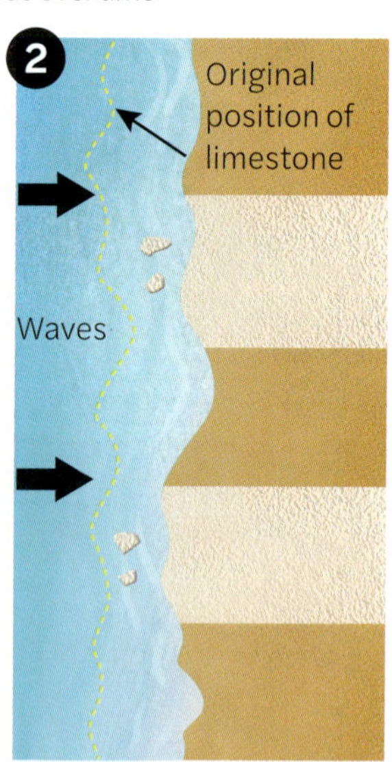

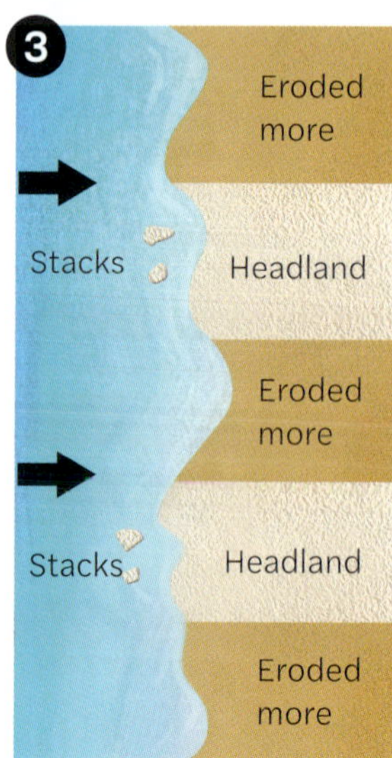

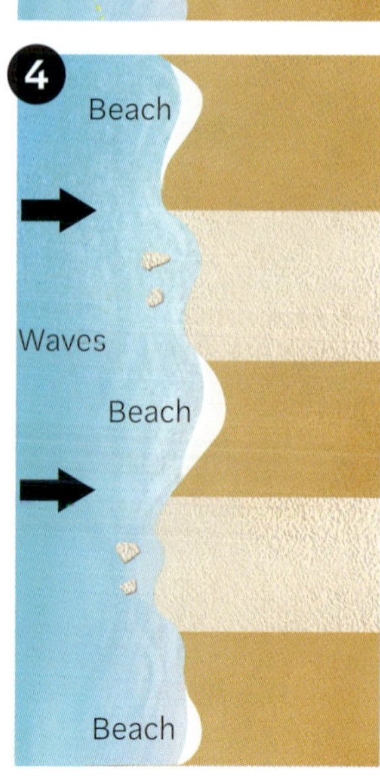

Learning ladder G2.3

Show what you know

1 Identify the difference between a bay and a headland.

2 What is meant by the terms 'hard rock' and 'soft rock'? What impact does rock type have on a coastline?

3 Using plasticine or clay, construct a headland, cave, arch and stack. Using these models, discuss the 'story' of coastal erosion with a partner.

Interconnections

Step 1: I can identify and describe interconnections

4 Source 1: Refraction is caused by the interconnection of waves and which coastal feature?

Step 2: I can explain interconnections

5 Source 3: How does the interconnection of waves and soft rock lead to the formation of bays?

Step 3: I can identify and explain the implications of interconnections

6 Source 3: Construct a flow diagram identifying how wave interaction with coastal landforms may have shaped Wineglass Bay (Source 2). Use the following key terms when constructing your diagram:

a hard rock
b soft rock
c erosion
d headland
e bay
f wave.

Step 4: I can evaluate the implications of significant interconnections

7 Evaluate the implications of the interconnection of headlands and waves.

HOW TO Flow diagrams, page 144

How are destructive waves eroding the coastline at Loch Ard Gorge?

Loch Ard Gorge at Port Campbell in Victoria is greatly affected by powerful destructive waves that travel long distances over the Southern Ocean. The soft limestone cliffs are eroded into erosional landforms such as gorges, headlands, caves, arches, stacks and blowholes.

Source 1

Eroding the limestone cliffs

Source 2

Oblique aerial view of Loch Ard Gorge

Learning ladder G2.4

Show what you know

1 Why is Loch Ard Gorge so prone to erosion?

2 Research the term abrasion. Describe how abrasion has helped erode the cliffs at Loch Ard Gorge.

Spatial characteristics

Step 1: I can identify and describe spatial characteristics

3 Using tracing paper, trace Source 2 and annotate at least five coastal landforms in this place.

Step 2: I can explain spatial characteristics

4 How are caves, arches and stacks associated with one another?

Step 3: I can explain processes influencing places

5 Using Sources 1 and 2, explain how destructive wave action has changed the headland marked 'A' on the oblique aerial photograph.

Step 4: I can predict changes in the characteristics of places

6 Looking at the oblique aerial image of Loch Ard Gorge in Source 2, describe how the headland marked 'A' will change in the future.

HOW TO

Sketches and annotating, page 142

How does sand move?

As waves hit beaches, the waves move sand along the coast. Constructive waves deposit sand to form other beaches, as the sand builds up against obstacles such as headlands or groynes. Dry sand on the beach can be blown by onshore winds to the back of the beach where it forms sand dunes.

How is sand moved by waves?

Most **constructive waves** hit the coast at an angle because of the direction of the wind and the shape of the land. Strong **swash** from the waves picks up grains of sand and moves them along the beach. Gravity takes the weak **backwash** and some sand directly back into the ocean ready for the next wave to move the sand further along the beach.

Along the beach, the sand moves in this zigzag fashion, known as **longshore drift**, until it meets an obstacle such as a **headland**, rocks or human constructions such as **groynes**. The sand builds up against the obstruction, widening the beach.

Longshore drift is also responsible for forming narrow bars of sand, called **spits**, that jut out into the ocean. Spits occur when the coastline changes direction or the beach meets the mouth of a river. The sand is deposited and builds up on itself to form a spit. Sometimes the spits join an island to form a **tombolo**.

How does wind move sand?

At low tide, sand particles on the beach dry out and can be picked up by onshore winds and moved further inland. The moving sand grains may be trapped by obstructions such as plants and mounds at the rear of the beach to form a **primary dune**.

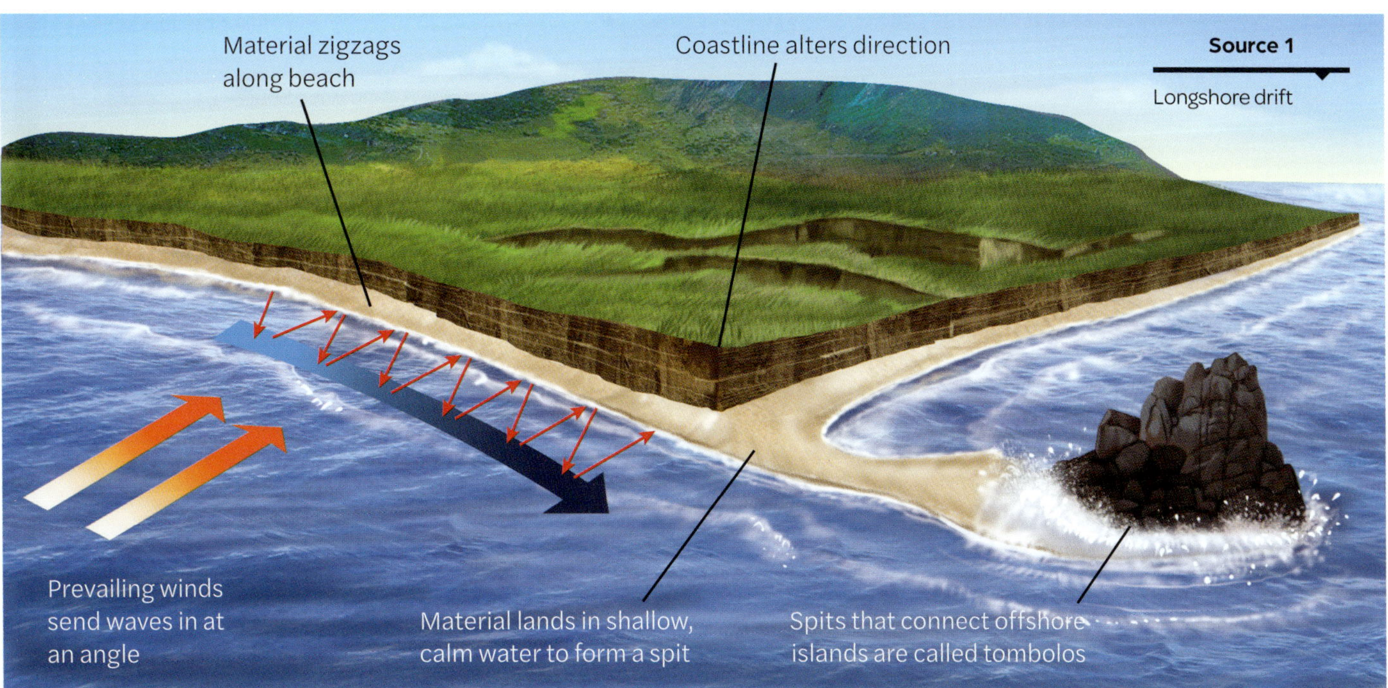

Source 1 Longshore drift

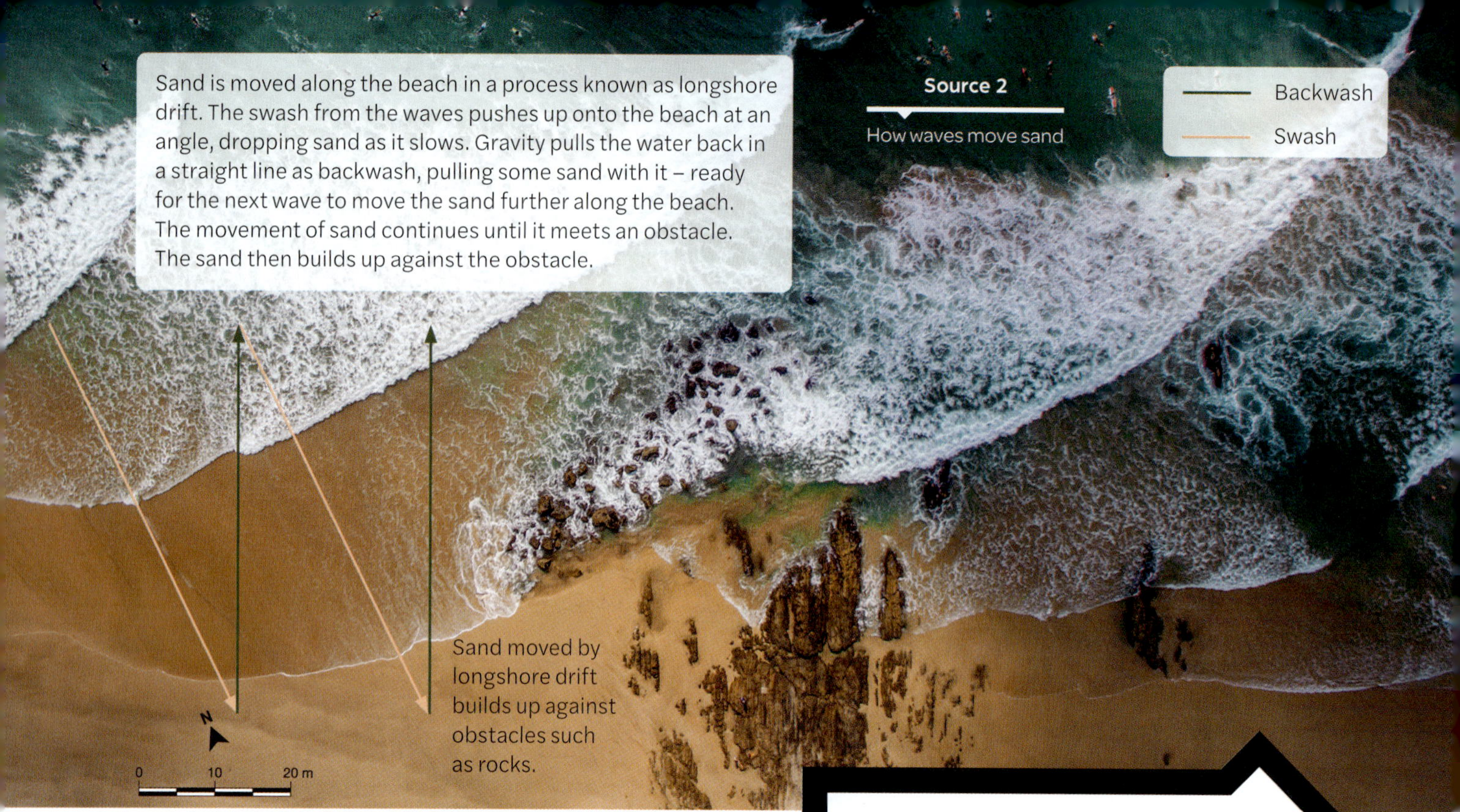

Sand is moved along the beach in a process known as longshore drift. The swash from the waves pushes up onto the beach at an angle, dropping sand as it slows. Gravity pulls the water back in a straight line as backwash, pulling some sand with it – ready for the next wave to move the sand further along the beach. The movement of sand continues until it meets an obstacle. The sand then builds up against the obstacle.
Backwash
Swash
Sand moved by longshore drift builds up against obstacles such as rocks.
N
0
10
20 m

Source 2

How waves move sand

There is little soil in a primary dune, so plants find it difficult to take root. Some grasses with shallow roots, such as marram grass, can grow in this very salty environment. Plants hold the sand dunes together and allow the dune to grow larger.

Wind can also erode bare patches on sand dunes, forming **blowouts**. The sand is blown further inland, eventually forming a **secondary dune**.

Onshore winds pick up dry sand from the beach and move it inland. The sand starts to mound.
Secondary dune
Blowout
Primary dune
A primary dune builds up, stabilised by plants. Wind scours sand to form a blowout.

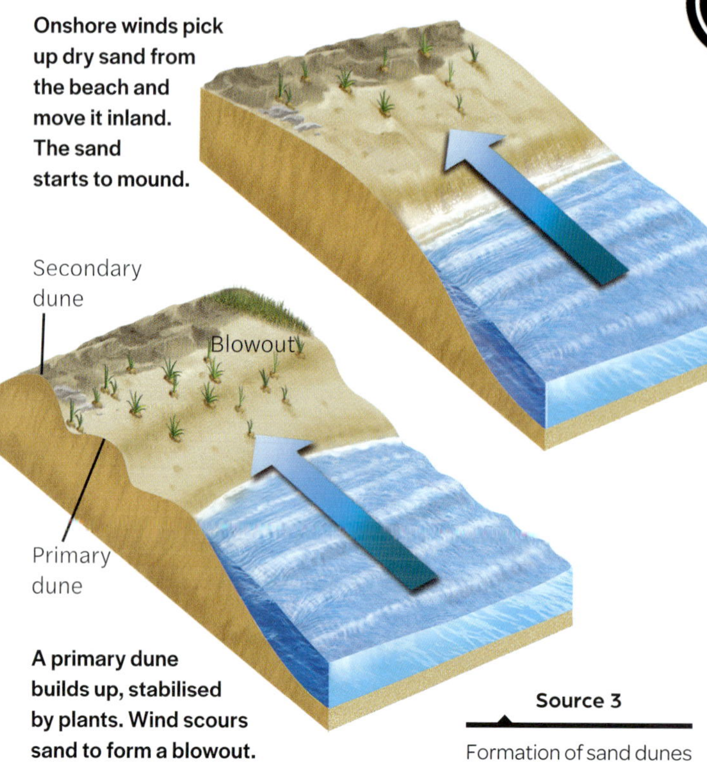

Source 3

Formation of sand dunes

Learning ladder G2.5

Show what you know

1 How does the movement of sand change coastal landscapes?

Interconnections

Step 1: I can identify and describe interconnections

2 Is longshore drift the only natural process that influences sand movement?

Step 2: I can explain interconnections

3 How is wind interconnected with the direction of longshore drift?

Step 3: I can identify and explain the implications of interconnections

4 Evaluate the implications of the interconnection of plants and wind on sand dunes. Why is it not wise to trample or remove vegetation from sand dunes?

Step 4: I can evaluate the implications of significant interconnections

5 Evaluate the implications of the building of Sandringham Harbour for Hampton Beach (pages 78–79). What has been done to repair the damage and why?

How is the Maroochydore beach constantly changing?

Like other beaches, the Maroochydore beach on Queensland's Sunshine Coast is made of loose sand. The beach is constantly changing as waves erode, transport and deposit material on the beach.

Rivers can also change coastlines where they meet the ocean. The Maroochy River mouth constantly changes and adds to coastal erosion issues. **Groynes** made of large sacks of sand have been placed to the south of the river mouth to help keep the sand on the beach. Note the groynes in Source 3 and how **longshore** drift (see pages 64–65) has pushed sand up against the groynes.

Source 2

One of the four sand bag groynes at the mouth of the Maroochy River, put in place in 2003.

Source 1

Plan aerial views of the Maroochy River from 1973, 1986 and 2003 show how the position of the river mouth changes and the impact it has on the beaches around it.

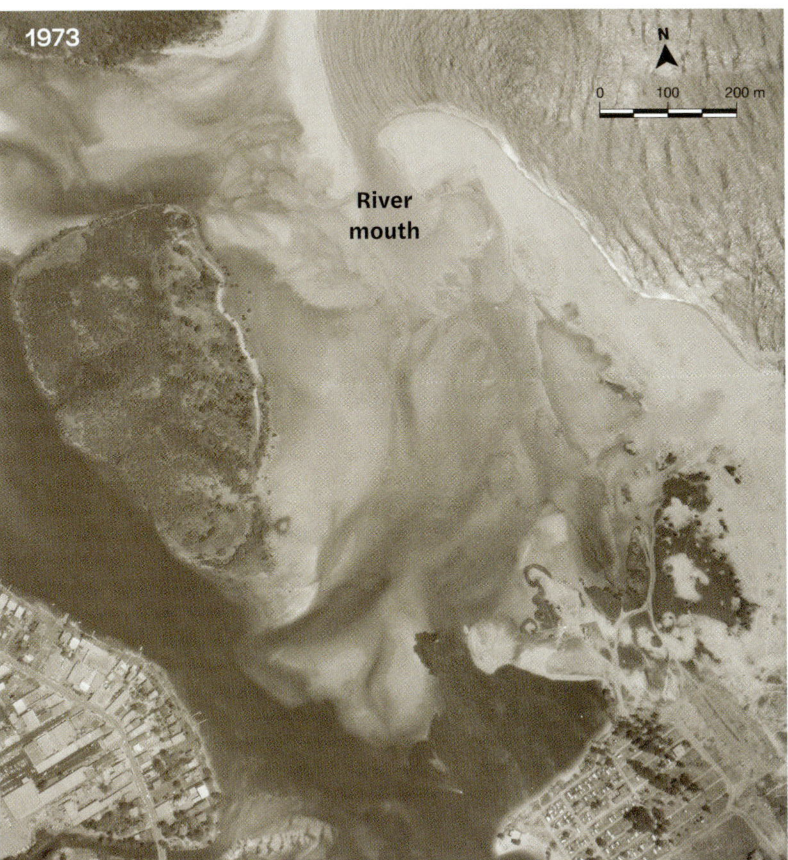

Source 3

Oblique aerial view of the mouth of the Maroochy River in 2018. Groynes have been positioned on the south bank of the river mouth to keep sand on the beach.

Source: State of Queensland, 2019

Learning ladder G2.6

Show what you know

1 What effects could longshore drift, wind and river movement have on the local communities near Maroochydore Beach?

Analyse data

Step 1: I can use geographic terminology to interpret data

2 What are groynes and how are they used to reduce the impacts of longshore drift?

Step 2: I can describe patterns and trends

3 Describe how the mouth of the Maroochy River has changed using the satellite images from Source 1.

Step 3: I can explain the reasons behind a trend or spatial distribution

4 Using the criteria below, provide a detailed description of at least one change identified in Question 3.
   a Relative location of *change*.
   b Name of the new coastal feature formed.
   c Probable cause of the *change*.

Step 4: I can analyse relationships between different data

5 Trace or sketch an outline of the scene in Source 3. Mark in the location of the groynes and describe their interaction with longshore drift of sand.

HOW TO
Sketches and annotating, page 142
Satellite images, page 149

How are coasts used?

In Australia, human activities are concentrated along the coast, where most people choose to live. Tourists are also attracted to Australia's beaches and businesses conduct international trade through major harbours along the coastline.

How do people use coasts?

Australia's coastline attracts heavy use from our population: more than 85 per cent of Australians live within 50 kilometres of the ocean, and more than 25 per cent live within 3 kilometres of the ocean. More and more people are choosing to live near the coast and coastal towns are growing, with new houses, shops, businesses and recreational facilities.

As well as being a popular place for people to live, Australia's coastline is also a major tourist destination. The Gold Coast attracts 13 million tourists each year, mainly from within Australia, in addition to gaining 12 500 new residents yearly. Sydney's Bondi Beach attracts 2.6 million tourists per year, with more than half coming from overseas.

Australia's coast is also important for overseas trade. Sheltered bays are transformed into harbours for large ships to export and import goods from overseas. These harbours are developed near major cities or near resources such as mines.

Source 1

The foreground of this photo shows Sea World, one of many Gold Coast theme parks. In the middle of the image is the Marina Mirage resort and boat marina. The background shows high-rise accommodation and the residential area of Surfers Paradise.

Source 2

Bondi Beach attracts thousands of local people and tourists each summer.

Australian population distribution

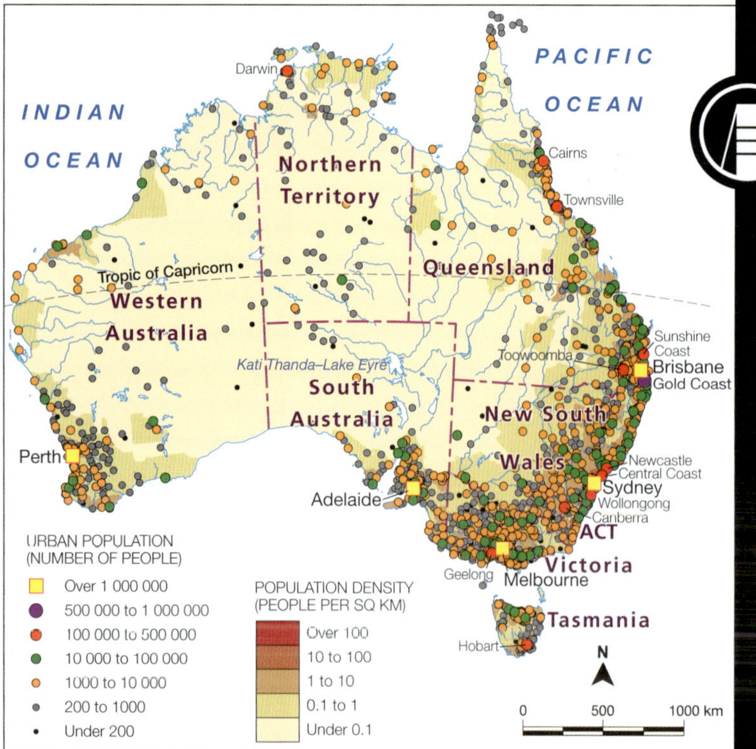

Source 3

This map shows the heavy distribution of Australia's population around the coastlines.

Source: Matilda Education Australia

Learning ladder G2.7

Show what you know

1 Where do most Australians live?

2 Why are sheltered bays important?

Analyse data

Step 1: I can use geographic terminology to interpret data

3 Use the terms foreground and background to describe the locations of tourist attractions on the Gold Coast in Source 1.

Step 2: I can describe patterns and trends

4 Describe, using PQE, the distribution of Australia's population using Source 3.

Step 3: I can explain the reasons behind a trend or spatial distribution

5 Using the results from Question 4, suggest reasons to explain the pattern. Classify your reasons into the SHEEPT categories.

Step 4: I can analyse relationships between different data

6 How are coastlines interconnected with our tourism industry?

HOW TO

PQE, page 138
SHEEPT, page 140

What are the effects of erosion?

Erosion Swallows Homes

Tuesday 7 June 2016 COLLAROY, SYDNEY

Huge waves and a king tide smash the New South Wales coast and swallow exclusive homes in Narrabeen and Collaroy.

Huge waves hit coast

Huge waves have battered the coast of New South Wales over the last four days – the deadly combination of a high spring tide and a violent storm. Narrabeen and Collaroy beaches have lost 50 metres of beach and many waterfront houses are now at risk of falling into the sea.

'The beach is 50 metres narrower now than it was on Saturday afternoon,' said Professor Ian Turner, the director of the Water Research Laboratory at the University of NSW.

'Also, in the area where we surveyed, for every metre along the beach we've seen up to 150 cubic metres of sand stripped off the upper beach, and it is now sitting out there in the surf zone.

'The storm is certainly starting to abate, but there are still very large waves, and still a large tide, which are continuing to cause damage along the beachfront.'

A plane with a laser scanner will be sent out today to survey the damage along the coastline.

'We will be able to very, very carefully detail the damage that has been caused by the storm and will be using that information to be able to predict the damage of future storms,' Professor Turner said.

Emergency response

Waterfront properties along Collaroy beach have been undermined by the large waves. Erosion to the beach has already claimed a swimming pool and the sunroom of one house has already fallen onto the beach. Coastal engineers inspected the damaged homes and the State Emergency Service sandbagged the beach to prevent further damage.

New South Wales Premier, Mike Baird, said, 'This was a ferocious storm that has devastated communities, washing away beaches, flooding inland areas and inundating businesses. Emergency services and volunteers have been doing an incredible job since the storm began last Friday. I want to ensure their efforts are coordinated in a way that allows communities to rebuild and recover as quickly as possible.'

Source: Author article based on combined press reports.

Source 1

Huge waves hit Bondi Beach

Source 2

Houses damaged along Collaroy beach

Learning ladder G2.8

Show what you know

1 What damage occurred to coastal properties in Narrabeen and Collaroy and what was the cause?

Geographic challenge

Step 1: I can identify responses to a geographical challenge

2 Source 1: What action do you think would have been taken by the operators of the pool in the foreground during this incident?

Step 2: I can compare responses to a geographical challenge

3 Source 2: Make a field sketch of the scene and provide labels to indicate the damage using information from the news article. What short-term responses are needed to make the area safe?

Step 3: I can compare strategies for a geographical challenge

4 Why was a plane with a laser scanner used to survey damage?

Step 4: I can evaluate alternatives for a geographical challenge

5 Source 2: What information in the photograph shows an attempt to stop erosion? What other action could be taken to protect the homes and the coastline?

HOW TO

Sketches and annotating, page 142

How do tourists value the coast?

Australia's coastline is the number one attraction for international visitors and it earns Australians more than $40 billion every year in local and international tourism.

Tourism

Tourism is a very important part of the Australian **economy**. It contributes 3.2 per cent of Australia's total wealth, also known as the country's **Gross Domestic Product (GDP)**. Tourism also employs nearly 5 per cent of all Australians. Australia welcomed 9.1 million international visitors in 2018 and they spent $42.5 billion.

Tourism Australia is the government agency responsible for attracting international visitors to Australia. To continue to grow tourism, Tourism Australia undertakes research to understand what tourists enjoy about Australia. This research underpins the planning and **marketing** activities that Tourism Australia undertakes.

Source 1

The 'There's nothing like Australia' marketing campaign puts the focus on exploring Australia's natural offerings, especially on the coast.

WELCOME TO THE WORLD'S MOST SCENIC RUNWAY

THERE'S NOTHING LIKE AUSTRALIA

Source 2

Tourism Australia's research shows that the Australian coastline is very popular with international visitors.

The economic value of the coast

Tourism Australia's research shows that 70 per cent of our international visitors enjoy aquatic and coastal experiences as part of their trip to Australia. The tourists rank Australia as the best place in the world to enjoy marine wildlife and coastal beaches. Approximately seven million international visitors enjoy Australia's coastline each year and they contribute $35 billion to the Australian economy.

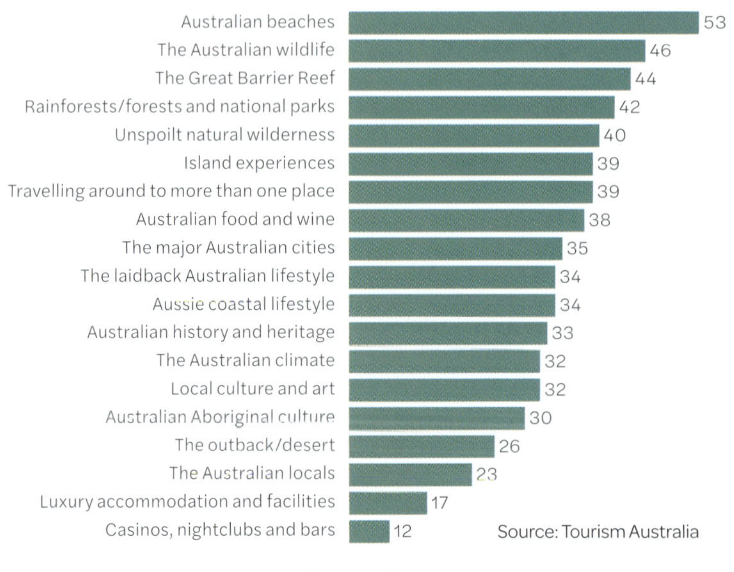

Source 3

What attracts international visitors – Tourism Australia 2020 report

Learning ladder G2.9

Economics and business

Step 1: I can recognise economic information

1 Using quantitative data, describe how tourism contributes to the Australian economy.

Step 2: I can describe economic issues

2 Outline two positive and two negative impacts to the environment and economy if tourism was banned from Australia's coastlines.

Step 3: I can explain issues in economics

3 Discuss the interconnection between Tourism Australia's marketing campaigns and the popularity of coastlines as a tourist destination.

Step 4: I can integrate different economic topics

4 Source 3: Using the data from the Tourism Australia 2020 report, create an advertising campaign that promotes other attractions within Australia, as an alternative to visiting coastlines.

Step 5: I can evaluate alternatives

5 How could we make tourism more sustainable to ensure our coastlines do not suffer for economic gain?

How do people manage coasts?

Natural processes such as longshore drift and events such as storms are constantly changing the shape of our coastlines. Settlements built in coastal areas are affected by these changes.

Managing coastlines

Humans have tried to manage coasts to ensure we have beaches to enjoy, that erosion doesn't threaten communities and that sand doesn't clog harbours. To stop or slow the movement of sand, communities construct walls from concrete, rock, timber and sandbags. Solid sea walls deflect wave energy down, scouring sand away and undermining the wall. Loose rock walls help disperse the energy of the wave.

The different types of walls are as follows:

- **Sea walls** – run parallel to the shore and are designed to stop erosion by waves and storms. They protect the paths, roads and houses built along the shore.
- **Breakwaters** – are rock walls parallel to the beach and located in the water. They are designed to direct the force of the waves at the wall and not at the beach and sand dunes behind it.
- **Groynes** – are walls that extend out from a beach into the sea. They are designed to trap the sand moving along the beach. Groynes also direct waves away from the shore, reducing erosion.
- **Training walls** – are built on either side of a river mouth to stop sand from blocking it.

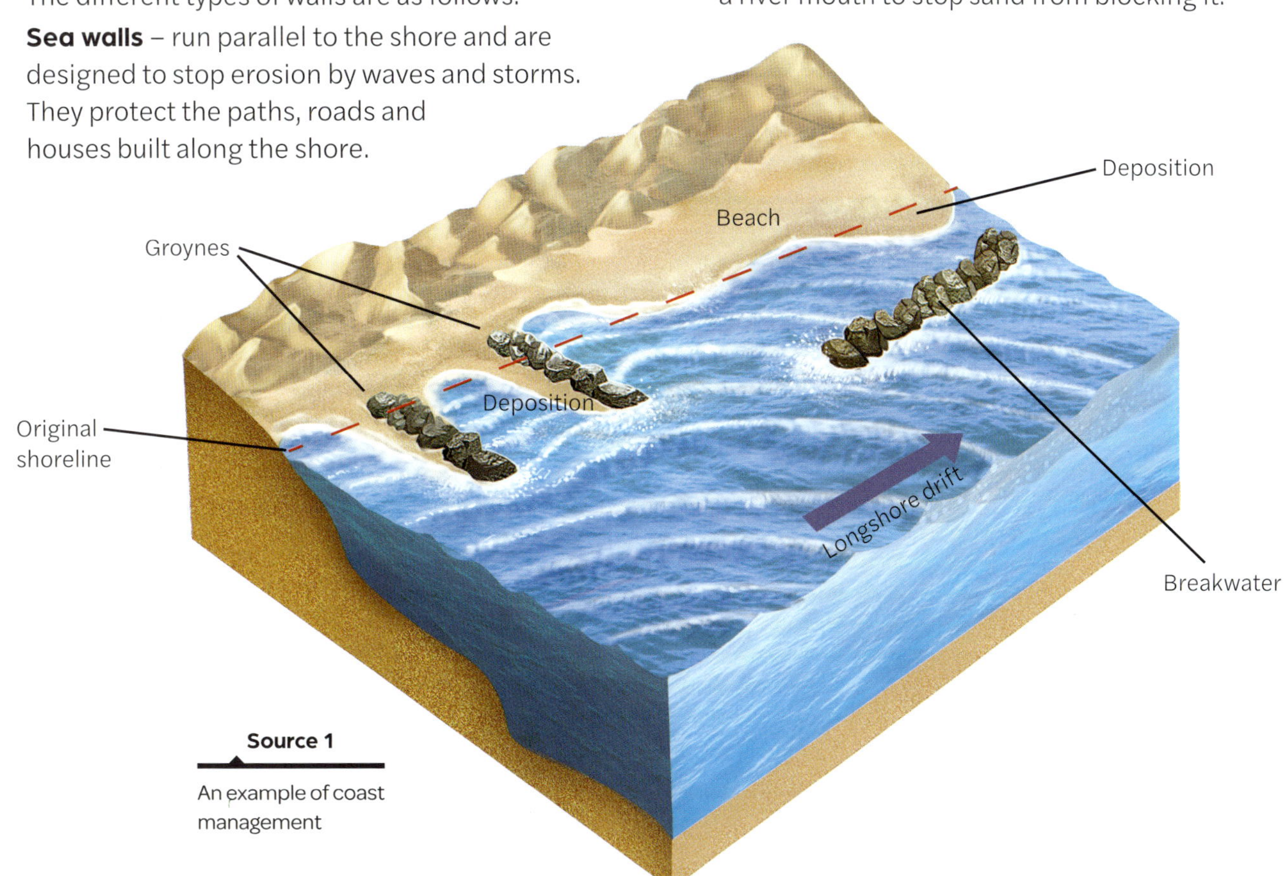

Source 1

An example of coast management

Source 2

A huge dredger pumps sand towards the Surfers Paradise coastline to replenish eroded beaches. Sand is dredged from kilometres offshore and pumped back about 50 metres away from the breaking waves to redistribute to the beaches.

Repairing damaged coastlines

Coastal engineers use a number of methods to repair damaged beaches. They can transport sand by truck or **dredge** it from offshore, and then pump it back just offshore to replenish eroded beaches. This process is known as **beach nourishment**. On the Gold Coast, millions of cubic metres of sand have been dredged from the sea floor or transported from other places to restock eroding beaches.

Sand dunes are stabilised and repaired by erecting fences and signs to keep people off the dunes. Replanting sand dunes with vegetation such as coastal grasses and native plants helps bind the sand together and holds it in place. Sand also mounds up against the plants.

Source 3

The North Cronulla sand dune revegetation program involved fencing and replanting.

Source 4

Plan view of Tweed Heads (NSW) and Coolangatta (Queensland)

Source: State of Queensland 2011

Source 5

The Tweed Heads and Coolangatta coastline

Managing the coast at Coolangatta

Coolangatta is a popular tourist resort on Queensland's Gold Coast, just north of Tweed Heads. After many attempts to restore them, the wide beaches along Coolangatta have returned to the way they were 100 years ago.

Changes made along the coastline on the border between Queensland and New South Wales greatly affected the beaches at Coolangatta. In 1962, the **training walls** on either side of the Tweed River were extended to protect the mouth of the river from a build-up of sand. Unfortunately, **longshore drift** was unable to move sand past the training walls and the Greenmount Beach quickly disappeared.

In 1972, a **groyne** was constructed at Kirra Point at the northern end of Greenmount Beach to help restore it. The beach expanded as sand accumulated against the groyne. **Beach nourishment** in 1974, 1989 and 1995 pumped sand onto the beach.

From 2001, sand has been pumped from south of the Tweed River training walls through pipes to Greenmount and Kirra beaches to match the natural longshore drift of sand that was in place before they were built.

Source 6

Changes in beaches at Coolangatta 1935 to 2004

Learning ladder G2.10

Show what you know

1 Source 6: How is the Coolangatta coastline now similar to the 1935 profile?

2 How has sand been pumped to Greenmount and Kirra beaches?

Geographical challenge

Step 1: I can identify responses to a geographical challenge

3 Source 3: Describe what coastal managers are attempting to do using these constructions.

Step 2: I can compare responses to a geographical challenge

4 Source 1: How are groynes and breakwaters similar and different in managing the coastline?

Step 3: I can compare strategies for a geographical challenge

5 What is the management strategy shown in Source 2? How is this strategy interconnected with the natural process of longshore drift?

Step 4: I can evaluate alternatives for a geographical challenge

6 Create a photo essay illustrating the use of key coastal management strategies in real-life situations. For each photo, describe the management strategy being used and what natural process it is trying to prevent.

HOW TO

Photo essays, page 145

G2.11

thinking locally

How did people destroy and then rescue Hampton Beach?

Hampton is a bayside suburb located 15 kilometres south of Melbourne on Port Phillip Bay, between the suburbs Brighton and Sandringham. Hampton Beach was nearly lost because of the construction of a local harbour.

Longshore drift moves sand northwest towards Green Point in summer and southeast towards Picnic Point in winter, as shown in Source 2. However, the building of Sandringham Harbour trapped sand in the harbour. In summer, the waves couldn't move the sand back to Green Point so, by the 1970s, Hampton Beach was gone. Sand has been pumped back onto the beach and rock and timber **groynes** were built to keep it there. The latest rock groyne was completed in 2012.

Seasonal longshore drift at Hampton Beach, Port Phillip Bay

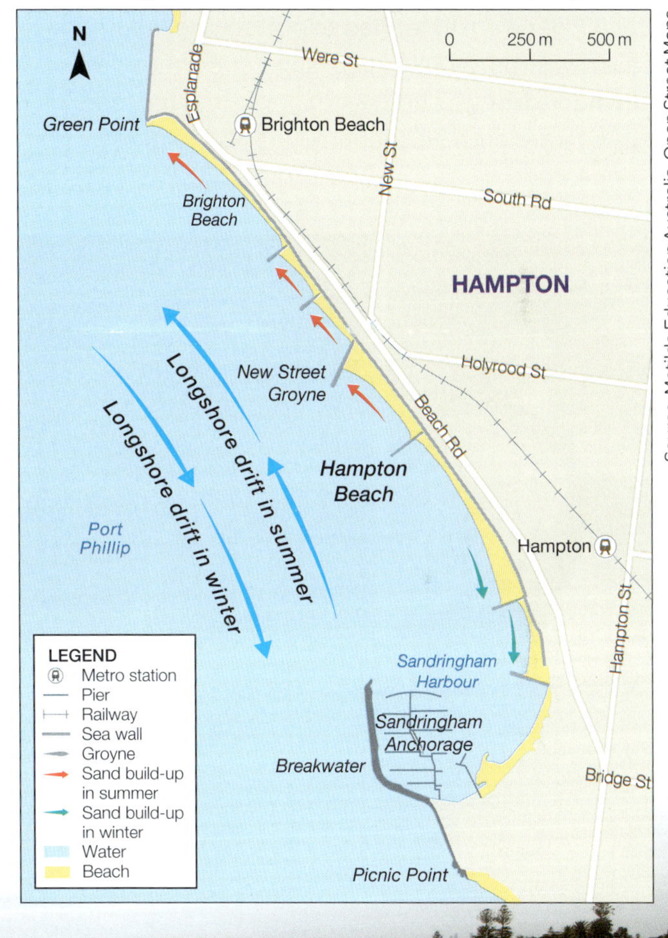

Source: Matilda Education Australia, Open Street Maps

Source 1

Groynes on Hampton Beach help retain the sand and control longshore drift.

Source 2

Longshore drift at Hampton Beach, Port Phillip Bay

Source 3

An early photo of Hampton Beach c. 1900

Source 4

Hampton beach in 2018

Learning ladder G2.11

Show what you know

1 Source 1: Why has the sand built up on only one side of the groyne?

2 Why did the building of Sandringham Harbour cause issues for Hampton Beach?

Interconnections

Step 1: I can identify and describe interconnections

3 Source 1: Describe the interconnection between sand and groynes on Hampton Beach.

Step 2: I can explain interconnections

4 Source 4: Trace or sketch the scene in the satellite image and annotate the following:
   a the various ways humans have changed this place over time.
   b natural characteristics and features of the coastal landscape.
   c the coastal management strategies that have been implemented.

Step 3: I can identify and explain the implications of interconnections

5 The direction of longshore drift at Hampton Beach depends on the season. Using your knowledge of this natural process, identify whether the aerial image in Source 4 was taken in summer or winter. Justify your response.

Step 4: I can evaluate the implications of significant interconnections

6 Using data from the map and images on this spread, explain why the marina would have a large impact on the movement of sand along the Hampton Beach coastline.

HOW TO

Sketches and annotating, page 142
Satellite images, page 149

How do tsunamis impact coastlines?

Tsunamis are **geohazards** that originate at or near Earth's surface, when water is pushed up by undersea earthquakes and volcanoes, as discussed in Chapter 1. They travel at speeds of up to 800 kilometres per hour and cause extensive damage when they crash onto coastlines.

How tsunamis work

When there is an undersea **earthquake** or volcanic eruption, the upward movement of the sea bed moves the water above it, creating a fast-moving **tsunami**. Tsunamis can cross the entire Pacific Ocean in less than a day.

The long wavelengths of tsunamis lose very little energy as they travel away from the focus of the earthquake, which is the location where the earthquake begins. In the deep ocean, tsunami waves may only be about 30 centimetres high. As the waves reach shallower water near the coast, they quickly slow down, and grow bigger. Tsunamis can grow up to 30 metres high.

The trough between the tsunami wave crests often reaches the shore first. This acts like a vacuum, sucking water in the shallows out to sea, exposing the sea bed. The retreating sea water is an important warning sign that a tsunami wave is coming.

Tsunamis hit the shore as a huge wave or a surging tide of water. They cause incredible damage to coastlines and in low-lying coastal areas can flow many kilometres inland, sweeping everything away.

Source 1

Formation of a tsunami

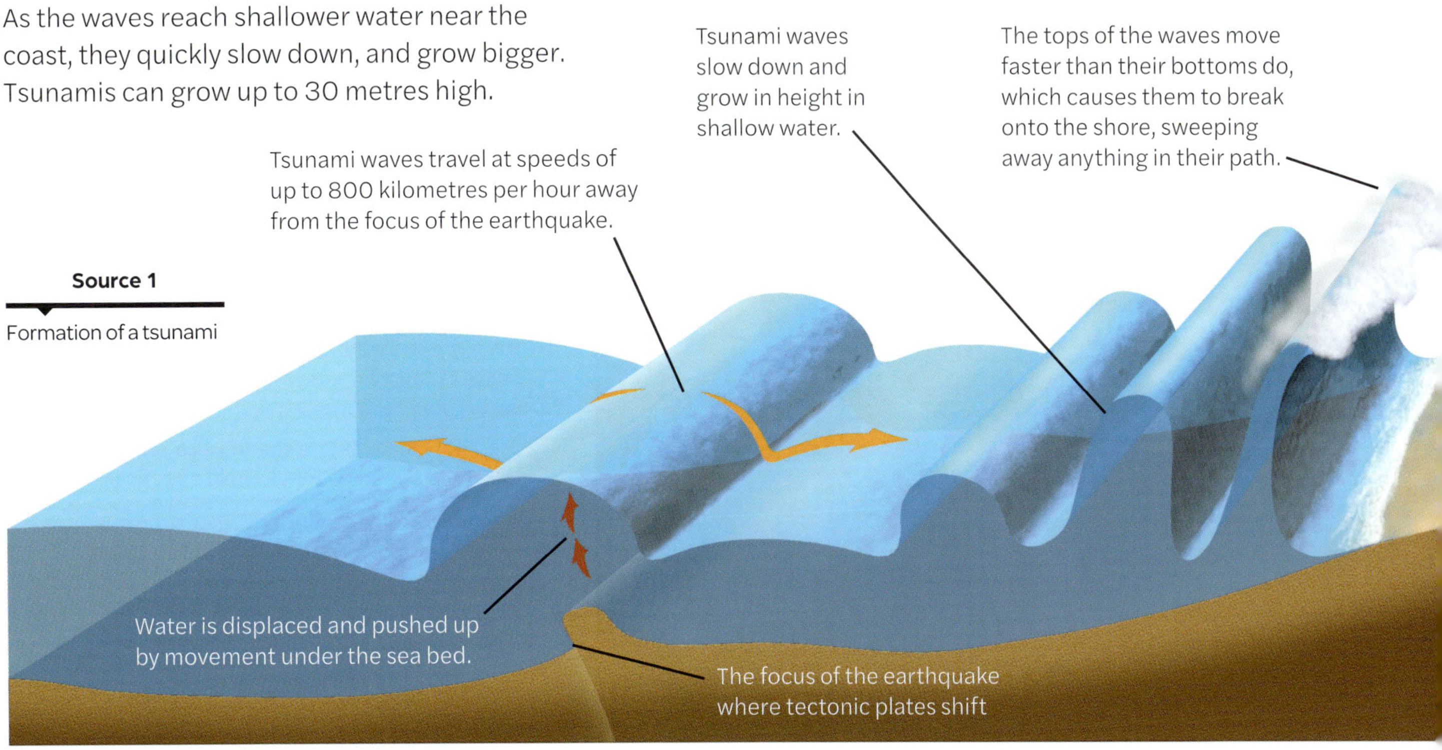

Source 2

The city of Miyako on the north-east coast of Japan's Honshu island was devastated by the 2011 tsunami. The 13-metre-high tsunami swamped the city's sea wall and tsunami shelters. Over 600 people were killed and 6000 buildings were destroyed in that city.

Japan Tsunami 2011

In 2011, a massive undersea earthquake, measuring 9 on the Richter scale, occurred about 80 kilometres off the east coast of Japan. Less than one hour later, tsunami waves up to 13 metres high crashed onto Japan's east coast.

The water flowed over the **sea walls** built to protect the coastline and washed away buildings, ships, cars, roads and people. More than 16 000 people were killed and 300 000 buildings were completely destroyed.

Learning ladder G2.12

Show what you know

1 What is a tsunami and how do they occur?

2 Why is Japan vulnerable to tsunamis? How does this vulnerability affect Japan's coastlines?

Collect, record and display data

Step 1: I can collect, record and display data in simple forms

3 With a partner, research five tsunamis. For each tsunami, summarise the following:
   a location and date
   b height of waves
   c impact to the coastline of the region
   d impact to human population of the region.

Step 2: I can recognise and use different types of data

4 Source 2: Give examples of damage caused by tsunamis from this oblique aerial image.

Step 3: I can choose, collect and display appropriate data

5 Consider the websites that you used to complete the research in Question 3. Were they reliable or appropriate for this research?

Step 4: I can use data to support claims

6 Explore http://mea.digital/Blyn
   a How does this data show Japan's tsunami vulnerability at any given time?
   b How could websites or data such as this be used to plan and prepare the coastal environment for a disaster event?

What was the impact of the Palu earthquake and tsunami?

On 28 September 2018, a large magnitude 7.5 earthquake occurred 77 kilometres from Palu in Indonesia. The earthquake caused major **soil liquefaction**, where the soil turned to mud, submerging buildings. It also created a tsunami.

What impact did the tsunami have on Palu?

The **tsunami** produced waves up to 12 metres high that struck Palu and other coastal settlements. It destroyed around 70 000 homes and pushed debris inland, damaging bridges and roads along its path. The combined impact of the earthquake and tsunami led to the deaths of at least 2000 people.

More than 200 000 people were made homeless by the natural disaster and relief camps were set up to give them shelter, supply food and clean water, and to treat injuries and infections.

Source 1

Disaster management map of Palu, October 2018

Source: European Commission, ERCC, Matilda Education, Copernicus EMS

Source 2

This ship was transported inland by the tsunami that struck Indonesia in 2018.

Source 3

An oblique aerial view showing the devastated city of Palu in Indonesia following an earthquake and tsunami in September 2018

Learning ladder G2.13

Show what you know

1 What was the impact of the tsunami on Palu?

2 What is soil liquefaction?

Geographical challenge

Step 1: I can identify responses to a geographical challenge

3 Source 1: International aid agencies needed to get to Palu. Describe the relative location of Palu in Indonesia using the inset map provided.

Step 2: I can compare responses to a geographical challenge

4 Source 1: Once aid arrived to Palu, which areas did they need to get to and why?

Step 3: I can compare strategies for a geographical challenge

5 Source 1: Imagine you are coordinating an aid team on the ground in Palu. List the different jobs needed to respond to the damage and threats.

Step 4: I can evaluate alternatives for a geographical challenge

6 Imagine your class is responsible for providing resources and shelter to people affected by the disaster in Palu. Based on the damage and outcomes of the tsunami, create an action plan to help them recover from this event.

Masterclass

Learning Ladder

Coastal population and shoreline degradation

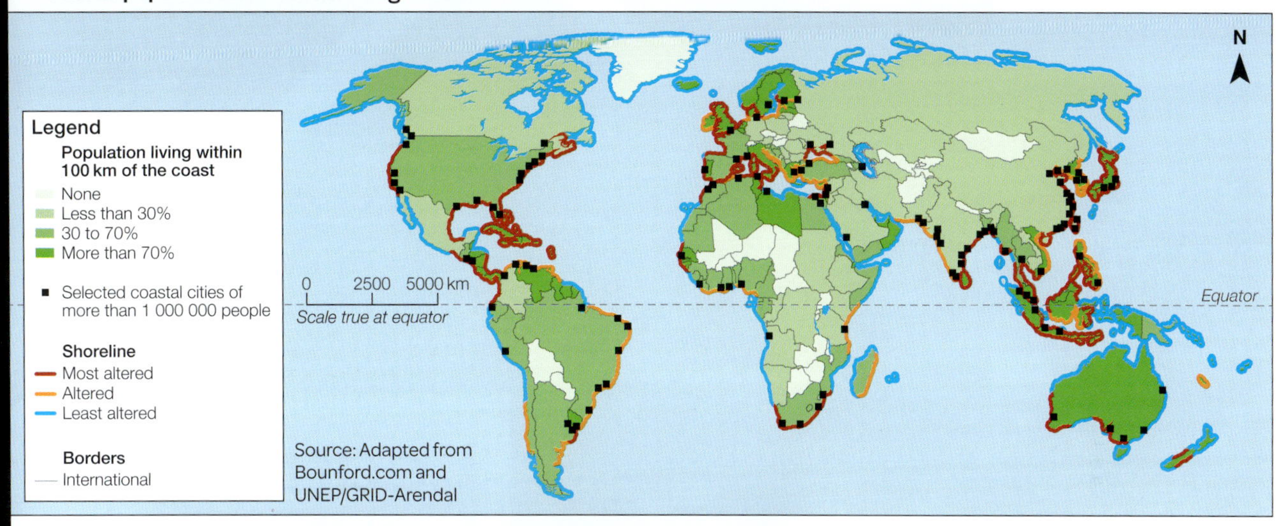

Source: Adapted from Bounford.com and UNEP/GRID-Arendal

Source: State of Queensland 2011

Source 1

Percentage of inhabitants that live along the coast

Source 2

Plan aerial view of Tweed Heads (NSW) and Coolangatta (Queensland)

Source 3

Human management of the coastline at Brighton in England.

Work at the level that is right for you or level-up for a learning challenge!

Step 1

a I can identify and describe spatial characteristics

Describe the place photographed in Source 2.

b I can identify and describe interconnections

Identify one key interconnection in Source 1.

c I can identify responses to a geographical challenge

Source 2: Why have groynes been installed on this beach?

d I can collect, record and display data in simple forms

Source 1: Identify what the black boxes represent on the map.

e I can use geographic terminology to interpret data

Source 1: Describe which sections of Australia's coastline have been most altered by people.

Step 2

a I can explain spatial characteristics

Source 2: How successful have the groynes in the north of the image been in keeping sand on the beaches?

b I can explain interconnections

Source 2: Why is sand been pumped from south of the Tweed River training walls through pipes to the beaches in the north of the image?

c I can compare responses to a geographical challenge

Source 2: Sketch this area as it might look if only the training walls at the mouth of the river were in place and no groynes were used on beaches.

d I can recognise and use different types of data

Sources 2 and 3 are both aerial views of the coast. Which is more like a map and why?

e I can describe patterns and trends

Refer to Source 1. Identify one key pattern that is evident on this map.

Step 3

a I can explain processes influencing places

Refer to Source 2. In which direction are the waves travelling as they hit the beach in the north of the image? Explain how this will impact on longshore drift in this area.

b I can identify and explain the implications of interconnections

Identify landforms mainly affected by an interconnection with the wind. Discuss why this interconnection is important.

c I can compare strategies for a geographical challenge

Why were training walls built at the mouth of the Tweed River and how else could this objective have been achieved (page 76)?

d I can choose, collect and display appropriate data

Imagine you were trying to collect data on the movement of sediment along the beach in Source 2. What data collection methods would be appropriate for this investigation?

e I can explain the reasons behind a trend or spatial distribution

Refer to Source 1. Explain, using SHEEPT, why so many people live along coastal regions.

Step 4

a I can predict changes in the characteristics of places

Source 2: Predict what would happen to Greenmount Beach (the large beach in the north-east of the photo) if it were to lose the groyne at the western end of the beach. Justify your response.

b I can evaluate the implications of significant interconnections

Describe the interconnection between wind and coastal vulnerability to erosion in Source 3.

Masterclass

c I can evaluate alternatives for a geographical challenge

Source 2: Compare the roles of the training walls and groynes in the image. What similarities and differences are there?

d I can use data to support claims

Referring to Source 3, discuss the conclusion 'This place has clearly been affected by longshore drift'.

Do you agree with this statement? Use evidence from the source to justify your answer.

e I can analyse relationships between different data

Compare Sources 1 and 3 on pages 82–83. Describe how useful each would be to aid workers before they arrive in Palu.

Step 5

a I can analyse the impact of change on places

Prepare a table to analyse the short and long-term impact on Palu following the 2018 earthquake and tsunami (pages 82–83).

b I can explore spatial association and interconnections

'A coastline is an environment that is highly interconnected with both land and marine processes.' Discuss.

c I can plan action to tackle a geographical challenge

A geographer conducting fieldwork on longshore drift completes a field sketch as their only source of data. Evaluate this method's likely success in helping understand longshore drift in this location. Suggest other methods that may be useful to the researcher.

d I can evaluate data

Justify whether you think Source 1 is a reliable data source given that populations are constantly changing and moving.

e I can draw conclusions by analysing collected data

Consider the coastal management strategies evident in Source 2. From the image, do you think they are successful? Suggest other management techniques that may also help preserve the coastline.

Capstone

How can I understand coastal landscapes?

In this chapter, you have learnt a lot about coastal landscapes. Now you can put your new knowledge and understanding together for the capstone project to show what you know and what you think.

In the world of building, a capstone is an element that finishes off an arch or tops off a building or wall. That is what the capstone project will offer you, too: a chance to top off and bring together your learning in interesting, critical and creative ways. You can complete this project yourself, or your teacher can make it a class task or a homework task.

mea.digital/GHV8_G2

Scan this QR code to find the capstone project online.

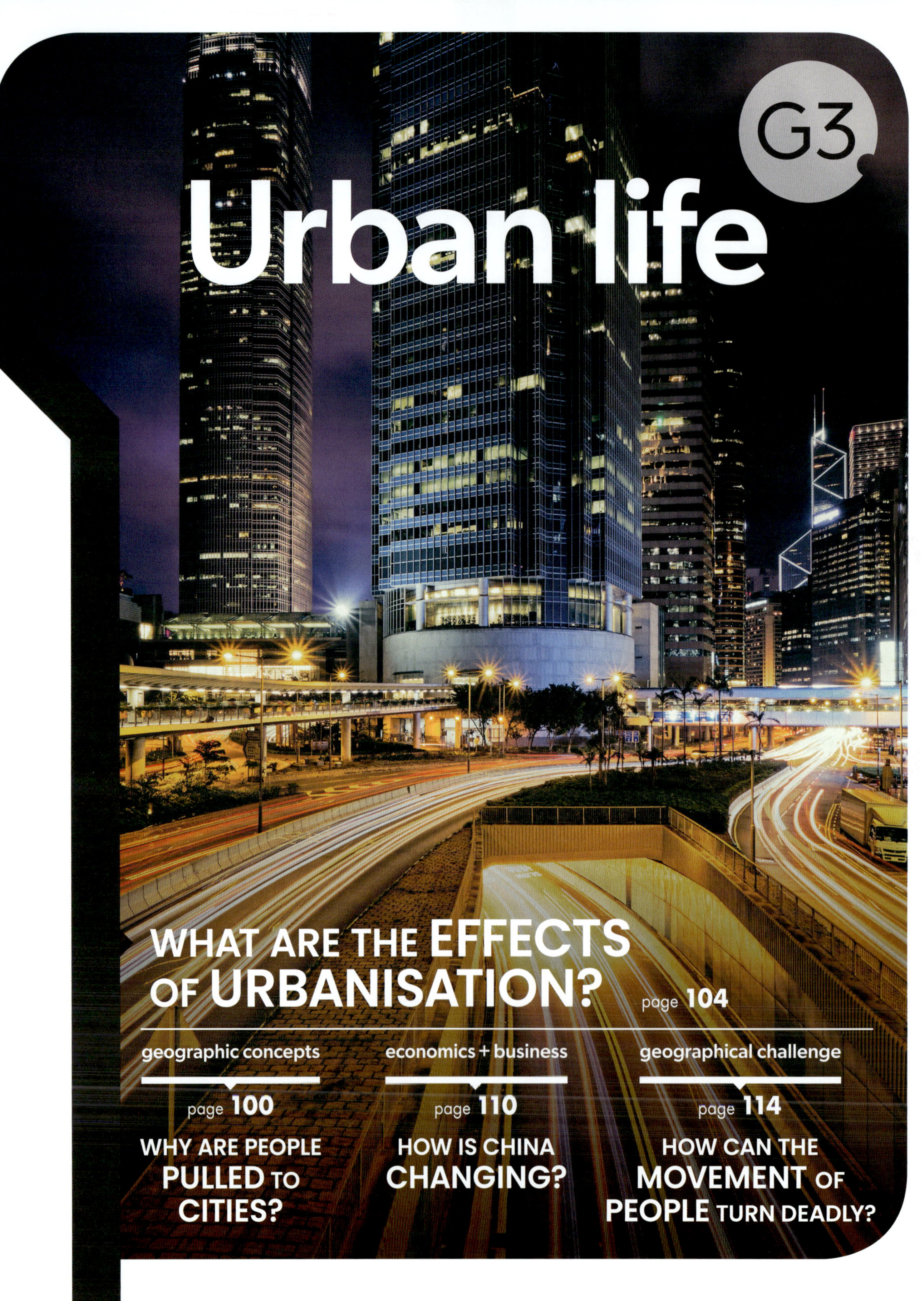
G3
Urban life
WHAT ARE THE EFFECTS OF URBANISATION?
page 104
geographic concepts
page 100
WHY ARE PEOPLE PULLED TO CITIES?
economics + business
page 110
HOW IS CHINA CHANGING?
geographical challenge
page 114
HOW CAN THE MOVEMENT OF PEOPLE TURN DEADLY?

How can I learn about urban life?

More than half the world's population now lives in urban areas, as large numbers of people move from rural areas to cities seeking job opportunities, higher pay, and better access to health care and education. However, urbanisation negatively impacts the environment through higher carbon emissions, deforestation and reduced farming areas, as well as straining existing infrastructure and services. Urbanisation leads to the growth of megacities, or to many cities spreading into one to form a conurbation.

learning ladder

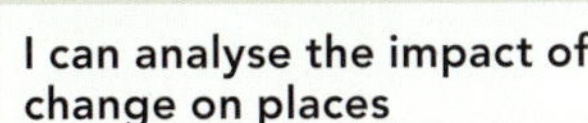

Step	Spatial characteristics	Interconnections	Geographical challenge
step 5	**I can analyse the impact of change on places** I can analyse and evaluate the implications of growth or decline of places and calculate the impact on people and environments.	**I can explore spatial association and interconnections** I can compare distribution patterns and the interconnections between them such as the growth of conurbations.	**I can plan action to tackle a geographical challenge** I can frame questions, evaluate findings, plan actions and predict outcomes to tackle a population-based geographical challenge.
step 4	**I can predict changes in the characteristics of places** I can predict changes in the characteristics of places over time due to urbanisation.	**I can evaluate the implications of significant interconnections** I can identify, analyse and explain key interconnections within and between places, and evaluate their implications over time.	**I can evaluate alternatives for a geographical challenge** I can weigh up alternative views and strategies on a population-based geographical challenge using environmental, social and economic criteria.
step 3	**I can explain processes influencing places** I can explain the series of actions leading to change in a place, such as rural-urban migration.	**I can identify and explain the implications of interconnections** I can identify, analyse and explain population-based interconnections and explain their implications.	**I can compare strategies for a geographical challenge** I can compare strategies for a geographical challenge, taking into account a range of factors and predicting the likely outcomes.
step 2	**I can explain spatial characteristics** I can identify concepts of Space, Place, Interconnection, Change, Environment, Scale and Sustainability (SPICESS) when I read about changing nations.	**I can explain interconnections** I can describe and explain interconnections and their effects, such as the growth of cities into one another to form conurbations.	**I can compare responses to a geographical challenge** I can identify and compare responses to a geographical challenge and describe its impact on different groups.
step 1	**I can identify and describe spatial characteristics** I can talk about spatial characteristics at a range of scales; e.g. urbanisation in the USA.	**I can identify and describe interconnections** I can identify and explain simple interconnections such as poorer people in rural areas being attracted to opportunities in large cities.	**I can identify responses to a geographical challenge** I can find responses to a geographical challenge such as urbanisation and understand the expected effects.

I can evaluate data
I can determine whether data presented about human settlements is reliable and assess whether the methods I used in the field or classroom were helpful in answering a population-based research question.

I can draw conclusions by analysing collected data
I can summarise findings and use collected data to support key patterns and trends I have identified for population-based research.

I can use data to support claims
I can select or collect the most appropriate data and create specialist maps and information using ICT to support investigations into urban settlements.

I can analyse relationships between different data
I can use multiple data sources, overlays and GIS to find relationships that exist in patterns of urbanisation on Earth.

I can choose, collect and display appropriate data
I can select useful sources of population data and represent them to conform with geographic conventions.

I can explain the reasons behind a trend or spatial distribution
I can identify Social, Historical, Economic, Environmental, Political and Technological (SHEEPT) factors to help me explain patterns in data.

I can recognise and use different types of data
I can define the terms primary, secondary, qualitative and quantitative data, and represent data in more complex forms.

I can describe patterns and trends
I can identify Patterns, Quantify them and point out Exceptions (PQE) to describe the patterns I see.

I can collect, record and display data in simple forms
I can identify that maps and graphs use symbols, colours and other graphics to represent data.

I can use geographic terminology to interpret data
I can identify increases, decreases or other key trends on a map, graph or chart about urban settlements.

Collect, record and display data

Analyse data

Spatial characteristics

1 Looking at Source 1, what geographic characteristics can you identify? Describe how they may have occurred using the SPICESS terms of space and place.

Interconnections

2 Source 1: Why do large cities such as London attract so many people?

Geographical challenge

3 What challenges do large cities present for planners and governments? Give three examples.

Collect, record and display data

4 Conduct a quick survey of your class, asking them the following question: Would you rather live in an area that was:
a urbanised?
b rural?
Record your data using a tally.

Analyse data

5 Look carefully at the graph in Source 3 on page 93. Is Australia an urbanised nation? Suggest reasons to support your answer.

G3.1

How do nations change?

Nations constantly change and develop as the population grows and technology improves. As the world population increases, we are seeing an increase in the urbanisation of environments. Urbanisation can be positive, as cities provide people with better access to employment, health care, education and higher living standards, but urbanisation also has negative impacts on people and place through higher carbon emissions, more competition for resources due to population density and a reduction in open space for recreation and farming.

Source 1

Changes in Dubai's urban landscape from 1990 to 2003 – look for the five square white buildings that appear in both photos to see the scale of the change.

1990

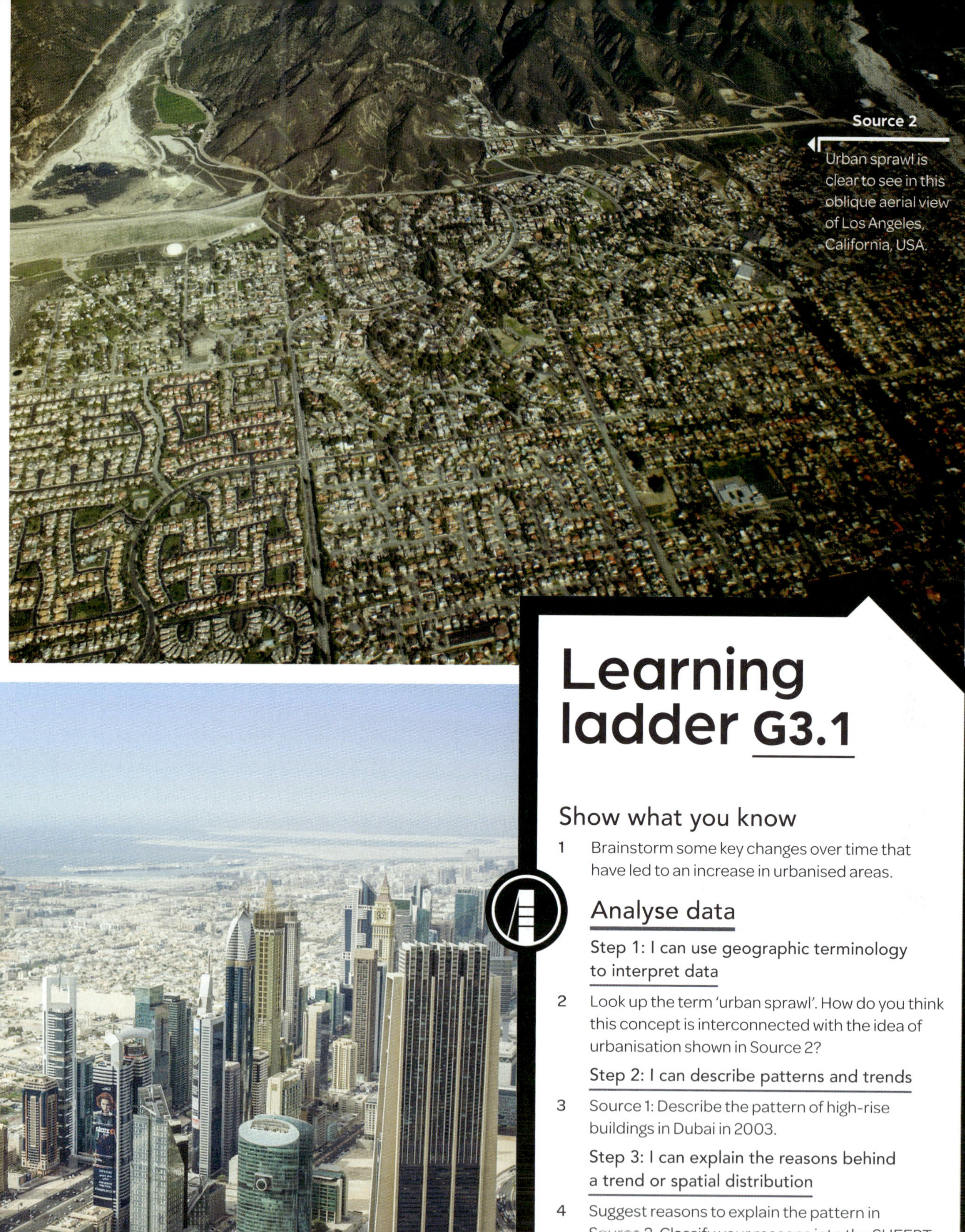

Source 2

Urban sprawl is clear to see in this oblique aerial view of Los Angeles, California, USA.

Learning ladder G3.1

Show what you know

1 Brainstorm some key changes over time that have led to an increase in urbanised areas.

Analyse data

Step 1: I can use geographic terminology to interpret data

2 Look up the term 'urban sprawl'. How do you think this concept is interconnected with the idea of urbanisation shown in Source 2?

Step 2: I can describe patterns and trends

3 Source 1: Describe the pattern of high-rise buildings in Dubai in 2003.

Step 3: I can explain the reasons behind a trend or spatial distribution

4 Suggest reasons to explain the pattern in Source 2. Classify your reasons into the SHEEPT categories.

Step 4: I can analyse relationships between different data

5 Identify the buildings from 1990 in the photo of Dubai in 2003. Describe what has changed and suggest what the scene may look like today.

HOW TO

SHEEPT, page 140

Where are the world's urbanised areas?

More than half of the world's population now lives in urban areas, mainly along coastal regions in the northern hemisphere. Urbanisation has seen cities in Asia grow rapidly.

What is an urban area?

An **urban** area is a densely populated human settlement. In Australia, urban areas are defined as centres of 1000 or more people, with a **population density** of at least 200 people per square kilometre. Urban areas such as cities and towns contain commercial and residential buildings, and **infrastructure** such as transport, power, water and sewage treatment.

Urbanisation refers to the growth of urban areas as more people move from **rural** areas to cities. Projections show 68 per cent of all people will live in urban areas by 2050.

Where are urban areas?

The world's urban areas are generally located close to coastlines and transport routes. They are found in areas of high rainfall and on flat land. Desert, mountain, forest and polar environments have some of the lowest population densities on Earth.

In addition, the **northern hemisphere** contains most of the world's largest urban areas and cities. Europe and the USA are two regions that have been highly urban for a long time and that feature **more economically developed countries (MEDC)** of the world.

Source 1

Gardens by the Bay in Singapore. Singapore is one of 11 countries in the world that are 100 per cent urbanised. With only about 700 square kilometres of land, Singapore has increased its land area by 23 per cent by reclaiming land from the sea and has relied on high-rise apartments to house its population.

Source 2

This composite satellite image shows the Earth at night. It highlights the city lights of the world's urban areas. The northern hemisphere is far more urbanised than the southern hemisphere.

The Asian countries of China, India, Indonesia and Singapore are examples of countries where urbanisation has been recent and rapid. The largest cities in the **southern hemisphere** are found along the coasts of South America.

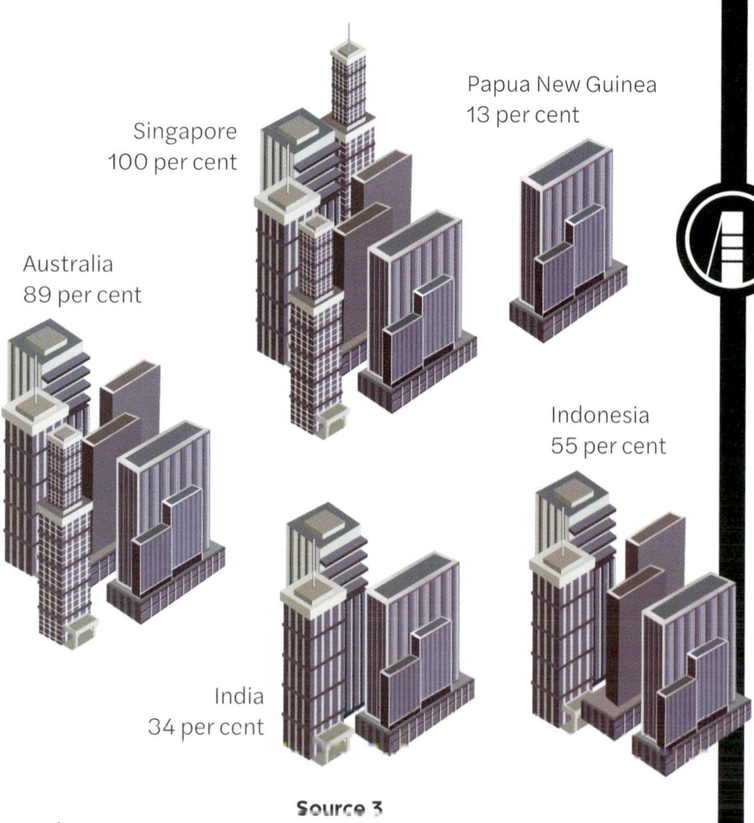

Source 3

In 1950, only 30 percent of the world's population lived in urban areas. Urbanisation occurred mainly in developed countries such as Australia. Today 55 per cent of people live in urban areas; however, some countries, such as Papua New Guinea, remain largely rural.

Learning ladder G3.2

Show what you know

1 Define the concept of an 'urban area' using at least four pieces of quantitative data.

Spatial characteristics

Step 1: I can identify and describe spatial characteristics

2 Source 2: Which hemisphere is the most urbanised? How can you tell from this image?

Step 2: I can explain spatial characteristics

3 Source 3: What impacts do you think urbanisation has had on the environment in Singapore?

Step 3: I can explain processes influencing places

4 As a class, discuss why most urbanised areas are located along coastlines. You may wish to use the SHEEPT factors to broaden your thinking.

Step 4: I can predict changes in the characteristics of places

5 Conduct some research and create a list of the top 10 fastest-growing cities on Earth. Discuss why these places are growing so quickly.

SHEEPT, page 140

How urbanised are Australia and the USA?

Highly urbanised and wealthy countries such as the USA and Australia have many similarities. Large cities develop along the coastline and continue to expand through a process known as urban sprawl.

Urbanisation

Although the USA has a population 13 times the size of Australia, the two countries have many similar urban patterns. Both countries are highly urbanised. In Australia, 89 per cent of the population lives in urban areas. In the USA, 82 per cent of people live in urban areas.

Urban sprawl

Australian and American cities are both characterised by **urban sprawl**. Land in city centres is expensive, so high-density offices and apartments are built for businesses and residents. Surrounding the city centre lies a broad area of medium-density residential housing called **suburbs**.

Settlement patterns

The populations of the USA and Australia are concentrated in large cities and towns along the east, west and south-east coasts. Both the USA and Australia have large areas of dry inland desert where virtually no-one lives. In the USA, the large desert city of Las Vegas is an exception to the pattern. The resort town has grown by 400 per cent since 1980 and now has a population of 640 000.

Australia's urban areas

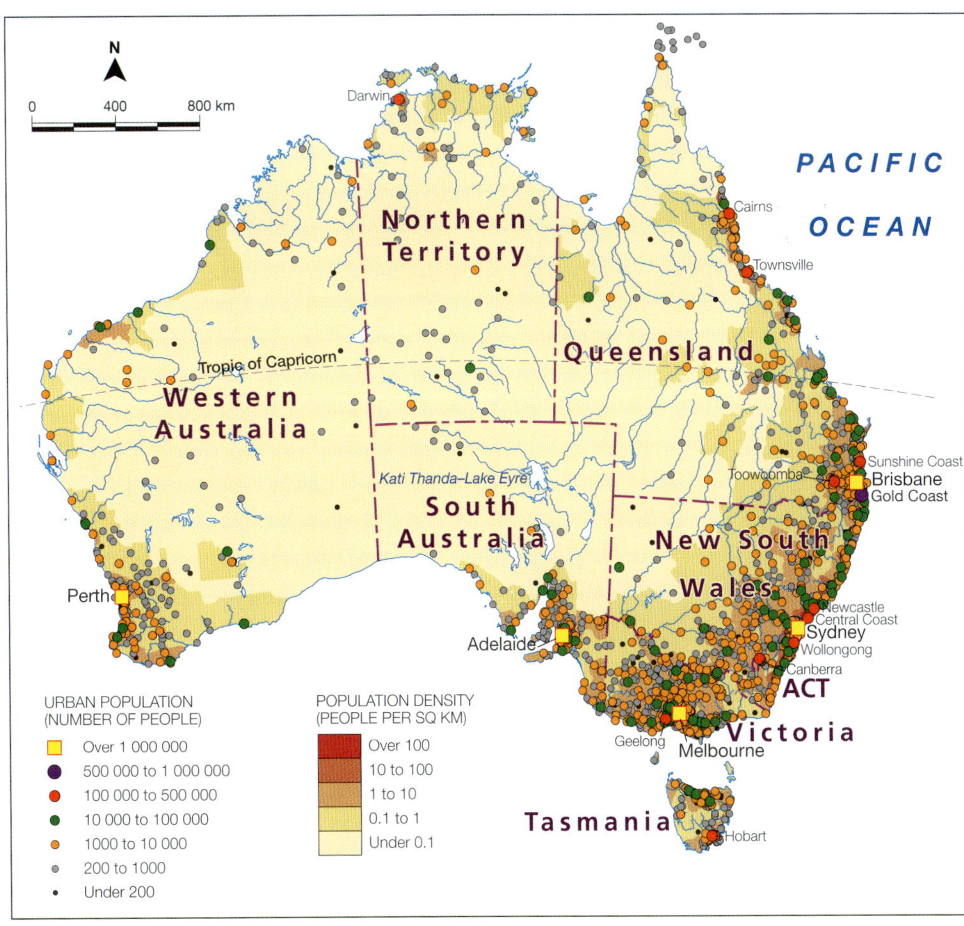

Source: Matilda Education Australia

Source 1

This map shows Australia's most densely populated urban areas

The USA's urban areas

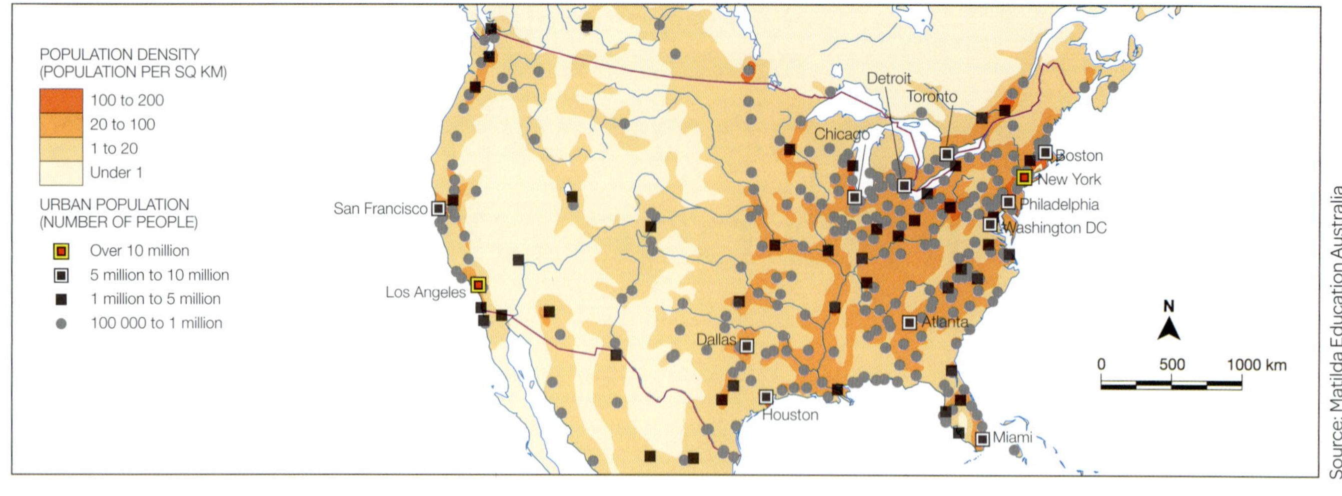

Source 2

This map shows the USA's most densely populated urban areas

Commuter towns are separated from suburbs by farmland and open space, but connected by roads and rail. People in these towns commute to the city for work. Over time, suburbs and commuter towns join together, increasing the urban sprawl.

Sometimes whole cities merge together into a continuous urban area known as a **conurbation**. The South East Queensland conurbation stretches for 200 kilometres joining the Sunshine Coast, Brisbane and the Gold Coast. In the north-eastern USA, the Boswash conurbation (see page 96) joins a number of major cities. More than 52 million people live in this conurbation – 16 per cent of the entire population of the USA.

Source 3

American and Australian cities are characterised by sprawling suburbs like Henderson in the Las Vegas metropolitan area. Las Vegas is an exception to the settlement patterns of the USA and Australia, where the urban areas are generally concentrated on the coast. Las Vegas is a large city in the middle of the American desert.

Learning ladder G3.3

Show what you know

1 Define urban sprawl using examples from Australia and the USA.

Spatial characteristics

Step 1: I can identify and describe spatial characteristics

2 Identify the most urbanised regions in the USA.

Step 2: I can explain spatial characteristics

3 Using Sources 1 and 2, discuss any similarities or differences between the location of urbanised areas in Australia and the USA.

Step 3: I can explain processes influencing places

4 Draw a sketch map of a place (either real or imaginary) that has urban sprawl. Include and annotate your sketch using the following key ideas:

a urban sprawl
b commuter town
c farmland
d open space
e roads
f rail
g suburbs
h city.

Step 4: I can predict changes in the characteristics of places

5 What are commuter towns and what is likely to happen to them in the future as urban areas expand?

G3.4

thinking globally

What happens when cities grow into one another?

Boswash population density

Manchester
Boston
Lowell
Worcester
Providence
Springfield
Hartford
Waterbury
New Haven
Bridgeport
Stamford
Newark
New York
Trenton
Philadelphia
Allentown
Reading
Wilmington
Harrisburg
Baltimore
Washington
Arlington
Alexandria

The USA includes the megacities of New York City and Los Angeles. As the cities around New York have expanded, they have joined together in what is known as a **conurbation**. Ranging from Boston to Washington on America's north-east coast, a number of cities have spread into one another to form one large conurbation referred to as Boswash or Bosnywash. One in six Americans live there.

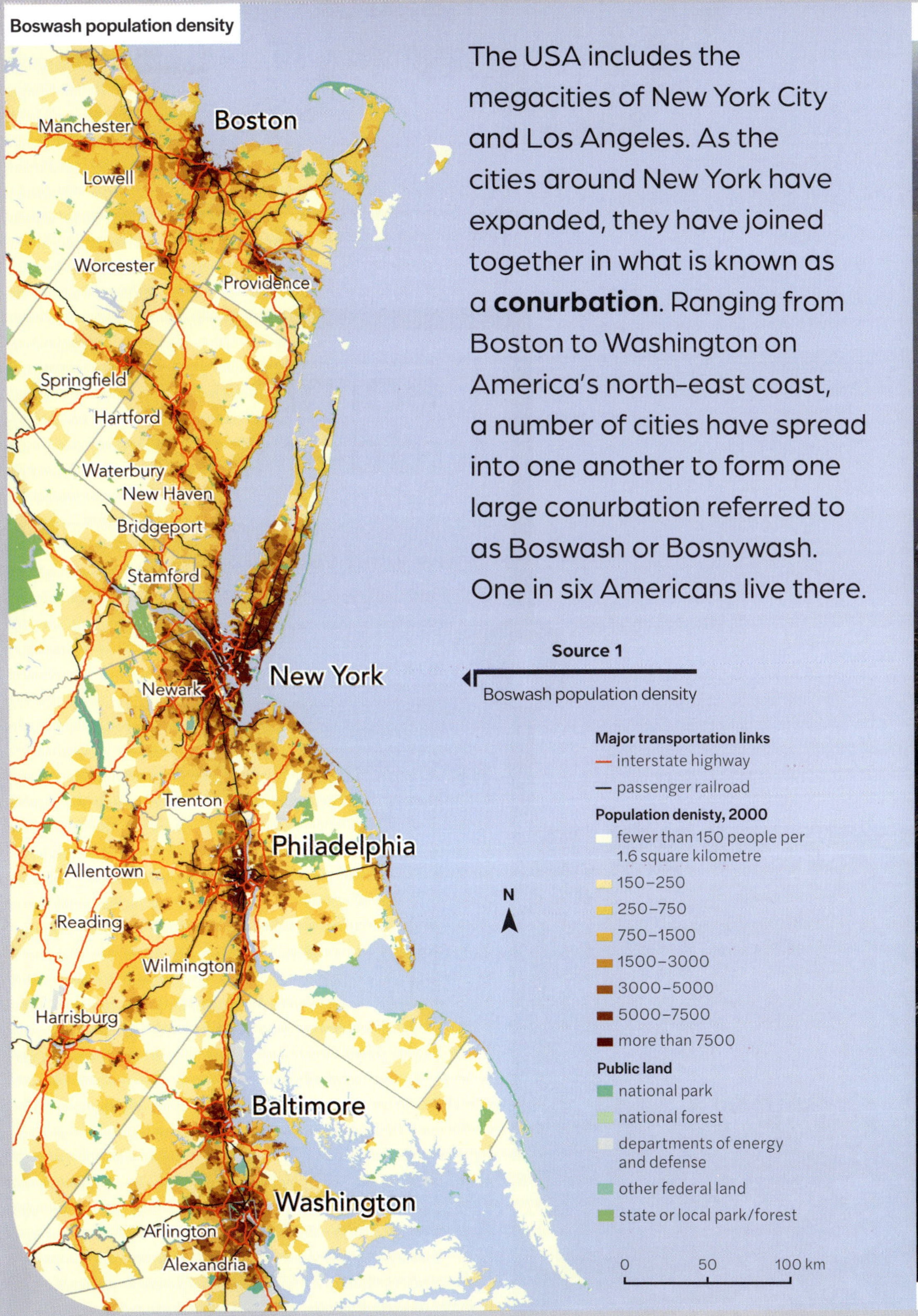

Source 1

Boswash population density

Major transportation links
- interstate highway
- passenger railroad

Population denisty, 2000
- fewer than 150 people per 1.6 square kilometre
- 150–250
- 250–750
- 750–1500
- 1500–3000
- 3000–5000
- 5000–7500
- more than 7500

Public land
- national park
- national forest
- departments of energy and defense
- other federal land
- state or local park/forest

N

0 50 100 km

Source: Radical Cartography

Growth of urban area for Boswash: 1974–2012

Source 2

Growth of urban area for Boswash

Forest area change 2000–2019 for Boswash area

Vermont
New Hampshire
Syracuse
New York
Albany
Massachusetts
Boston
Springfield
Hartford
Connecticut
Scranton
New Haven
Pennsylvania
New Jersey
New York City
Rhode Island
Harrisburg
Philadelphia
ATLANTIC OCEAN
Maryland
Baltimore
WASHINGTON, DC
Delaware
Virginia
Richmond
0 100 200 km
Legend
Major cities
Areas of forest loss between 2000–2019

Source: Matilda Education Australia, USGS, Hansen/UMD

Source 3

Boswash forest area change 2000–2019

Source 4

With a population of 22 million people, New York City is the largest and most densely populated city in the USA, with 72 000 people per square kilometre. The city is home to over 6500 high rise buildings, including the tallest building in the USA, the 541-metre One World Trade Center. The tower was built on the site of the former World Trade Center that was destroyed in the terrorist attacks of 11 September 2001.

Learning ladder G3.4

Show what you know

1 Define the term conurbation and provide an example.

2 Source 3: When did the greatest growth between New York and Philadelphia occur?

Interconnections

Step 1: I can identify and describe interconnections

3 Source 1: Which American cities are interconnected as part of the Boswash conurbation?

Step 2: I can explain interconnections

4 Explain how conurbations develop using the term interconnection.

Step 3: I can identify and explain the implications of interconnections

5 Source 4: Why is it difficult for New York to expand further? How has the city responded to this problem?

Step 4: I can evaluate the implications of significant interconnections

6 Compare Source 1 and Source 4. As a class, discuss how the urbanisation of Boswash has impacted the region's natural environment. Do you believe that urbanisation should be prioritised over environmental sustainability? As a class, justify your answer with data.

How are populations changing?

Rural to urban migration has seen large numbers of people move from farming areas to cities. Today, 54 per cent of all people in the world live in urban areas.

The lure of the city

The promise of job opportunities and better services pulls people from **rural** areas to cities. In 1900, just 16 per cent of people lived in **urban** areas. By 2050, this percentage is likely to grow to two-thirds of the population, as more African and Asian rural dwellers move to cities.

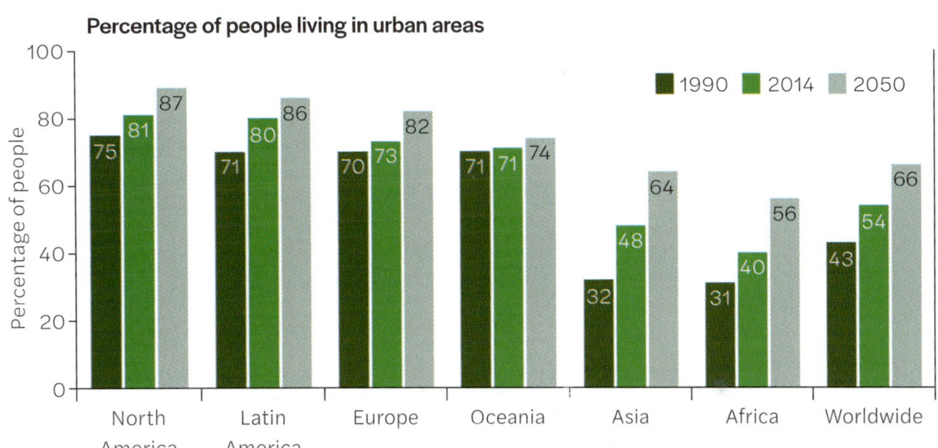

Source 1

Seoul was the destination for most rural migrants in South Korea. Between 1960 and 1970 many farmers left for opportunities in the city. In 1960, just 28% of all South Koreans lived in cities. By 1970, 41% lived in cities. Today, South Korea's urban population is 82% and it has quickly grown from one of the poorest to one of the wealthiest nations in the world.

Source 2

Percentage of people living in urban areas

Global patterns of urbanisation, 1995

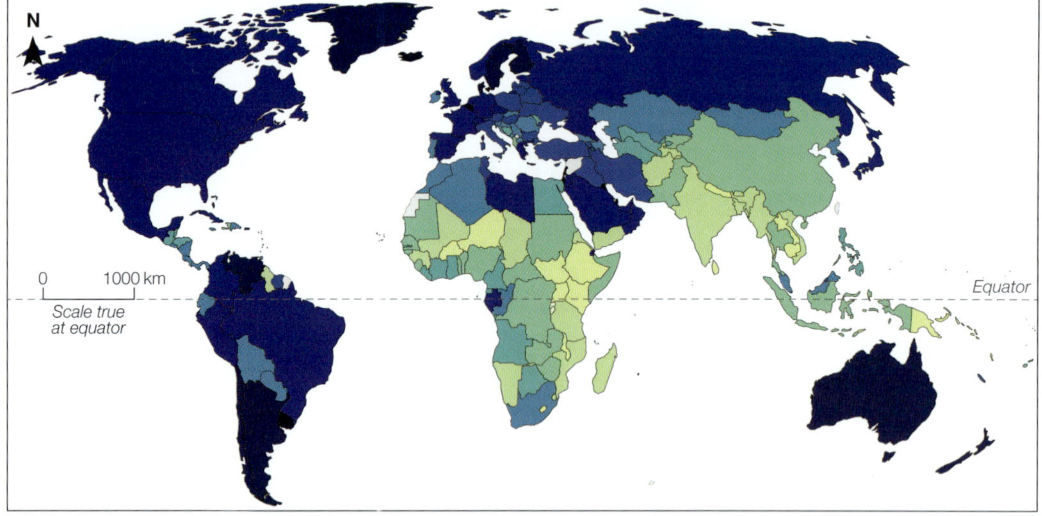

Global patterns of urbanisation, 2015

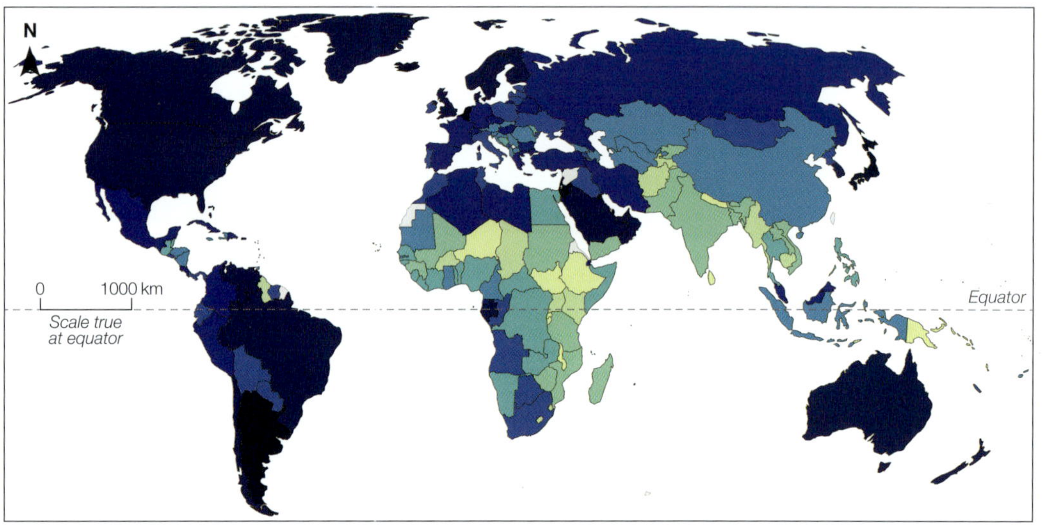

Source: Our World in Data (CC BY 4.0)

Source 3

Global patterns of urbanisation, 1995

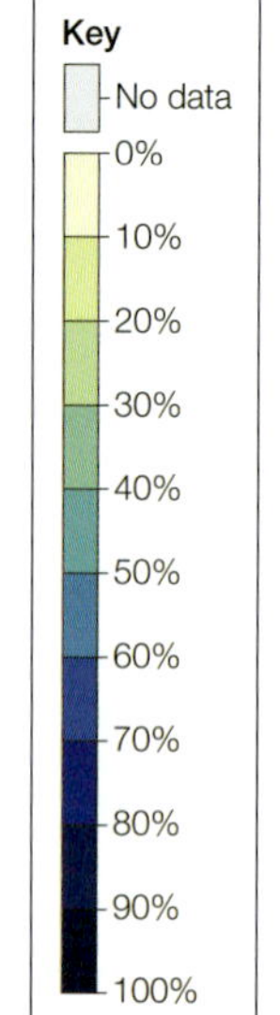

Source 4

Global patterns of urbanisation, 2015, this map shows that many countries, particularly in Asia and Africa, are becoming more urbanised.

Learning ladder G3.5

Show what you know

1 Source 2: Identify which regions have the largest predicted increase in urbanisation from 1990 to 2050.

2 Consider what you know about each of the world regions you identified in Question 1. Why would these places have such large changes over time?

3 Create a comparison table highlighting the pros and cons of living in an urban or rural environment (consider the SHEEPT factors).

Spatial characteristics

Step 1: I can identify and describe spatial characteristics

4 Source 2: Using PQE, describe the pattern of change in urban areas between 1990 and 2014.

Step 2: I can explain spatial characteristics

5 Explain why people from rural areas are drawn to cities.

Step 3: I can explain processes influencing places

6 Source 4: Using SHEEPT, explain why we have seen increased urbanisation occur over time on a global scale. You may wish to reference specific regions to quantify your response.

Step 4: I can predict changes in the characteristics of places

7 Source 2: Where is the greatest and least urbanisation set to occur between 2014 and 2050? Suggest reasons to support the trend identified.

PQE, page 138
SHEEPT, page 140

Why are people pulled to cities?

Over the last 150 years, there has been a clear pattern of **migration** from rural to urban areas. This urbanisation of the world's population is driven by people seeking better jobs and services.

Push and pull factors

There are a number of factors pushing people from rural areas and pulling people towards urban areas. Push factors can include droughts, floods, famine, lack of employment opportunities, and even **civil war**. Many young people today also see farming the land as a difficult, unglamorous and unrewarding career.

Pull factors encouraging people to move to cities include the chance of better jobs with higher salaries; greater access to education, goods and services; and a higher living standard.

Urbanisation

The push factors from rural areas and the pull factors to urban areas have seen millions of people migrate to cities in a process called **urbanisation**. Rapid urbanisation can lead to issues such as overcrowding and shortages in jobs, housing and services. Many new arrivals in cities in **less economically developed countries (LEDC)** cities are forced to live in tightly packed **slum** communities.

Source 1

The crowded streets of Kolkata in India. Rural to urban migration in India has seen cities like Kolkata become some of the most densely populated areas in the world. More than 24 000 people live in each square kilometre of the city. One-third of the city's population lives in slums and shantytowns, which house up to 28 000 people per square kilometre.

Push factors

- Few services
- Lack of job opportunities
- Poor transport links
- Natural disasters
- Wars
- Shortage of food

Pull factors

- Access to services
- Better job opportunities
- More entertainment facilities
- Better transport links
- Improved living conditions
- Hope for a better way of life
- Family links

Source 2

The perceived advantages of urban life versus some perceived disadvantages of rural life

Learning ladder G3.6

Show what you know

1 Outline the difference between push and pull factors.

2 Source 1: What issues are faced by rural migrants when they arrive in cities such as Kolkata?

Interconnections

Step 1: I can identify and describe interconnections

3 Source 2: Describe the interconnection between migration and better job opportunities.

Step 2: I can explain interconnections

4 How might push and pull factors differ for people living in more or less economically developed countries?

Step 3: I can identify and explain the implications of interconnections

5 What are the implications of large numbers of rural migrants arriving in large cities in less economically developed countries?

Step 4: I can evaluate the implications of significant interconnections

6 Write a short narrative of a person who is wanting to move from a rural to an urban area in an LEDC. In the narrative, highlight the push and pull factors for the place they want to leave and the location they want to move to and how the change of *place* will impact them physically, emotionally and financially.

What is a megacity?

Urbanisation is leading to the development of more megacities – cities with 10 million people or more. Most megacities are found in Asia, where fast-growing cities are causing headaches for planners.

The world's largest cities

Although measurements of urban boundaries can vary, in 2011, the United Nations stated there were 22 **megacities** in the world. By 2025, the UN predicts there will be 35. One in eight people lived in one of the world's 33 megacities in 2018.

Where are megacities?

Nearly 60 per cent of all megacities are found in Asia, including the five largest. Australia is the only continent with no megacities.

Tokyo in Japan is the world's largest city, with 37 million people – if it were a country, the city would be the 31st largest in the world. It is growing slowly and will only increase its population by 5 per cent from 2001 to 2025.

Source 1

Tokyo is currently the world's largest megacity but the United Nations predicts that it will be overtaken by New Delhi in 2028.

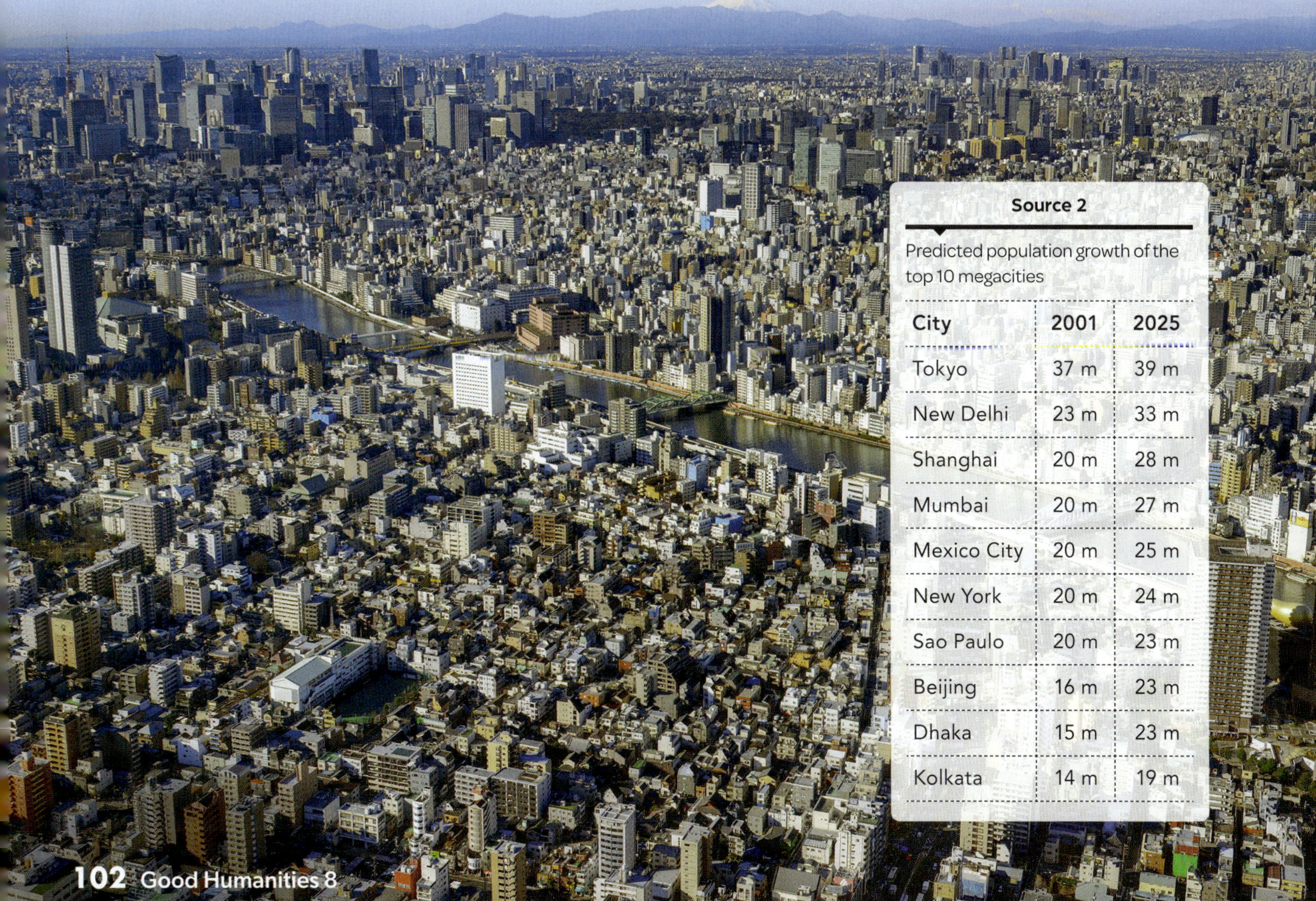

Source 2

Predicted population growth of the top 10 megacities

City	2001	2025
Tokyo	37 m	39 m
New Delhi	23 m	33 m
Shanghai	20 m	28 m
Mumbai	20 m	27 m
Mexico City	20 m	25 m
New York	20 m	24 m
Sao Paulo	20 m	23 m
Beijing	16 m	23 m
Dhaka	15 m	23 m
Kolkata	14 m	19 m

Megacities of the world

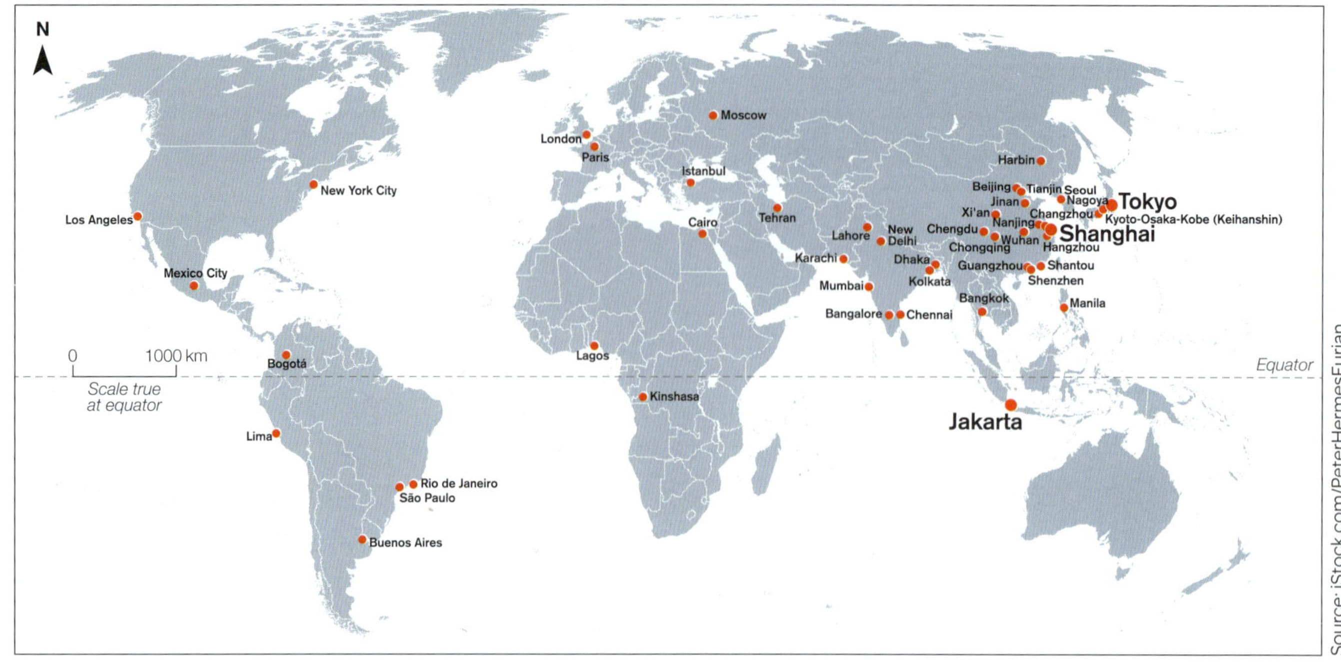

Source: iStock.com/PeterHermesFurian

Source 2

Megacities of the world – including three of the biggest: Tokyo, Shanghai and Jakarta.

Challenges for New Delhi

New Delhi in India is the world's second largest city, with 29 million people. It is growing rapidly and based on current growth will increase to 33 million people by 2025. New Delhi is projected to become the world's largest city in 2028, as more migrants move to the city.

It is difficult to keep pace with the massive increase in population in New Delhi. Each year its population grows by 700 000. Since 2005, the government has managed to build 43 000 new homes for the poor, but in the same time the population rose by more than 10 million people. Many new rural to urban migrants are forced to live in slums without clean drinking water or toilets.

Source 3

A crowded street in Dhaka, the world's most densely populated megacity

Learning ladder G3.7

Show what you know

1 Define the term 'megacity' using real-life examples and quantitative data.

Analyse data

Step 1: I can use geographic terminology to interpret data

2 Source 4: Define the term 'density' with reference to the population of Dhaka.

Step 2: I can describe patterns and trends

3 Rank the megacities in Source 2, according to the largest population increases from 2001 to 2025.

Step 3: I can explain the reasons behind a trend or spatial distribution

4 Based on your rankings from Question 3, draw some conclusions about the location, growth rate and change in megacities over time.

Step 4: I can analyse relationships between different data

5 Compare Source 1 to a satellite image of Melbourne. How do they differ in scale and geographic characteristics?

What are the effects of urbanisation?

As the pace of urbanisation increases, urban areas in less economically developed countries (LEDCs) often struggle to cope with large numbers of additional people, leading to problems supplying housing and basic services to the new arrivals.

Positive effects of urbanisation

Businesses and industries are mainly located in **urban** areas to take advantage of the availability of a workforce, power, water, communication, transport and other services.

People are attracted to cities for the employment opportunities and the convenience of access to education, health, entertainment and social services. Urban areas also enjoy more advanced communication and transportation networks than rural areas.

Negative effects of urbanisation

Urbanisation brings with it a number of problems, particularly if it occurs over a short period of time, as rapid change means the government does not have the opportunity to plan for large numbers of extra people.

Rapid urbanisation can create problems with meeting the cost of living and providing adequate housing as well as difficulties around obtaining access to clean water and proper sanitation.

Housing problems

The increased population in cities often leads to a shortage of space for housing, particularly affordable housing. In many of the cities of LEDCs, large numbers of **rural to urban migrants** live in **slums**. Conversely, when rural communities are abandoned, this can create the problem of ghost towns.

Source 1

As people leave rural areas in great numbers, they can leave ghost towns behind, such as Novo-Tishevoye on the outskirts of Moscow, Russia. Ghost towns are now found all around Moscow, as the rural inhabitants look for greater economic opportunities in cities.

Source 2

Mumbai is one of the world's megacities, with a population of 20 million people. More than half of the population is made up of migrants who come to the city in search of a better life. Around 42 per cent of Mumbai's population lives in slums. Half of Mumbai's slums are not connected to water, sewerage or electricity.

Cost of living

Incomes in urban areas are usually higher than in rural areas, but the cost of living in urban areas is much higher. As more migrants arrive in cities, there is more competition for jobs, which can make it even harder for people to live.

Water and sanitation problems

Water shortages can occur in cities with rapidly increasing populations. Sewerage facilities also struggle to cope and human waste increases the pollution of local rivers, lakes and seas. Poor water quality leads to increases in diseases such as typhoid and diarrhoea.

Learning ladder G3.8

Show what you know

1 List the negative and positive effects of urbanisation.

2 How are 'ghost towns' a result of rural migrants moving to cities for a better life?

Geographical challenge

Step 1: I can identify responses to a geographical challenge

3 Refer to Source 2. Identify why people would move to an urbanised area, only to live in a slum environment.

Step 2: I can compare responses to a geographical challenge

4 Imagine you are an immigrant who has recently moved into a new urbanised region. List the challenges you may face given the existing high-density population in that place.

Step 3: I can compare strategies for a geographical challenge

5 Brainstorm the stakeholders involved in planning urban areas. Take on the role of one of these stakeholders and write a short narrative explaining the three most important features to improve local wellbeing in an urban environment.

Step 4: I can evaluate alternatives for a geographical challenge

6 As a class, discuss the following statement: 'Given high levels of population density in urban environments, we need to revitalise rural places rather than expand megacities'.

How is urbanisation changing Indonesia?

Just 50 years ago, Indonesia was mainly a rural society, with just 17 per cent of Indonesians living in urban areas. Rapid urbanisation has seen massive growth in Indonesian cities, which has led to problems such as housing shortages, congestion and pollution.

Indonesia's growing cities

Indonesia is the world's fourth most populous country and 60 per cent of Indonesia's 268 million people live on the island of Java. Indonesia has quickly transformed from a rural society to an urban one. In 2018, 55 per cent of all Indonesians lived in urban areas.

In 1950, only Indonesia's capital city, Jakarta (see pages 108–09), had more than one million people. Today Indonesia is home to 11 cities with populations over one million as well as two megacities – Jakarta and Surabaya.

Indonesia's city populations are growing at a rate of 4.1 per cent per year – faster than any other Asian country. By 2025, Indonesia is estimated to have 68 per cent of its population living in cities.

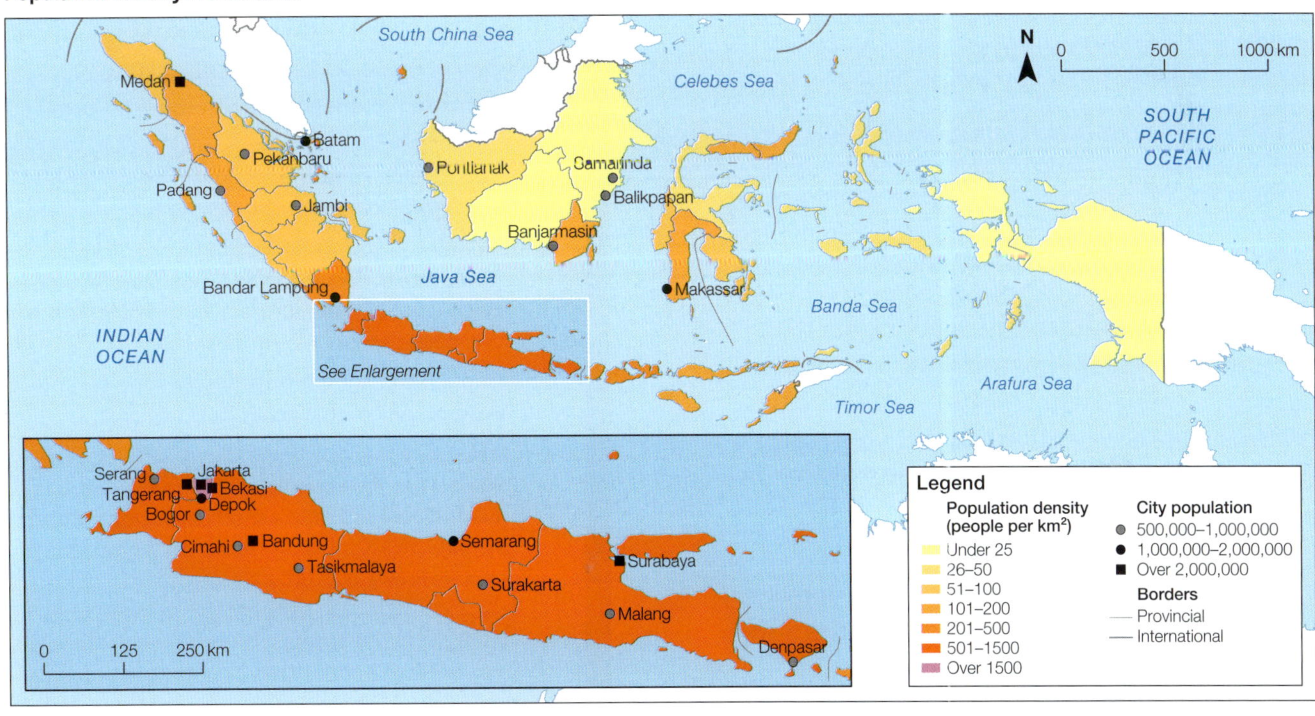

Source: Matilda Education Australia, Statistics Indonesia

Source 1

Population density in Indonesia

Source 2

A heavily polluted river runs through a Jakarta slum. Twenty-six per cent of all Indonesians live in slums.

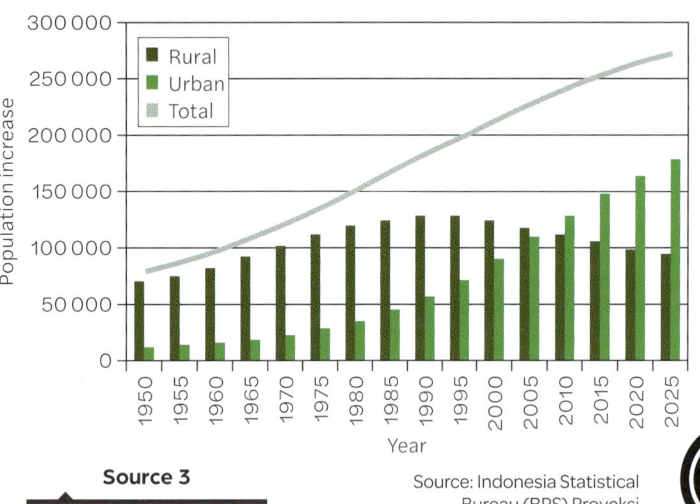

Source 3

Urbanisation of Indonesia

Source: Indonesia Statistical Bureau (BPS) Proyeksi Penduduk, 2005

Rural to urban migration in Indonesia

The Indonesian government has placed few restrictions on rural to urban migration. Most internal migration in Indonesia consists of the rural poor moving into cities for better wages, education and healthcare access. Most rural migrants are young people seeking stable jobs and a more modern life in Indonesia's large cities.

Almost 80 per cent of Indonesia's 140 million farmers are now aged 45 or older. Officials worry that future food supplies might be affected as the rural population continues to age.

The infrastructure in Indonesia's cities is not keeping pace with growing populations. New arrivals to cities will often live in poor housing in city slums, battle traffic jams and suffer high levels of air and water pollution.

Learning ladder G3.9

Show what you know

1 Source 1: What is the population density of Jakarta?

2 Source 1: Imagine you are a rural migrant from near Pekanbaru who moves to Jakarta. Describe the different environments in terms of density. How might the rural migrant feel when they arrive to live in Jakarta?

3 Source 3: Quantify the change in rural and urban dwellers in Indonesia between 1950 and 2020.

4 What problems is urbanisation causing for Indonesia's rural areas?

Collect, record and display data

Step 1: I can collect, record and display data in simple forms

5 Look at the comparative bar chart in Source 3. When did there become more urban dwellers than rural dwellers in Indonesia?

Step 2: I can recognise and use different types of data

6 Is Source 2 a primary or secondary source of data? Justify your response.

Step 3: I can choose, collect and display appropriate data

7 Imagine you are a geographer conducting a social study in Indonesia. What primary data would be useful to understand the growing population density and urbanisation?

Step 4: I can use data to support claims

8 What does Source 2 indicate about living conditions in this Jakarta slum? Why do people choose to live here?

What is the impact of Jakarta's population growth?

Jakarta is the capital city of Indonesia, and a megacity with a population of more than 30 million. The sprawling city has expanded into all usable space.

Large areas of Jakarta sit below sea level, so flooding is a constant problem. The rapid pace of Jakarta's urbanisation has introduced multiple problems such as a lack of affordable housing, traffic congestion, and health problems associated with polluted air and water. Poor rural migrants are forced to settle in slums along riverbanks. The new arrivals cannot access health, education and other public services because they don't have official Jakarta identity cards.

Life in the slums

Five million people live in slums in Jakarta, and 60 per cent of the population suffers breathing problems from the polluted air. Ninety-six per cent of Jakarta's river water is severely polluted and dangerous to drink. The government provides some clean water, but it can be too expensive for the poor to afford, so they use river water.

Source 1

Satellite images of Jakarta from 1976 to 2004. The green area shows the expanding urban area, while the red shows areas of vegetation such as farms and forest.

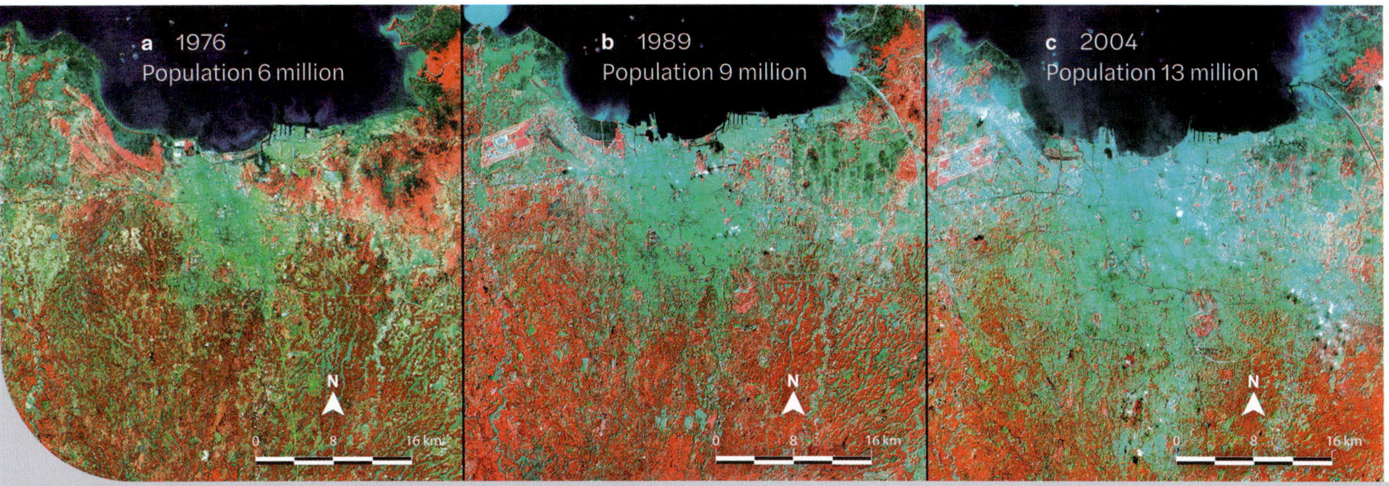

Source: Science Photo Library

Source 2

Each day, 15 million motorbikes and 5 million cars try to use Jakarta's gridlocked roads. Just 18 per cent of commuters use public transport.

Source 3

Children play and collect materials in a downtown slum in Jakarta.

Learning ladder G3.10

Show what you know

1 Describe the relative and absolute location of Jakarta.

2 Why is Jakarta a megacity?

3 Outline the positive and negative impacts of urbanisation in Jakarta.

Geographical challenge

Step 1: I can identify responses to a geographical challenge

4 Source 3: Why are slums an example of a response from rural migrants to survive in their new urban environment?

Step 2: I can compare responses to a geographical challenge

5 Source 1: Use the satellite images to describe the growth of Jakarta between 1976 and 2004.

6 Source 1: The northern part of the city on Jakarta Bay is constantly flooding and below sea level. What response can the government make to protect the city from flooding?

Step 3: I can compare strategies for a geographical challenge

7 Source 2: Suggest different strategies to help ease Jakarta's traffic problem.

Step 4: I can evaluate alternatives for a geographical challenge

8 In 2019 the Indonesian government announced it would move the administrative centre of Indonesia from overcrowded Jakarta to a new site on the island of Borneo. Evaluate the merits and problems associated with this plan.

HOW TO

Satellite images, page 149

How is China changing?

The Chinese government controls where its people work through the *hukou* system. Workers must obtain a permit to work outside their home village.

Internal migration in China

The largest **internal migration** in human history is occurring in China. Its people are migrating from inland **rural** villages to work in large coastal manufacturing cities such as Dongguan in Guangdong province. In 2015, there were 278 million migrant workers in China – 36 per cent of China's total workforce.

Under China's *hukou* system everyone must be registered in their home town or village to be able to access education, housing and welfare. Individuals are categorised as either rural or **urban**, and they are required to work within their designated area. China allowed people to move to other provinces to work in 1984 to provide the cheap labour required for China's booming manufacturing industries.

Under the *hukou* system, people moving from rural areas to cities must buy a work permit. Most permits only allow for temporary work of up to 12 months. There is great demand for these permits and they can be difficult to obtain.

Different living standards

Differences in living standards are so great between rural and urban areas in China that **migration** is a very attractive prospect for rural workers. Rural incomes are less than 40 per cent of urban ones. Farmers move to the large industrial centres where they can earn far more money than they could growing crops or tending cattle.

Problems for migrant families

Migrant workers often leave their children behind so they can work in urban areas to earn a better income. Good quality housing is difficult to find and rents are expensive. Migrants also struggle to get access to services such as education and health.

Migrants can spend many years away from their family. More than one-third of children in rural China are left behind in their villages under the care of their grandparents or other relatives.

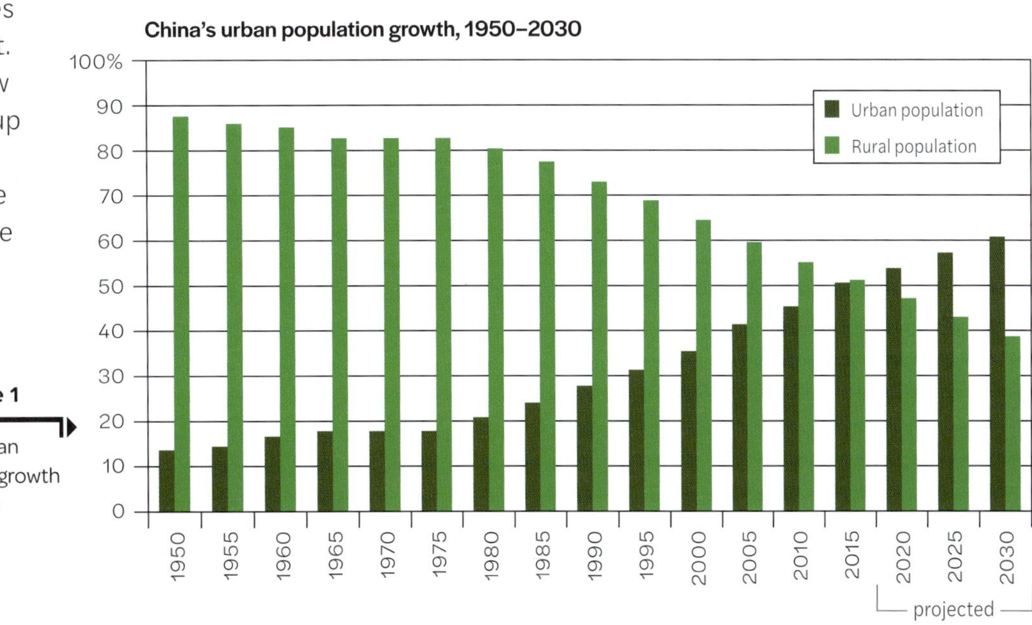

Source 1

China's urban population growth 1950–2030

Source 2

China's population 23 May 2019. Guangdong province, in the south, accounts for 27 per cent of China's temporary migrants.

People's Republic of China Population

Source: iStock/Rainer Lesniewski

Source 3

Smartphone assembly line in Dongguan, Guangdong province. China produces 60 per cent of all mobile phones and about half of them are manufactured in Guangdong province.

Learning ladder G3.11

Economics and business

Step 1: I can recognise economic information

1 Consider your existing knowledge of China's trade and economy. Along with some research, create a list of China's main imports and exports.

Step 2: I can describe economic issues

2 Comment on how internal migration in China can affect the local economy of a *place*.

Step 3: I can explain issues in economics

3 Discuss the statement: 'The *hukou* system is a successful method of ensuring that all Chinese people gain equal access to resources and employment opportunities.'

Step 4: I can integrate different economic topics

4 Relative scarcity is where we do not have enough resources to satisfy all our wants and needs. Discuss the concept of relative scarcity in terms of China's population growth and associated resource demand.

Step 5: I can evaluate alternatives

5 Suggest an alternative system allowing Chinese residents to move freely but not overpopulate and compete for resources in megacities.

How is migration changing Australia?

Half of Australia's population growth comes from immigration. Permanent migrants come to Australia to join family members, or as skilled workers or **refugees**. Most migrants come from China, India or New Zealand, and settle in Sydney or Melbourne.

Growth through migration

Australia's population is growing much faster than most countries in the world. Between 2006 and 2016 Australia's population grew by 1.7 per cent each year compared to a growth rate of just 0.3 per cent in other **MEDCs**. Half of the growth in Australia's population came through **international migration**.

In 2018, immigration into Australia was around 330 000 people. Most immigrants come to Australia as students or **skilled workers**. The Department of Immigration publishes lists of skilled occupations Australia requires and people from overseas can apply for them.

Following the end of World War II, most Australian migrants came from the United Kingdom and Europe. Today, the fastest growing populations of migrants come from China, India and New Zealand. Around half of new arrivals to Australia settle in Sydney or Melbourne, which can add to urbanisation problems such as a lack of housing, traffic congestion and greater demands on infrastructure such as schools and hospitals.

Source 1

Migration to Australia, 1990–2017. The relative thickness or thinness of a country represents the proportion of migrants from that country.

Migration to Australia, 1990–2017

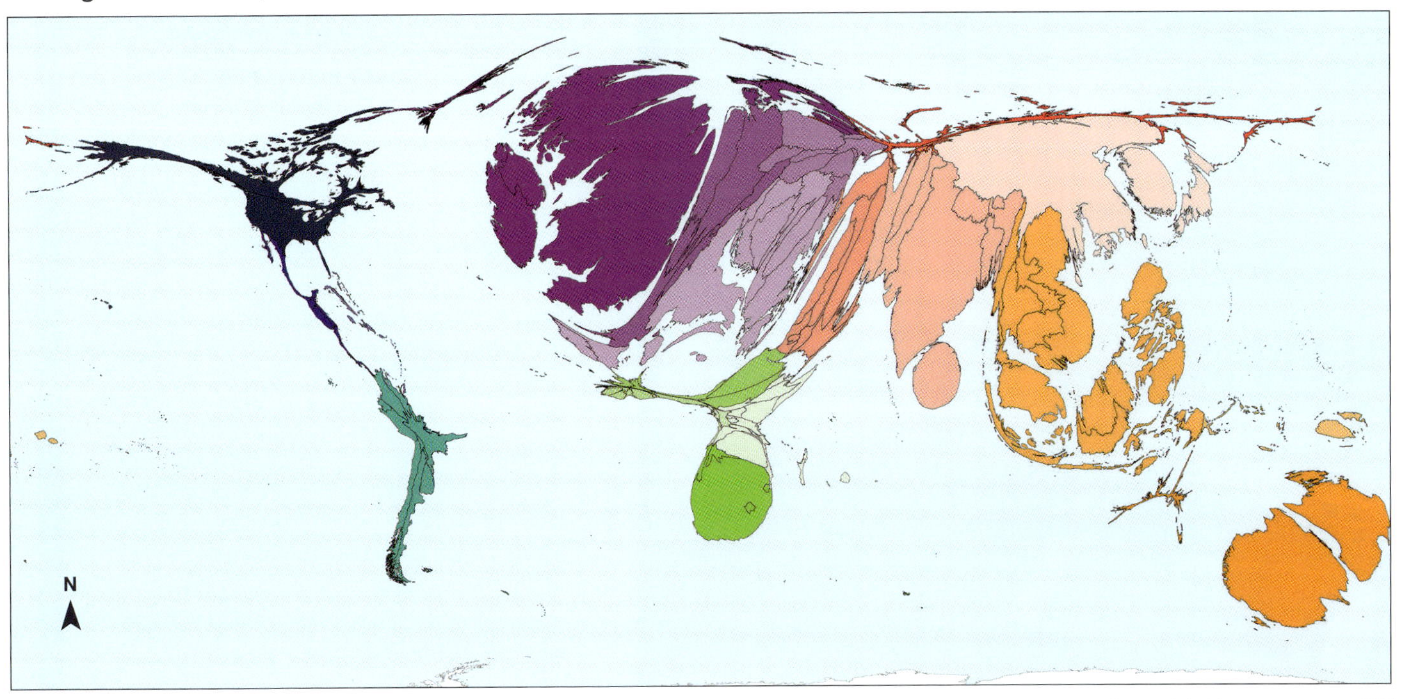

Source: Worldmapper, map no. 367

Source 2

Asylum seekers arrive by boat on Christmas Island, 8 July 2011. Refugees who arrive by boat are sent to offshore detention centres in countries other than Australia.

Internal migration

Internal migration within Australia is important to relocate skills and labour resources from one part of the country to another. During Australia's mining boom from 2005 to 2015, large numbers of internal migrants moved to mining centres in Western Australia and Queensland. Between 2006 and 2016, Queensland and Victoria were the only states to record net gains of internal migrants, with other states recording net losses.

Internal migration in Australia 2011–2016

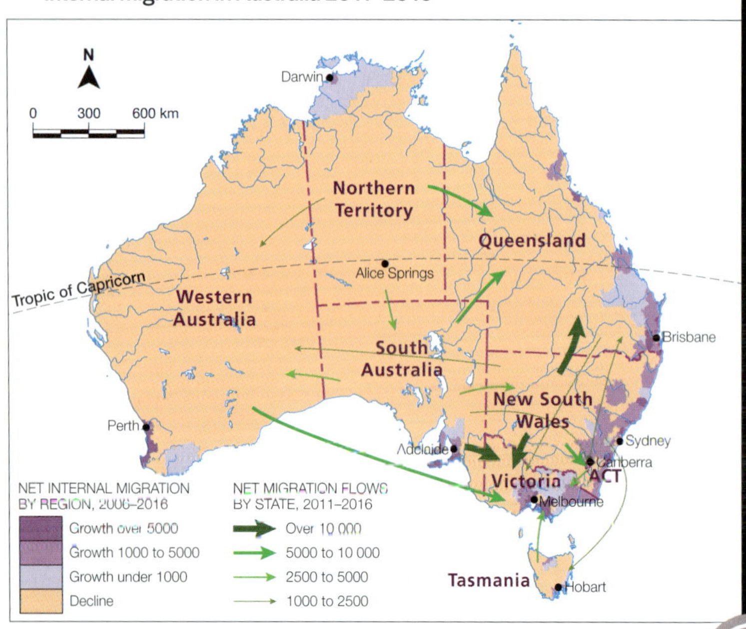

Source 3

Internal migration in Australia 2011–2016

Source: Matilda Education Australia

Learning ladder G3.12

Show what you know

1 How does migration contribute to Australia's population growth?

2 What are the pull factors for immigrants moving to Australia?

3 Source 3: Comment on how people migrate within Australia.

Collect, record and display data

Step 1: I can collect, record and display data in simple forms

4 Source 1: Use an atlas to help identify five countries where most Australian migrants come from. Remember that the shapes in Source 1 are distorted like blown-up balloons.

Step 2: I can recognise and use different types of data

5 Conduct a survey of your class or broader school community and identify the percentage of Australian-born students compared with those born internationally. Display the data you collect in a table or graph.

Step 3: I can choose, collect and display appropriate data

6 Were you surprised by the results of the survey conducted in Question 5? Explain whether you think a survey is a relevant method of primary data collection to investigate the rate of migration. How could the government use surveys to collect national scale data on population change?

Step 4: I can use data to support claims

7 Source 2: What dangers do refugees face when trying to come to Australia by boat? Research the government policies that apply to people who arrive by boat. Then research the case against these policies.

HOW TO

Surveys, page 132

How can the movement of people turn deadly?

The consequences of living in a highly connected world turned deadly in 2020 when the COVID-19 pandemic swept around the world, killing hundreds of thousands of people. Many governments moved quickly to shut down borders and restrict people's movement internationally and domestically.

The COVID-19 virus has spread to more than 180 countries since it was first identified in Wuhan, China, at the end of 2019. By January 20 in 2020, the virus had spread from China to neighbouring countries of Japan, South Korea and Thailand. In just two months, it spread to nearly every country in the world.

The World Health Organization (WHO) declared the coronavirus COVID-19 a **pandemic** on 11 March 2020. A pandemic is a disease that spreads in multiple countries around the world at the same time, usually affecting a large number of people.

As the pandemic spread, different **epicentres** for the disease sprung up. As the rise in new COVID-19 cases slowed in China, outbreaks in other countries soared. By mid-March 2020, Europe became the epicentre for the spread of the virus, with Italy and Spain recording tens of thousands of cases and thousands of deaths. At the end of March 2020, the US had become the new epicentre with more than 100 000 cases recorded, most of these in New York. By mid June 2020, the USA had 2.1 million confirmed cases and almost 120 000 deaths. Brazil was another epicentre, with 920 000 total infections and 45 000 deaths.

Slowing the spread of coronavirus

The COVID-19 pandemic sparked a public health emergency around the world, with severe restrictions placed on the movement of people. By 20 March 2020, only Australian citizens could travel to Australia. All travellers arriving in Australia were required to undertake a mandatory 14-day quarantine period at a hotel. The Australian government also restricted all cruise ships from entering Australia, following a number of outbreaks on cruise ships.

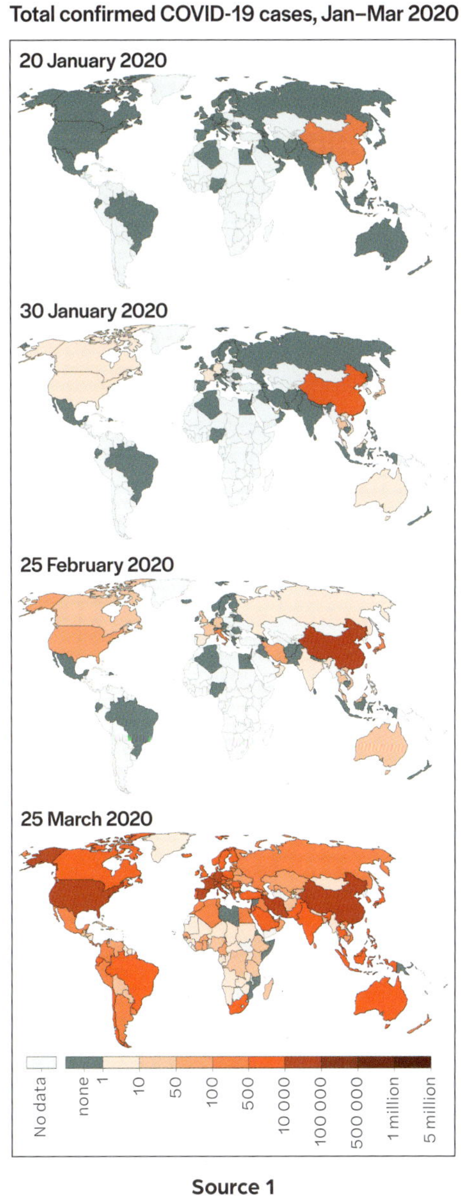

Source 1

Global spread of COVID-19 virus in the early months of 2020

Total confirmed COVID-19 cases: how rapidly were they increasing?

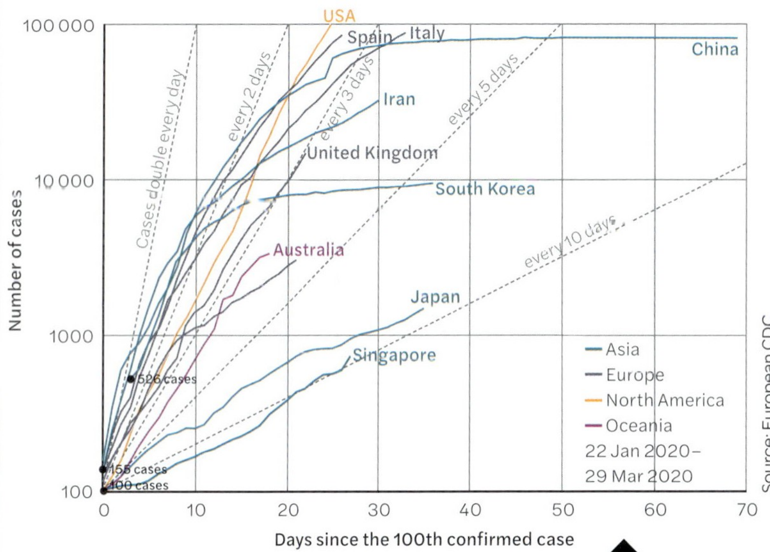

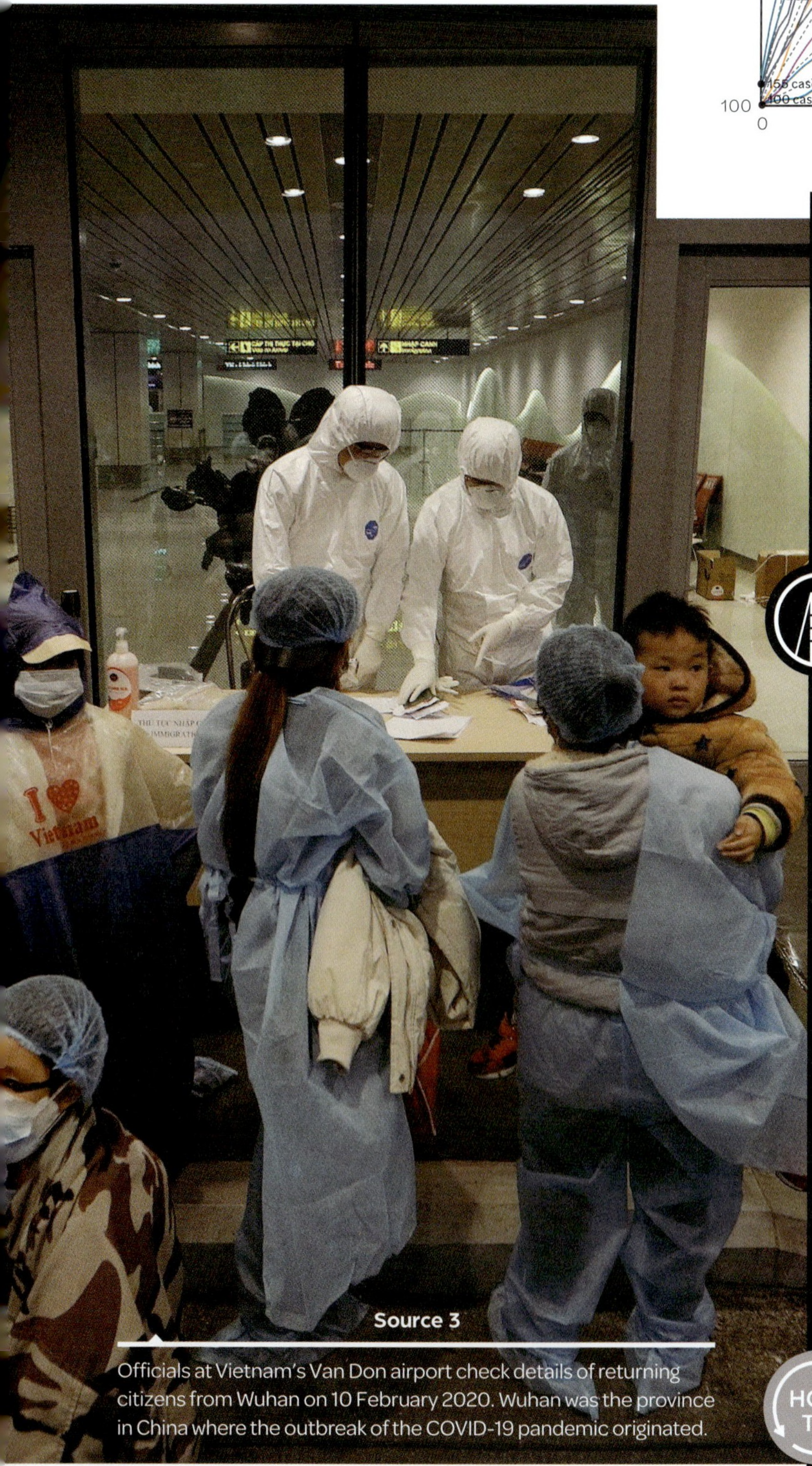

Source 2

Confirmed cases of COVID-19 for select countries and regions on 29 March 2020. This chart uses a logarithmic scale to compare the widely ranging values. The scale is nonlinear, so numbers 100 and 1000 and 10 000 are equally spaced. The graph also begins on the first day total confirmed deaths reached 0.1 per million for each country to make the growth of the pandemic easier to compare.

Source 3

Officials at Vietnam's Van Don airport check details of returning citizens from Wuhan on 10 February 2020. Wuhan was the province in China where the outbreak of the COVID-19 pandemic originated.

Learning Ladder G3.13

Show what you know

1 What is an epicentre during a pandemic?

2 Source 1: Use PQE to describe the global spread of coronavirus.

Geographical challenge

Step 1: I can identify responses to a geographical challenge

3 Source 3: What is happening in this scene and why?

Step 2: I can compare responses to a geographical challenge

4 Source 2: Which country do you think was most successful at controlling the spread of the virus by the end of March 2020 and why?

Step 3: I can compare strategies for a geographical challenge

5 Australia restricted travel from coronavirus hotspots before eventually closing its border to international visitors. What other action could it have taken?

Step 4: I can evaluate alternatives for a geographical challenge

6 Research the action taken by the Australian government to restrict the movement of people within Australia. Why did the government take this action?

HOW TO

PQE, page 138

How do we plan for urbanisation?

Governments need to plan far into the future to successfully manage Australia's growing urban populations. The government wants to ensure we have enough affordable housing and that infrastructure for health, education, transport and communications keeps pace with changes. Strategies to redevelop underused areas and to encourage people to move to regional areas also helps authorities deal with rapid urbanisation.

Planning for the future

Australia is one of the world's most highly urbanised countries, with nearly four in every 10 people living in Sydney or Melbourne. Half of all jobs growth is concentrated within two kilometres of these two cities. In 2018, Australia's population reached 25 million.

By 2050 it is projected that Australia will have a population of 36 million people. Nearly all of the growth will come from **international migration** and more than 70 per cent of this growth is likely to be in the major cities. Melbourne and Sydney will each have populations of eight million and Brisbane and Perth will each have populations of more than four million.

Future Melbourne

Melbourne hit the five million population mark in 2018 and on average is growing by 1800 people a week because of a high level of international migration. Melbourne's population growth is the equivalent of adding the population of Darwin to Melbourne each year.

Based on current growth rates, Melbourne will overtake Sydney as Australia's largest city in 2026. Careful planning needs to be undertaken to ensure that population growth is **sustainable**. Already planners are suggesting that by 2050 Melbourne will need:

- 1.6 million new homes
- 8 new hospitals
- 67 new secondary schools
- 222 new kindergartens.

Source 1

At 2500 square kilometres, Melbourne is one of the world's largest cities by area.

Melbourne's urban growth, from before 1880 to after 1960

Source 2

Melbourne's urban growth, from before 1880 to after 1960

LEGEND

Urban area

- Before 1880
- 1880 – 1919
- 1920 – 1960
- After 1960
- Non urban
- Parkland or reserve
- Major road
- Other road
- Major railway
- Mountain
- River
- Lake

1 : 760 000

Melbourne was founded by European free settlers on the lands of the Wurundjeri, Boonwurrung and Wathaurong peoples in 1835 by John Batman and others. Initially the settlement was called Dootigala, then Batmania and finally Melbourne in 1837.

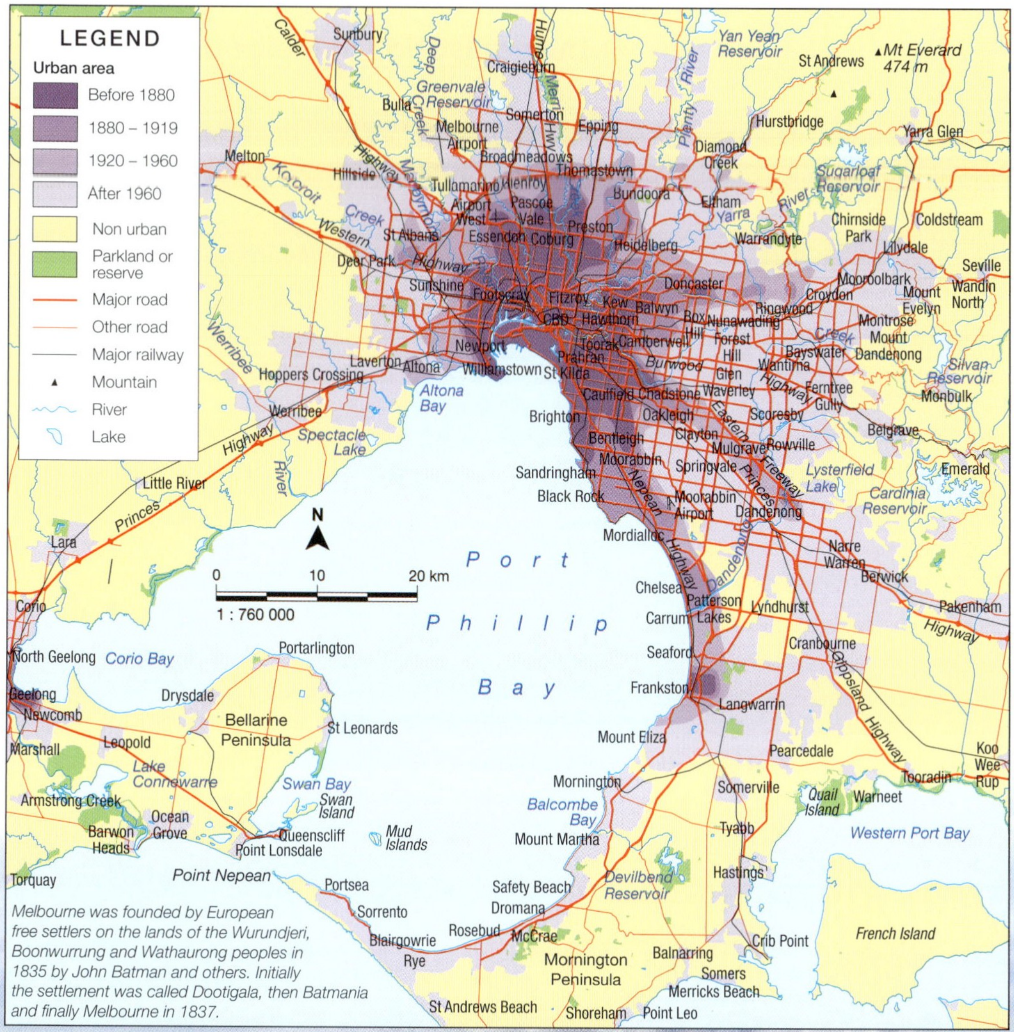

Source: Matilda Education Australia

When planning for Melbourne's future growth, governments consider three key trends:

1 **Suburbanisation** – where land on the city's outer edge is made available for housing. The suburbs grow outwards as new houses and services are constructed. Expanding cities through suburbanisation replaces natural environments or land previously used for farming.

2 **Counterurbanisation** – when people move from an **urban** area to a **rural** or coastal area to escape urban problems such as expensive housing, traffic and pollution. The **decentralisation** of the population helps take pressure off the urban centre.

3 **Reurbanisation** – when people move back into inner city areas that have been improved through **urban renewal** to attract people who enjoy the advantages of being near the city centre.

Source 3

Melbourne's growth corridors

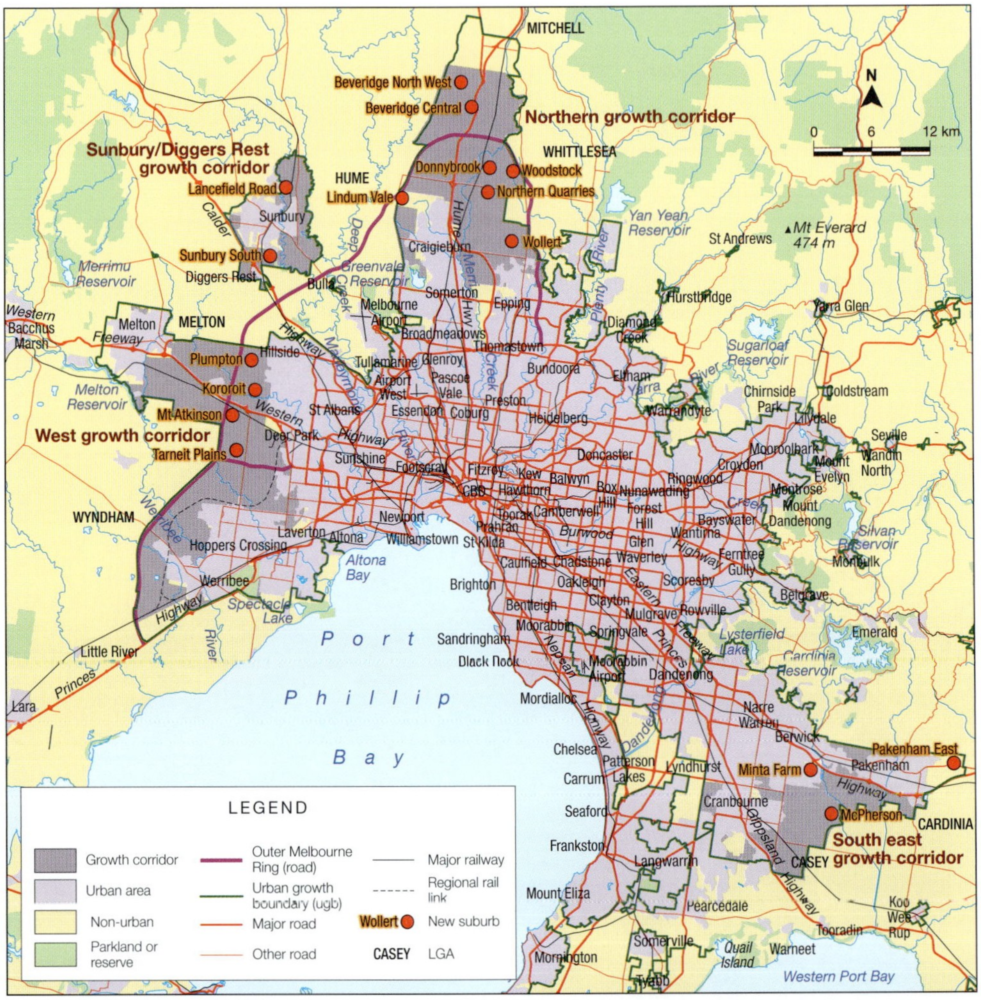

Source: Matilda Education Australia

Suburbanisation in Melbourne

At 2500 square kilometres, Melbourne is already one of the world's largest cities by area – and it is still expanding. More than 50 suburbs have been added to Melbourne since 2005.

In 2017, the Victorian government attempted to provide more affordable housing by rezoning areas on the edge of the city to create 100 000 new housing blocks and 17 new suburbs in Melbourne's key growth corridors:

Corridor	Houses	Population
West	170 000	480 000
North	117 000	330 000
South east	103 000	110 000
Sunbury	32 000	90 000

The major advantage of suburbanisation as a strategy is that affordable housing estates can be quickly established and designed.

The disadvantages of suburbanisation are that open land is sacrificed to build new urban areas and expensive new **infrastructure** has to be built. With most jobs available near the city centre, many residents in outer suburbs face long commutes to get to work.

Counterurbanisation in Melbourne

A total of 77 per cent of Victoria's population lives in Melbourne and about 90 per cent of population growth occurs in the state capital. Apart from capital cities, coastal communities are the fastest growing regions in Australia. Victoria's Surf Coast, centred on Torquay, is also rapidly growing in size and population. The region is 90 minutes from Melbourne, enabling residents to commute for work.

Source 4

Sky Park takes up 25 per cent of the new Melbourne Quarter commercial, retail and residential project and is an example of a reurbanisation project. The park provides city residents and workers with new green space. Designers have also added high-speed public wi-fi, power outlets and USB charge points to the park.

Governments encourage population growth and job creation in areas outside major cities in a process known as **decentralisation** or counterurbanisation. Businesses are given financial benefits to move from capital cities to regional centres. Governments also establish or relocate departments and authorities to towns and cities outside of Melbourne.

Reurbanisation in Melbourne

As Melbourne continues to grow, many areas are being reimagined as 21st-century suburbs. Through the process of urban renewal, old industrial areas and docks are being redeveloped to provide new homes, offices and entertainment spaces.

The advantage of redeveloping existing areas is that infrastructure such as power, water and transport already exists. Higher density developments of townhouses and apartment blocks allow more people to be housed in established suburbs. New green spaces can also be created for entertainment and leisure. Melbourne's Docklands and new Fisherman's Bend project (see pages 120–21) are examples of urban renewal projects.

Learning ladder G3.14

Show what you know

1 Source 1: In which direction did Melbourne expand after 1960?

2 Source 3: In which direction will Melbourne grow in the future? Why do you think it is expanding in these areas?

Geographical challenge

Step 1: I can identify responses to a geographical challenge

3 What are the three key trends that governments need to consider when planning for Melbourne's future growth?

Step 2: I can compare responses to a geographical challenge

4 What new infrastructure will Melbourne need by 2050 to support its growing population? How is reurbanisation one solution to providing infrastructure more quickly and cheaply?

Step 3: I can compare strategies for a geographical challenge

5 Source 4: Look at the photograph of Sky Park. Compare this to the usual land use in high-rise buildings in the centre of cities. How does it differ and what benefits might this bring to the community?

Step 4: I can evaluate alternatives for a geographical challenge

6 What is counterurbanisation and what benefits does it offer? Why might counterurbanisation be difficult to introduce?

G3.15

thinking locally

How is Fisherman's Bend changing?

The industrial region of Fisherman's Bend once hosted large car and aircraft factories, but is now the site of Australia's largest urban renewal project. The 4.8 square kilometres of underused land just a few kilometres from Melbourne's central business district (CBD) will be transformed to cater to the city's growth.

The Victorian government has rezoned the former industrial land to provide inner-city homes for more than 80 000 people by 2050. The employment precinct is projected to generate jobs for up to 40 000 people.

Fisherman's Bend will be served by train and tram links to the city and linked by a new bridge across the Yarra River. Other important **infrastructure** to be built includes a new secondary school, primary schools, child care centres, parks and shopping centres to support the growing population.

A key focus of the Fisherman's Bend **urban renewal** is on creating a **sustainable** environment. One quarter of the project area will be dedicated to open space, which will be no more than a 200-metre walk away for any resident or worker. The planners project that 80 per cent of transport movements will be made by walking, cycling and public transport. Waste water will be recycled for use in homes and businesses to reduce the impact on the environment.

Fisherman's Bend urban renewal area

Source: Victorian Department of Jobs, Precincts and Regions.

Source 1

Fisherman's Bend urban renewal area

Source 2

You can see the current Fisherman's Bend district in the foreground of this oblique aerial photograph with the nearby Melbourne CBD in the background.

Learning ladder G3.15

Show what you know

1 Outline three key changes that have occurred in Fisherman's Bend over time.

2 What is urban renewal and how will it affect Fisherman's Bend?

3 Source 2: Why do you think the developers of the project used this oblique aerial photograph to attract investors to the project?

Analyse data

Step 1: I can use geographic terminology to interpret data

4 Source 2: Locate the Fisherman's Bend Urban Renewal Area using the map in Source 1 and describe what the area looks like today.

Step 2: I can describe patterns and trends

5 Use Source 1 to answer the following questions.

a What are the four residential precincts in the Fisherman's Bend Urban Renewal Area?

b How tall can apartment buildings be built in most of the Lorimer precinct?

c If you lived in a 40-storey apartment building on Munro Street in the Montague precinct, how far would you travel to reach Marvel Stadium at Docklands?

d Imagine you live on the top floor – the penthouse – of the tallest building in Sandridge. In which direction will you have views? Explain why.

Step 3: I can explain the reasons behind a trend or spatial distribution

6 Give three examples of features of the Fisherman's Bend urban renewal project that will lead to a more sustainable environment.

Step 4: I can analyse relationships between different data

7 Use Sources 1 and 2 to support claims that the Lorimer residential precinct will offer the most expensive apartments. Justify your decision.

Masterclass

Learning Ladder

Work at the level that is right for you or level-up for a learning challenge!

Satellite images of San Antonio, Texas, 1991 (left) and 2010 (right)

Source 1

Satellite images of the city of San Antonio in Texas, USA

Changes in Australia's population growth

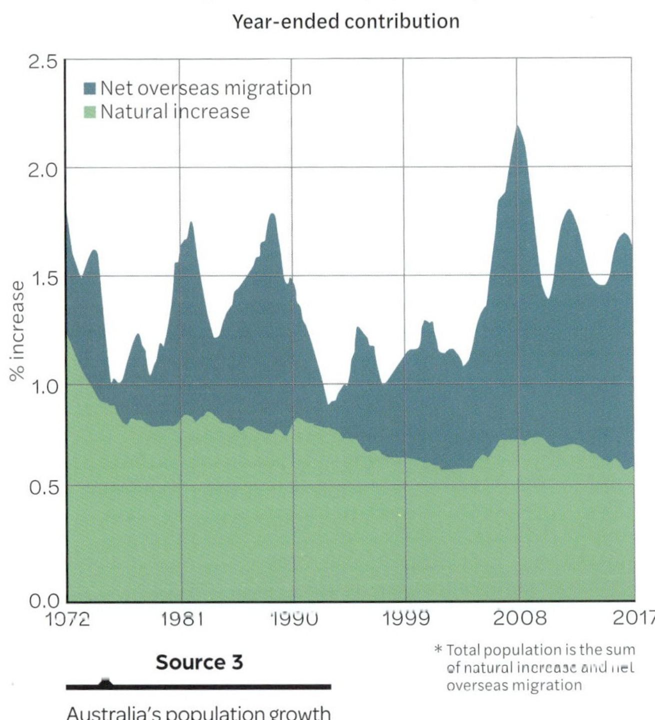

Source 3

Australia's population growth

Change in migrant origins in Australia over time

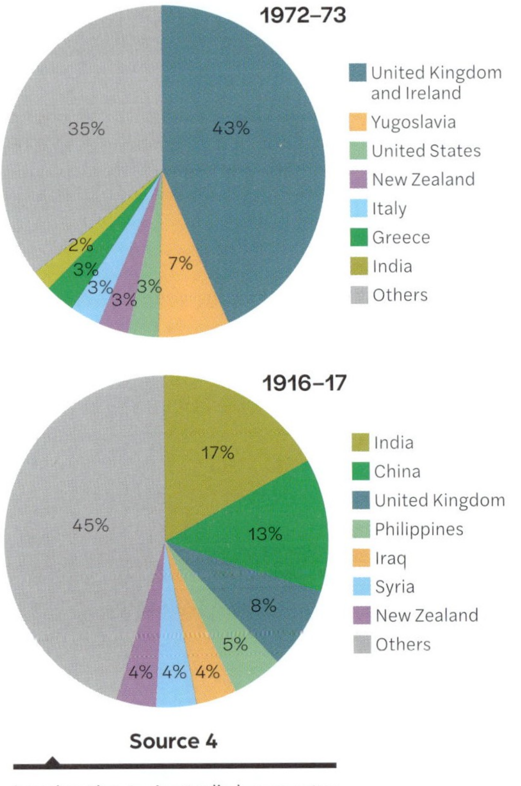

Source 4

Immigration to Australia by country

Source 2

A downtown slum in Jakarta.

Step 1

a I can identify and describe spatial characteristics

Describe the region your school is in. Why would they build a school in that *place*?

b I can identify and describe interconnections

How are rural migrants interconnected with job and education opportunities in cities?

c I can identify responses to a geographical challenge

Refer to Source 1. Describe how urban sprawl has changed this place over time.

d I can collect, record and display data in simple forms

Refer to Source 1. Identify what the different colours represent on these satellite images.

e I can use geographic terminology to interpret data

Source 1: In which directions has this city grown the most?

Step 2

a I can explain spatial characteristics

Source 1: Discuss how land use has changed in this region over time.

b I can explain interconnections

Source 2: Why are rural migrants interconnected with slums in many less economically developed countries?

c I can compare responses to a geographical challenge

Source 2: How do rural migrants with little money survive in large cities?

d I can recognise and use different types of data

A census is a survey conducted within a state or country to collect data on population dynamics. Suggest other ways governments could collect data on human population movement and changes.

e I can describe patterns and trends

Refer to Source 3. Using PQE (page 138), describe how net international migration has changed over time in Australia.

Step 3

a I can explain processes influencing places

Refer to Source 2. What push and pull factors do you think might have operated to attract these rural people to the city?

b I can identify and explain the implications of interconnections

What are the implications of large numbers of rural migrants arriving in large cities in less economically developed countries?

c I can compare strategies for a geographical challenge

Suggest different strategies to help ease traffic jams in large cities.

d I can choose, collect and display appropriate data

Refer to Source 4. Using two blank world maps, illustrate the origins of migrants that moved to Australia between 1972–73 and 2016–17. What advantage do the maps have over the pie charts in Source 4?

e I can explain the reasons behind a trend or spatial distribution

Refer to Source 4. Explain why the countries of origin of migrants to Australia have changed over time. Hint: You may need to discuss or research world events, MEDC, LEDC or geographical locations to help you answer this question.

Masterclass

Step 4

a I can predict changes in the characteristics of places

Megacities are going to become more common in the future. Discuss with reference to the concepts of environment, place and sustainability.

b I can evaluate the implications of significant interconnections

Increased migration is *interconnected* with urban sprawl. Do you agree with this statement? Justify your response with examples.

c I can evaluate alternatives for a geographical challenge

Refer to Source 4. What changes to towns, infrastructure and facilities would town planners need to consider in 2016–17 with the changing origins of migrants to Australia?

d I can use data to support claims

Referring to Source 4, discuss the conclusion that 'Australia's migrant profile has changed dramatically in the last 50 years'. Use evidence to support or refute this statement.

e I can analyse relationships between different data

Is there a relationship between Sources 3 and 4? Explain.

Step 5

a I can analyse the impact of change on places

Source 2: Why would this slum be located next to a river and what are the health dangers associated with this location?

b I can explore spatial association and interconnections

What type of land has been replaced by urban land use in Source 1? Identify an area on the satellite image where urban expansion will be more difficult and explain why.

c I can plan action to tackle a geographic challenge

Source 1: This is an example of suburbanisation to cater for a growing population. What other strategies could planners use to cater for an increasing number of people?

d I can evaluate data

Referring to Source 1, discuss whether you think satellite images are an appropriate method of displaying urban sprawl and change over time in a location. What other image or method may be more successful?

e I can draw conclusions by analysing collected data

Source 3: What is the trend for natural population increase in Australia? What is the trend for international migration to Australia? How can the government increase the rate of growth of the Australian population?

Capstone

How can I understand changing nations?

In this chapter, you have learnt a lot about changing nations. Now you can put your new knowledge and understanding together for the capstone project to show what you know and what you think.

In the world of building, a capstone is an element that finishes off an arch or tops off a building or wall. That is what the capstone project will offer you, too: a chance to top off and bring together your learning in interesting, critical and creative ways. You can complete this project yourself, or your teacher can make it a class task or a homework task.

mea.digital/GHV8_G3

Scan this QR code to find the capstone project online.

G4
Fieldwork
WHAT IS THE PURPOSE OF FIELDWORK? page 126
fieldwork task 1
page 128
HOW DO WE DEGRADE OUR LOCAL ENVIRONMENT?
fieldwork task 2
page 130
WHAT ARE THE CHALLENGES OF PLANNING FOR URBANISATION IN YOUR PLACE?
fieldwork
page 132
HOW DO GEOGRAPHERS USE SURVEYS?

What is the purpose of fieldwork?

Fieldwork is an important part of Geography. We can use fieldwork to discover new things, learn about patterns and relationships, and explore the world around us. In the last three chapters you have been exploring landforms and landscapes, as well as how places change over time. Now you are going to conduct your own research in the local area to answer a research question and write a mini-fieldwork report.

Quantitative and qualitative data

When conducting fieldwork, we collect data. The two main kinds of data are **qualitative** and **quantitative**. Quantitative data tends to be recorded with numbers. Examples of quantitative data would be the height of a building, the number of people in a population and the flow rate of a river. Qualitative data tends to be more observational. Examples of qualitative data include descriptions of a place, a field sketch or photos you may take. Both aspects of data collection are important and you should try to collect both quantitative and qualitative data in your fieldwork.

Research question

A **research question** is an overarching big idea that we want to investigate. Once you develop a research question, you will need to write a hypothesis. A **hypothesis** is an 'educated guess' about what the answer to the research question may be.

Choosing a fieldwork location

Choosing the right location to conduct your fieldwork is very important. The place you choose will depend upon your question, hypothesis and topic. Ensure you decide the location of your fieldwork before you begin to conduct research and collect data.

Pre-fieldwork research

Before you complete your fieldwork, you need to do some background research so you have an understanding of the field site, its characteristics and other relevant information. Using the internet to start your research is usually the most obvious and accessible source of information, keeping in mind that you must only use reliable websites.

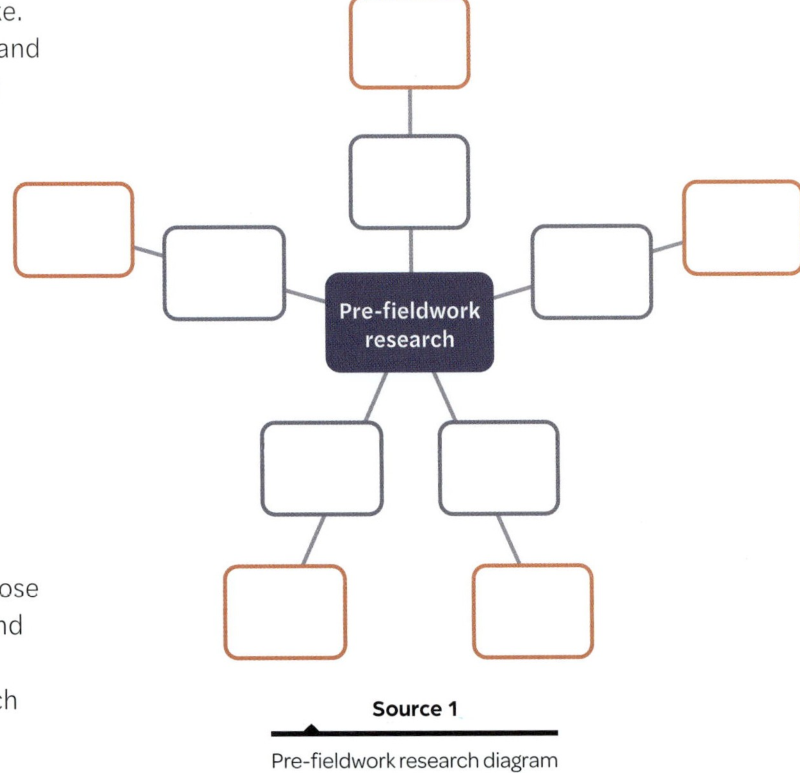

Source 1

Pre-fieldwork research diagram

Government, educational and other approved sites are most useful and usually contain reliable data. Blogs, social media and other public-based sites may be **biased** and not provide you with correct evidence. Books, handouts and other paper resources are also useful when conducting research. Using a template similar to the Source 1, you can record some pre-fieldwork research in the blue bubbles and note the source of your information to make a bibliography in the red bubbles.

Collecting data

As a class, you need to develop a list of **primary** and **secondary methods**. Primary methods are things that you will do in the field to find evidence to answer your research question and prove or disprove your hypothesis. Secondary methods are ways you can collect data back in the classroom to help you answer the question; for example, researching websites, books or other publications.

Some primary methods are:

- observations
- photographs
- note taking
- surveys (questionnaire or observational)
- field sketches.

Source 2

An observational checklist

Using the table below as a template, as a class discuss some that may be successful in helping you answer your research question

Primary method	This data will help me answer my research question by ...	Secondary method	This data will help me answer my research question by ...

Preparation for fieldwork

Before you go out in the field, it is important to create a checklist of things you need to bring and data you need to collect. Add to the checklist below according to what you plan to do in the field.

- Clipboard
- Paper (blank and lined)
- Pencils
- Camera
- Survey questions

Be prepared for any weather – sometimes doing fieldwork in the rain gives you a good insight into how water is acting to degrade your local area.

Source 3

Water-resistant notepads can help with wet-weather fieldwork.

Analysing data

Once you have been in the field and collected data using your primary and secondary methods, you need to communicate your findings. You can communicate your research via a report or display it as a visual poster or presentation. When you communicate findings, you need to ensure you highlight any patterns, *interconnections* or significant data that allows you to answer the research question and prove or disprove the hypothesis.

Analysing your results

Once you have collected and presented the data, you need to analyse your results. Use the information in the next section to help you complete your fieldwork analysis.

G4.2

fieldwork task 1

How do we degrade our local *environment*?

Understanding how landforms and landscapes form and change is vital so we know how to sustainably manage our environment for the future. Unfortunately, much of the change we make to the environment tends to affect it negatively or is unsustainable. This means we are not ensuring that the landscapes we enjoy now will be the same for future generations. Fieldwork is an important step in furthering our understanding of the environment and also in helping develop successful and sustainable management strategies.

Research location

Our research question for this fieldwork investigation states that we are considering environmental degradation on the local *scale*. Thus, you first need to decide what location you will work with for this fieldwork task. You may choose a part of your school, a local park or even the street where you live. Once you have decided on a location, you are ready to undertake some pre-fieldwork research.

Source 1

Locate an area that has clear erosion or has been impacted negatively by humans

Research question

Research question: What can we do on a local scale to reduce environmental degradation?

Hypothesis

Even small actions can increase the sustainability of our environment.

Source 2

Air pollution is an example of environmental degradation

Writing your report

Defining your topic and method

1 State your research location, question and hypothesis.
2 Provide a brief outline of the location, characteristics and history of your local region, using around four or five sentences.
3 State two primary fieldwork techniques you used to complete this fieldwork and describe how they were helpful in answering the research question.
4 State two secondary resources you used in the lead-up to the fieldwork and describe how they were helpful in answering the research question.

Presenting your data

As discussed previously, present the data you collected using tables, graphs, sketches with annotations, photographs and so on.

Analysing the data

Answer your research question: What can we do on a local scale to reduce environmental degradation?

Consider:

- What evidence of *environmental* degradation was present in this location?
- Was there large-*scale* or small-*scale* degradation?
- What was causing the *environmental* degradation?
- What evidence of this was there?
- Will the *environmental* degradation at this location have long-term impacts?
- How could the *environmental* degradation be managed sustainably?
- Could small changes to our behaviour reduce the impacts of the *environmental* degradation at this *place*?

If you completed a survey:

1 What were your main findings?
2 How does this help us answer the research question?
3 Were you surprised by these results? Why or why not?

Drawing conclusions

1 Summarise the key findings of your research.
2 Was the hypothesis correct (supported)? Why or why not?
3 Record five key pieces of evidence (such as data, photos, sketches or surveys) that support your answer above.

Evaluating your research

1 Which methods were the most successful for collecting data to answer the research question? Why?
2 Which methods were the least successful for collecting data to answer the research question? Why?
3 What changes to your fieldwork approach would you suggest to future Year 8 students completing the same task?

Handing in your report

Once you have completed all the steps of your fieldwork report, you are ready to collate your work into a folio and take action! What small changes will you make to ensure that your local environment is used sustainably for the future?

G4.3

fieldwork task 2

What are the challenges of planning for urbanisation in your *place*?

As our national population continues to grow, our communities are under pressure to change in order to meet the new demands and needs of their residents. This creates challenges for urban planners, who need to ensure that these areas have access to enough resources and sustainable living practices for both the locals and the new arrivals. Whether you live in a rural or an urban community, it is likely that your *place* has changed significantly over time. Fieldwork is an excellent way to review, monitor and predict change and challenges your area has faced or may face is the future.

Research location

As a class, you will need to determine a field site that has undergone some change over time. This change should be linked with an increase in population and may be obvious, such as a growth in urbanisation, or may be more subtle, such as a reduction in the size of a standard housing block or increased resources or infrastructure such as the appearance of major shopping centres or freeways.

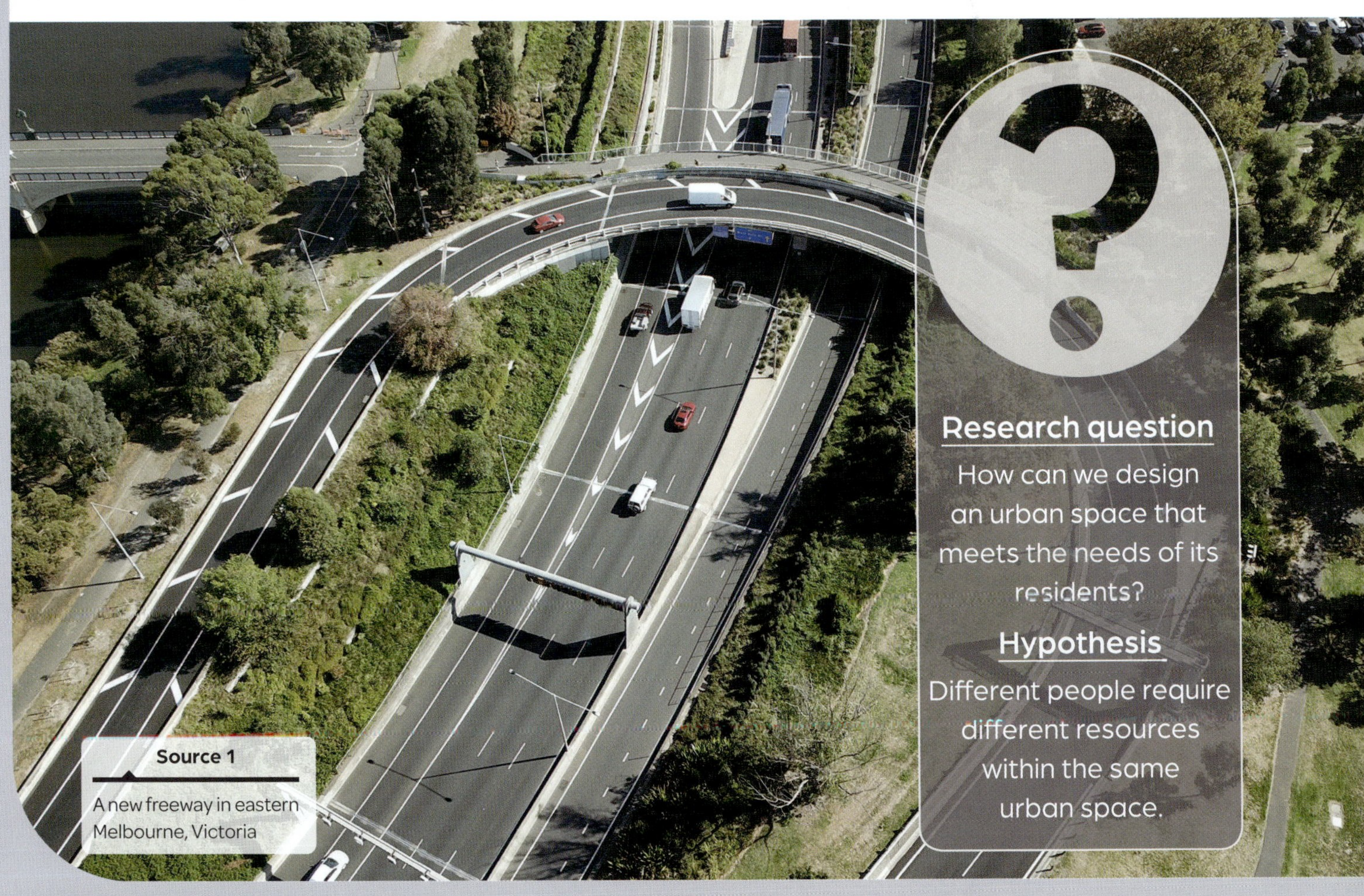

Research question

How can we design an urban space that meets the needs of its residents?

Hypothesis

Different people require different resources within the same urban space.

Source 1

A new freeway in eastern Melbourne, Victoria

Writing your report

Defining your topic and method

1. State your research location, question and hypothesis.
2. Provide a brief outline of the location, characteristics and history of your local region, using around four or five sentences.
3. State two primary fieldwork techniques you used to complete this fieldwork and describe how they were helpful in answering the research question.
4. State two secondary resources you used in the lead up to the fieldwork and describe how they were helpful in answering the research question.

Presenting your data

Present the data you collected using tables, graphs, sketches with annotations, photographs and so on.

Analysing the data

Answer your research question: How can we design an urban space that meets the needs of its residents?

Consider:

- What is urbanisation and why does it occur?
- When did or will urbanisation occur in your area?
- What new infrastructure will the urbanisation of your community provide?
- What are the different needs of the community?
- How can urbanisation help meet these needs?
- What are some negative impacts of urbanisation?

If you completed a survey:

1. What were your main findings?
2. How does this help us answer the research question?
3. Were you surprised by these results? Why or why not?

Drawing conclusions

1. Summarise the key findings of your research.
2. Was the hypothesis correct (supported)? Why or why not?
3. Record five key pieces of evidence (such as data, photos, sketches or surveys) that support your answer above.

Evaluating your research

1. Which methods were the most successful for collecting data to answer the research question? Why?
2. Which methods were the least successful for collecting data to answer the research question? Why?
3. What changes to your fieldwork approach would you suggest to future Year 8 students completing the same task?

Handing in your report

Once you have completed all the steps of your fieldwork report, you are ready to collate your work into a folio and take action! What changes would you make to your local area so it can plan for the challenges it may face in the future as population grows?

G4.4

fieldwork

How do geographers use surveys?

Surveys are useful in Geography to collect data in the field. Your surveys could be in the form of questionnaires to help gain an understanding of people's perspectives.

You could also do a vehicle survey, where you tally the number of vehicles that pass in a particular amount of time. Surveys provide you with a sample of what is occurring in a natural or human *environment*.

Surveys can be written by hand or prepared on software such as Survey Monkey or Google Forms.

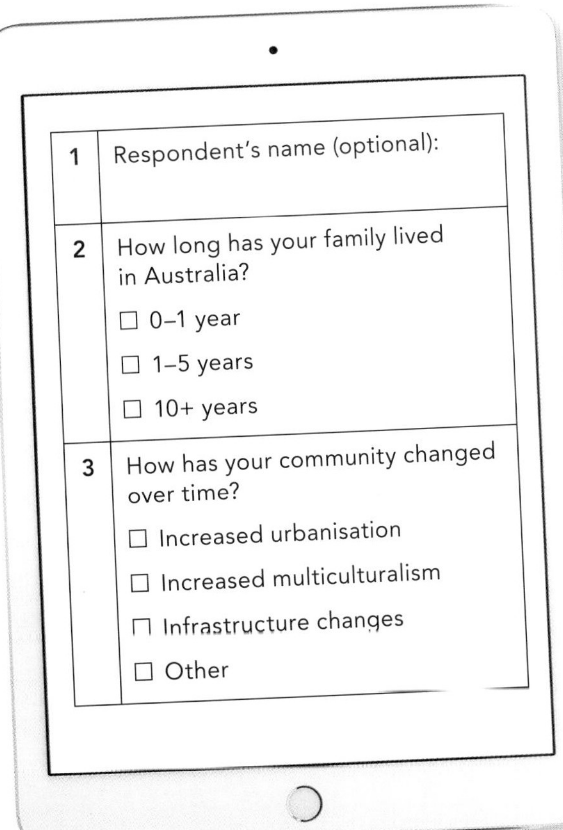

1	Respondent's name (optional):
2	How long has your family lived in Australia? ☐ 0–1 year ☐ 1–5 years ☐ 10+ years
3	How has your community changed over time? ☐ Increased urbanisation ☐ Increased multiculturalism ☐ Infrastructure changes ☐ Other

Source 1

A sample questionnaire

In the field

1. Are surveys a primary or secondary data collection method?
2. The stages of fieldwork include defining the research question, collecting data and analysing that data. In which stage would a survey be a useful tool?
3. Write a survey question that would gather quantitative data about landscapes and urbanisation.
4. Write a survey question that would gather qualitative data about urbanisation.
5. Why are multiple-choice questions often used in surveys?
6. Follow these steps to design a survey on a topic you are passionate about:
   a. Write an objective for your survey. Think about who will read the data and what decisions might be made based on the data you collect. Make sure your objective is short and clear.
   b. Write a title for your survey.
   c. Write two or three sentences to introduce and explain your survey to the people who will be completing it. Decide whether or not your survey will be anonymous.
   d. Choose how you will create your survey: on paper, on your computer using Word or Excel, or using an online platform such Survey Monkey or Google Forms.
   e. Using the method you chose in part d, write 10 survey questions. Keep your questions short and specific. Try to use multiple-choice questions to make analysing your data easier.
   f. Swap your survey with someone else in your class. Proofread and evaluate each other's surveys. Give your classmate appropriate feedback on how they could make their survey more effective.
   g. With your teacher's permission, conduct your survey in your class, school or community.

G5

Geo How-To

Knowing the locations of countries, their capitals – and even their flags – can be useful geographic knowledge. However, the key to success in Geography is understanding key skills and being able to apply them in different situations. This chapter will walk you through some of the key skills for Year 8 and provide you with examples of how to use them.

Mapping with BOLTSS (NA)

Every map requires BOLTSS in order to be understood by the geographic audience, and some people add (NA) to the acronym as well to remind them to be neat and accurate. When you construct maps, use the following checklist to ensure you have completed your mapping tasks correctly.

B Border

A border is important to show the edges of the mapping field. It provides a clear area for you to construct your map and makes it appear clear and neat to readers.

O Orientation

An orientation, or compass, helps us understand direction when reading the map. Orientations should be drawn as a 16-point compass.

N
nnw nne
nw ne
wnw ene
W E
wsw ese
sw se
ssw sse
S

Source 1

A 16-point compass

Street map of Williamstown, Victoria

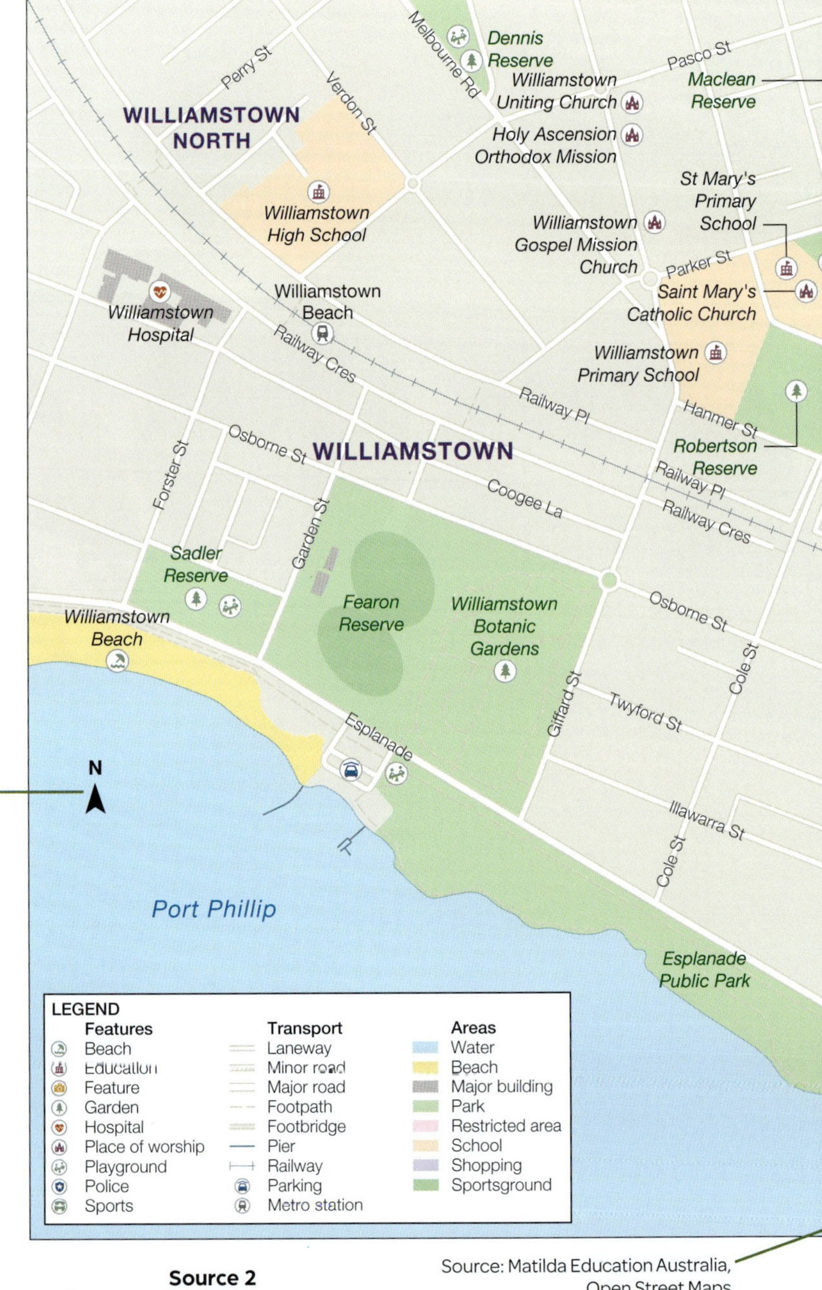

Source: Matilda Education Australia, Open Street Maps

Source 2

This map features all aspects of BOLTSS

HIGHWAY
MAJOR ROAD
SECONDARY ROAD
RIVER
PROPOSED ROAD
BRIDGE
HOUSING WITH NUMBER BESIDE ROAD
1
20
21
25
26
LIGHT RAIL & STATION
TRAIN RAIL & STATION
STATE BOUNDARY
CITY BOUNDARY
PETROL STATION
RESTING AREA
SCHOOL
POLICE STATION
BUS STATION
LIGHT RAIL STATION
TRAIN STATION
AIRPORT
SEAPORT
FIRE STATION
LIBRARY
POST OFFICE
PLACE OF INTEREST
INFORMATION KIOSK
HOSPITAL

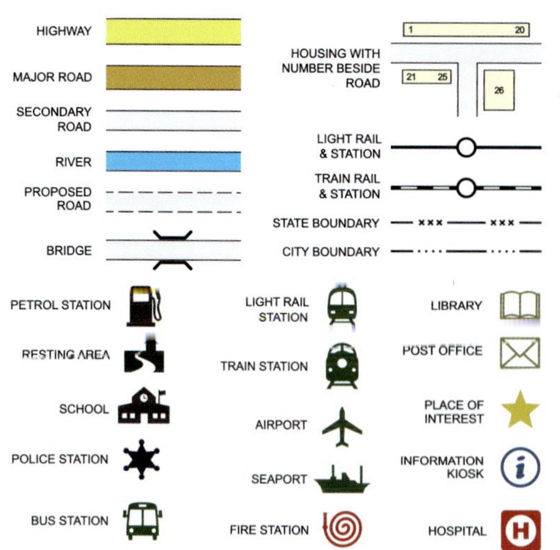

Source 3

A sample legend

L Legend

A legend, or key, is vital to understanding the map. Without a legend, colours and symbols would not make sense and we would not be able to interpret patterns or distributions.

T Title

A title gives us an understanding of what the map is showing. If you are drawing a sketch map, you should also provide a date and time. This allows you to monitor change in a location over time.

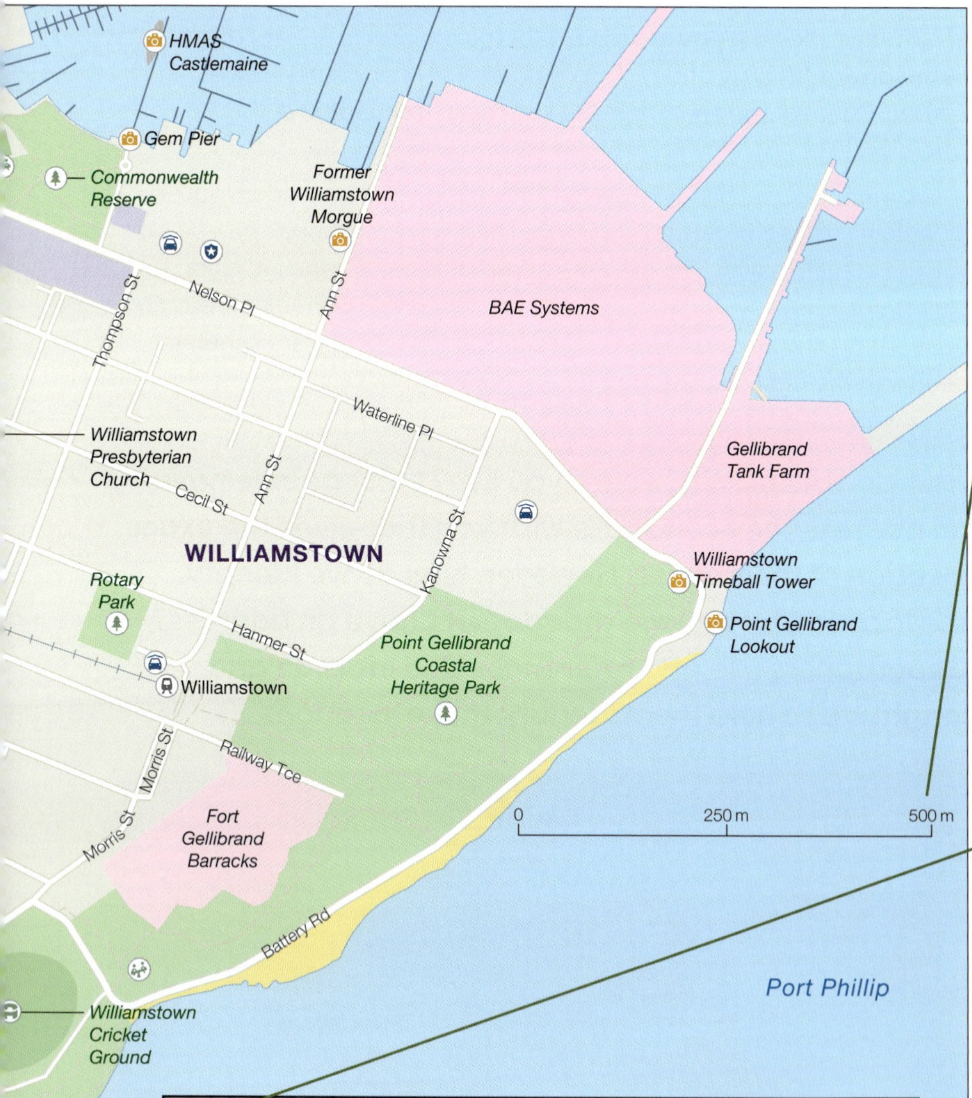

HMAS Castlemaine
Gem Pier
Commonwealth Reserve
Former Williamstown Morgue
Nelson Pl
Thompson St
Ann St
BAE Systems
Waterline Pl
Williamstown Presbyterian Church
Cecil St
Ann St
Gellibrand Tank Farm
Kanowna St
WILLIAMSTOWN
Rotary Park
Hanmer St
Williamstown Timeball Tower
Point Gellibrand Lookout
Point Gellibrand Coastal Heritage Park
Williamstown
Railway Tce
Morris St
Morris St
Fort Gellibrand Barracks
0
250 m
500 m
Battery Rd
Port Phillip
Williamstown Cricket Ground

S Scale

A scale provides us with information on how big something is in real life. While a house on a map may be 2 centimetres across, it may be representing a 15-metre wide three-bedroom home. Scales can come in many forms: linear, ratio or fraction.

0
10
20 km
1
1 000 000
1:1 000 000

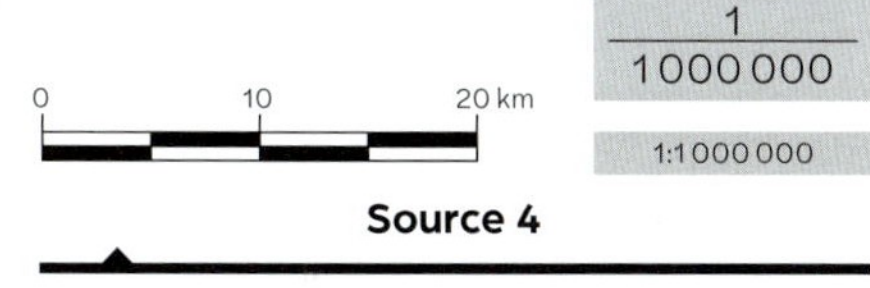

Source 4

Map scales: linear, fraction and ratio

S Source

Always acknowledge the source of the information that you illustrate on your map. The source can also indicate whether the map is reputable or not.

(NA) Neatness and Accuracy

Some people add the final letters 'NA'. When we read a map, we rely on it being neat and legible and we expect that its data is displayed accurately. Therefore, when you construct a map it is important you take the time to correctly illustrate the patterns and distributions you see in the data.

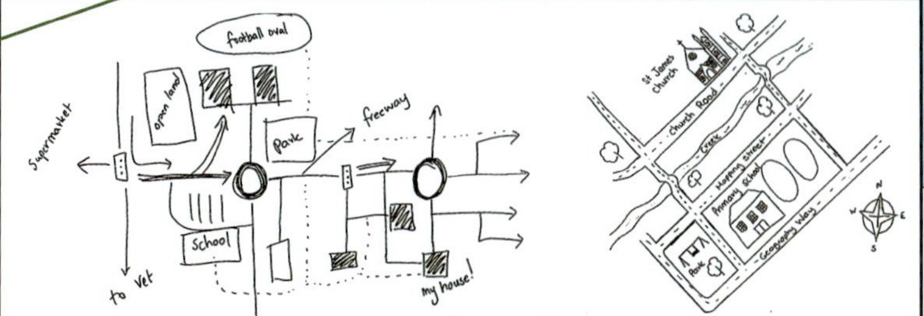

football oval
open land
Supermarket
freeway
Park
School
to Vet
my house!
St James Church
Church Road
Creek

Source 5

Which sketch map is easier to read?

HOW TO

Mapping skills

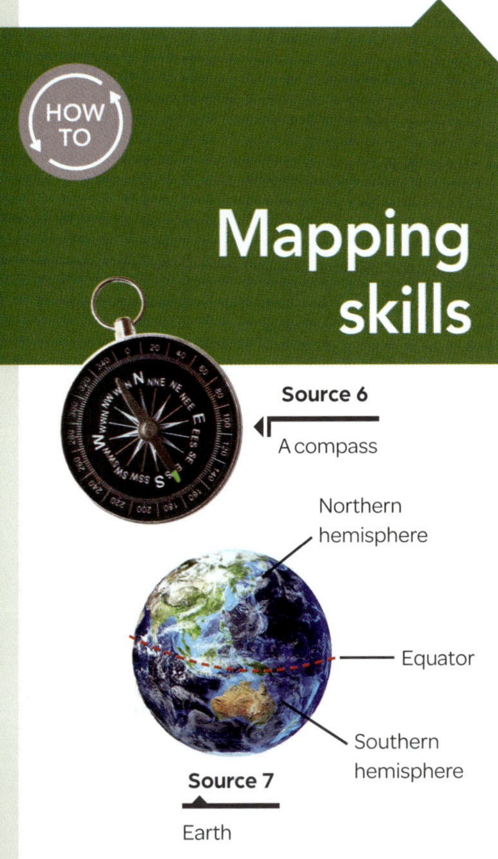

Source 6

A compass

Source 7

Earth

Direction

Up until now, direction may have been described to you as 'left' and 'right', 'above' or 'below'. While these words are helpful, in Geography we also need to use compass points: north, south, east and west.

Consider the world globe in Source 7. Around the centre there is a line of latitude called the **equator**. North of the equator is the **northern hemisphere** which includes the continents of North America and Europe. South of the equator is the **southern hemisphere**; this is where we live!

Sometimes, we mistakenly use the word 'above' instead of 'north' or 'below' instead of 'south'. If you say something is 'above' the equator, you are actually saying it is floating in the air over the top of it! If you say something is 'below' the equator, you are describing something buried beneath it! Use directional terms carefully to ensure you are sending people in the right direction.

Latitude and longitude

How do you identify where you live? Most people have an address they can share with you that gives the exact location of their house. However, how do we identify the location for a place that doesn't have an address? Geographers use grid references and latitude and longitude to help them identify these locations.

Source 8

World map with lines of latitude and longitude shown.

Latitudinal lines are shown running around Earth like a series of belts. We use the equator as a reference point, so it represents 0° latitude. Lines north of the equator are numbered 1, 2, 3 etc. degrees north (°N), all the way to 90°N at the North Pole. Lines south of the equator are 1, 2, 3 etc. degrees south (°S), all the way to 90 °S at the South Pole.

Longitudinal lines are like jail bars wrapping around the globe from east to west. These lines are measured starting from the Prime Meridian (the line that passes through Greenwich, in the UK), which is 0° longitude. Lines of longitude east of the Prime Meridian are 1, 2, 3 etc. degrees east (°E) and lines of longitude west of the Prime Meridian are 1, 2, 3 etc. degrees west (°W). On the map in Source 8, Manila can be described as having an absolute location of 15°N and 120°E, as it is north of the equator on the 15° line of latitude and 120° east of the Prime Meridian.

Scale

Look at Source 7 again. Logically, we know that this image of Earth is not to scale. In reality, the Earth is over 12 700 km in diameter!

Scale is important in Geography as it allows us to shrink maps and other images down to a size where we can see patterns, distributions and changes. Typically, scale is displayed using a line that acts like a ruler, showing you how many centimetres on the map represent the real distance.

Scale can also be used to describe changes that occur over time. A 'large-scale change' indicates that something has caused major alterations to a region. A 'small-scale change' indicates that little has been altered. Review the images in Source 10. What is the scale of the city's change?

Source 9

Obviously, this map is not to scale. It is a large area that has been shrunk down to fit on this page. The scale on this map helps us determine how big things are in real life.

Source 10

Change over time to Surfers Paradise, 2007 to 2019

In geography we use maps and graphs to understand what is happening around us. In many cases, these resources provide us with information on patterns and spatial distributions of phenomena.

The formula PQE helps us describe these patterns and distributions. P stands for pattern; Q stands for quantify and E stands for exception.

P Pattern

A pattern is a trend in the data. When looking for a pattern, read the legend and interpret what the colours or symbols mean. On a graph or map, you may notice that all the data points tend to be clustered in one spot or that there is an uneven distribution of data points. You may need to use the names of places or even your compass points to describe where on the map these clusters appear. For example, when observing Source 11 at the right, we notice that most countries in the northern hemisphere have higher literacy rates than those in the southern hemisphere.

Descriptive words that may help you describe patterns are: *clustered, even, uneven, highly distributed, north, south, east, west, increase, decrease and fluctuate.*

Q Quantify

When we quantify our pattern, we need to use numerical data to provide evidence of what we see. You could gain data by using the legend, measuring using the scale or conducting a count. You need to ensure that the data you provide relates directly to the pattern you recorded earlier. For example, we noticed that countries in the northern hemisphere tend to have higher literacy rates than those in the southern hemisphere: the US, Europe and China all have high literacy levels. To quantify, only seven countries in the southern hemisphere have more than 95 per cent literacy.

E Exception

An exception is a trend on the map or graph that doesn't 'fit in' with our original pattern statement. When we observe an exception, it is also good to quantify it to provide a comparison to our original statement. For example, previously we noticed in Source 11 that countries in the northern hemisphere tend to have higher rates of literacy than those in the southern hemisphere. However, Australia is in the southern hemisphere and has a population with over 95 per cent literacy. This is an exception to the pattern.

What is the difference between qualifying data and quantifying data?

PQE helps us describe patterns and distributions. When we describe, we 'say what we see'. In a PQE analysis, we do not explain or give a reason why we see patterns; this is done in a SHEEPT analysis (page 140).

To quantify means to use percentages, counts, ratios or data to provide details about the patterns you are describing. To qualify a statement means to use general describing words such as 'large', 'many', 'broad' or 'small' to describe a pattern or change.

By using quantifiable data, we can more easily see key differences between locations or monitor change over time. Imagine that your PQE analysis stated: 'The ability to read varies worldwide'. Does this sentence help you show exactly what is happening around the world? Or does this quantified statement provide more detail: 'Seven countries in the southern hemisphere have populations with over 95 per cent literacy'?

How do I start my PQE sentences?

When writing a PQE analysis, start sentences with the following key terms:

Pattern:

Overall ...

For example:

Overall, countries in the northern hemisphere tend to have higher literacy rates than those in the southern hemisphere.

Quantify:

To quantify ...

For example:

To quantify, only seven countries in the southern hemisphere have more than 95% literacy.

Exception:

However ...

For example:

However, Australia is in the southern hemisphere and has a highly literate population.

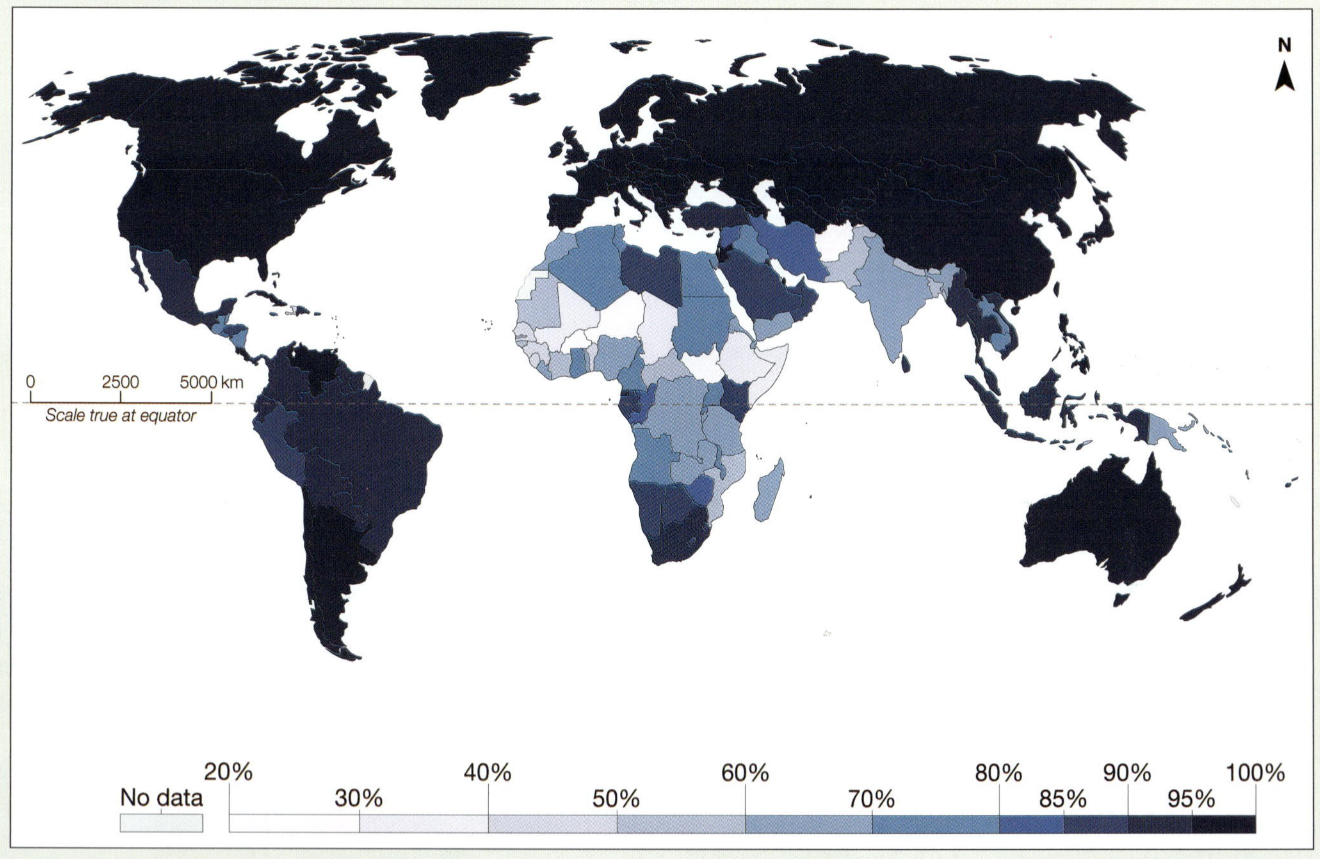

Source: Map created by Our World in Data with data from CIA Factbook (2016)

Source 11

World map of literacy rates

SHEEPT

SHEEPT is an acronym that helps you remember the reasons why a spatial pattern occurs. It stands for: Social, Historical, Economic, Environmental, Political and Technological.

S Social

1

Social factors are anything to do with people. Social factors include population, culture, language and religion. Here we see social differences between living conditions in India.

H Historical

Historical factors are anything to do with our past. Historical events, buildings, people and changes to climate all influence what we see in our world today. India's urban landscape has changed significantly over time.

E Economic

2

Economic factors are those relating to money. In Geography, income, costs of things and how much money is spent can provide us with information on a place. There are large scale economic differences between communities in this place.

How do I write a SHEEPT analysis?

SHEEPT is usually used to explain why patterns or distributions may occur in a particular region. It can also be used to expand our thinking when annotating images or considering new geographical content. The following is an example of how we can write a SHEEPT analysis for an image. The highlighted terms indicate the use of a SHEEPT term. Can you identify all of them? The analysis does not need to include every term.

Source 12 is an image of slums in Mumbai, an Indian megacity of more than 12 million people. Rural-urban migrants come in the hope of better economic conditions but find it difficult to afford housing. In Mumbai, 42 per cent of the population live in slums, built of recycled materials on any available open land. Lack of infrastructure and basic amenities such as a safe water supply and sanitation produce dangerous health conditions for the people living in slums. These are among the most overcrowded and densely populated places to live. Historically, governments have tried to improve services in slums and more recently there have been plans to relocate people to millions of affordable homes away from the city centre.

Source 12
Mumbai, India

E Environmental (3)

Environmental factors are those relating to the natural or human environments on Earth. Humans can manipulate the environment to suit their needs. Here we see how parklands have been urbanised to support a growing population.

P Political

Political factors are those to do with the government or leading groups. When we consider political factors, we refer to laws and policies.

T Technological

Technological factors relate to the different kinds of technology that we have access to. This could be in the form of gadgets, **spatial technology**, medical technology or even roads, transport or basic machines used in the home.

Sketches and annotating

Field **sketches** are an excellent way of recording data when you are investigating a research question.

Sketches allow you to annotate movement, patterns or any interconnections you see. Field sketching is not a test of your artistic skills – the idea is to record a simplified version of what you can see.

T Title (1)

Provide a heading for your sketch that is clear and provides information on the location. You may wish to record both the **absolute** and **relative locations**.

O Orientation (2)

An orientation shows the direction that you were facing when you conducted your sketch. In order to record a correct orientation, you need to use a compass.

A Annotations (3)

Annotations are the most important thing to complete when drawing a field sketch. Annotations allow you to record details about what you see and explain how elements of your drawing relate back to the research question. Ensure that lines pointing to your annotations are completed with a ruler and do not overlap.

TROUPS CREEK WEST WETLAND
(1) NARRE WARREN NORTH:
7.6.19 12.50pm (5)

(3) A path allows individuals to walk around the wetland and enjoy the outdoor environment

When you are making a field sketch, you need to remember TOASTIE!

TOASTIE will help you remember the key skills when making a field sketch.

It stands for Title, Orientation, Annotations, Scale, Time, Information and Edge.

Source 13

An annotated field sketch

S Scale 4

Most sketches are not to scale. However, all geographical maps and sketches require a scale to give the reader some indication of size. To estimate a scale, use a metre-rule or pace out an area that you have sketched. Then, using a small ruler, identify how large the same area is on your drawing. For example, you may estimate that the path you are looking at is 1 metre wide and when you measure your drawing of the path it is 1 centimetre wide. Therefore, your rough estimated scale is 1 m = 1 cm.

T Time

By recording the time your sketch was completed, you can analyse how the environment changes over the course of a day, a month or even years!

I Information

Provide more than just one-word annotations on your sketch. Annotations should be at least one sentence and help your reader to identify any patterns.

E Edge

Draw a border so it is clear where your sketch starts and ends.

Visual communication

Flow diagrams

Flow diagrams are helpful to show processes or how different parts of the environment are *interconnected*. The arrows represent the movement between stages or the connection between things.

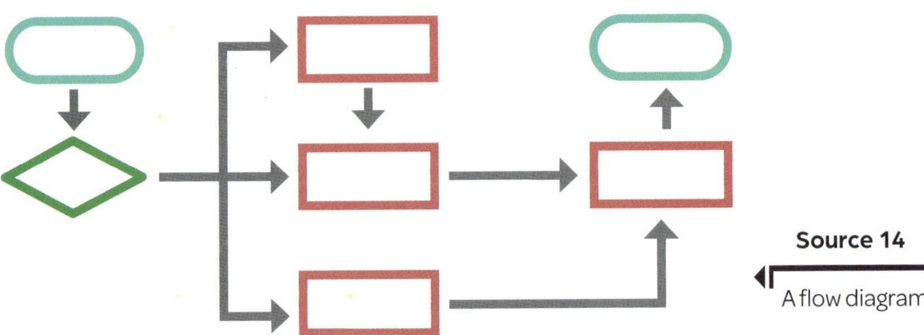

Source 14

A flow diagram

Block diagrams

Block diagrams are a useful way of drawing landscapes to show what is happening both above and below ground. For example, when we consider the movement of water, block diagrams help us visualise how water can move both on the surface of the Earth through rivers, oceans and streams, as well as under the ground through water tables and pipes.

When you create a block diagram, you need to ensure that you draw it three dimensionally, with equal amounts of drawing room both above and below the surface of Earth. Annotations are important when drawing a block diagram as this is how your audience can interpret your illustration.

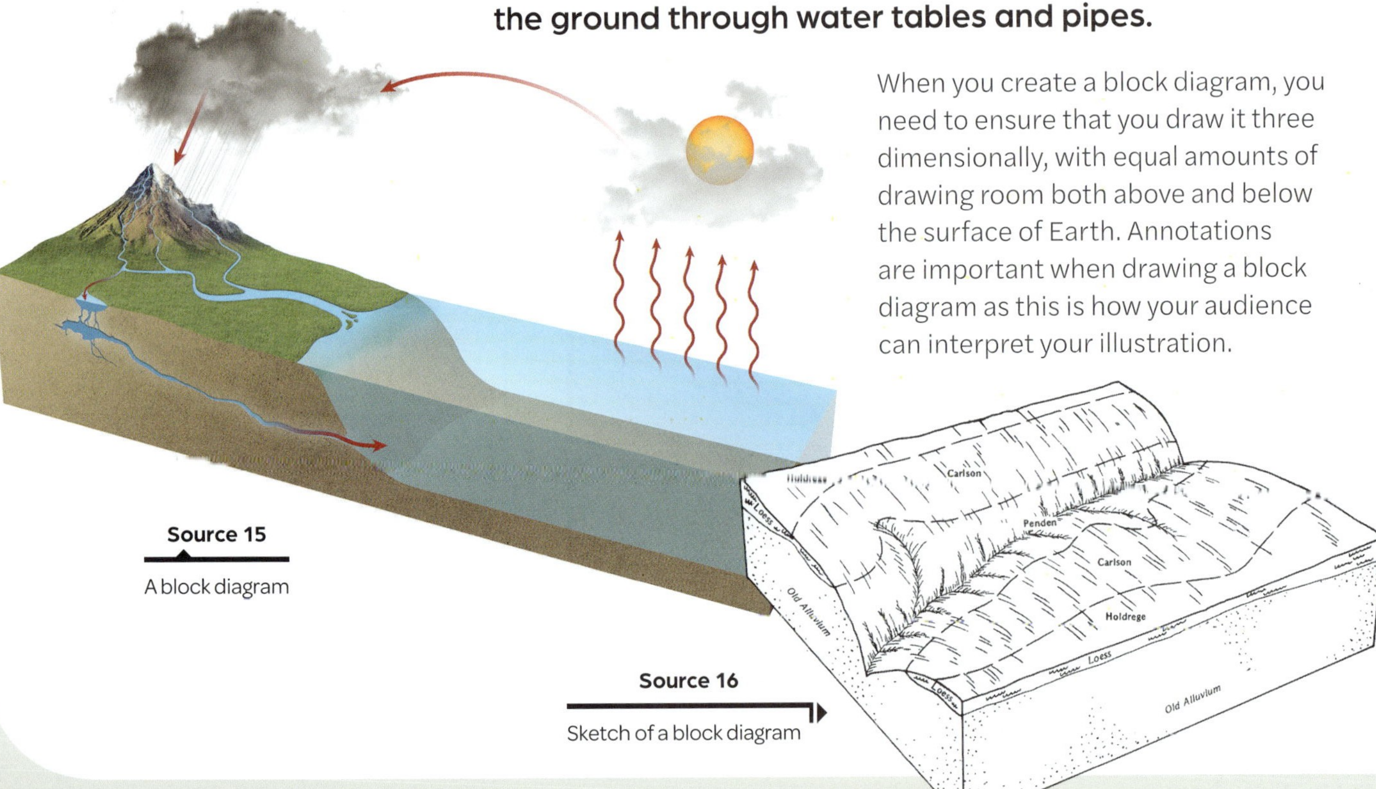

Source 15

A block diagram

Source 16

Sketch of a block diagram

Photo essays

A **photo essay** is a way of presenting information visually to show characteristics of a place or process. A photo essay usually includes a series of photos with specific annotations or captions that provide a brief background into the key features of the image or the meaning behind the image choice.

Source 17

A photo essay

Historical Hampton Beach

Modern-day Hampton Beach

Department of Primary Industries sign at Hampton Beach, which indicates that the intertidal zone is protected to a depth of two metres.

Early boating at Hampton Beach

Hampton Beach bathing boxes and pier c. 1900

Aerial view of Port Phillip Bay, including Hampton Beach

Graphing

Simple graphs

Graphing is an important way to display geographic information. It shows us patterns and changes over time. When creating a graph, think SALTS!

S Scale

The scale of your graph will depend on the data you are trying to visualise. The easiest way to create a graph axis scale is to identify the lowest and highest value (range) and then fill in the numbers in between so you can mark your data points easily.

A Axis

Each graph has an *x*-axis and *y*-axis. The *x*-axis is the horizontal axis and the *y*-axis is the vertical. Make sure you label each axis!

L Legend

A graph often uses colours to represent data. A legend indicates to the audience what these colours mean and how to read the data.

T Title

A title lets your audience know what your graph is showing.

S Source

When you graph information, it is important to acknowledge where you gained the data from. Maybe you collected it yourself or maybe you sourced it from a website. By stating the source of your information, the reader knows how reliable it is.

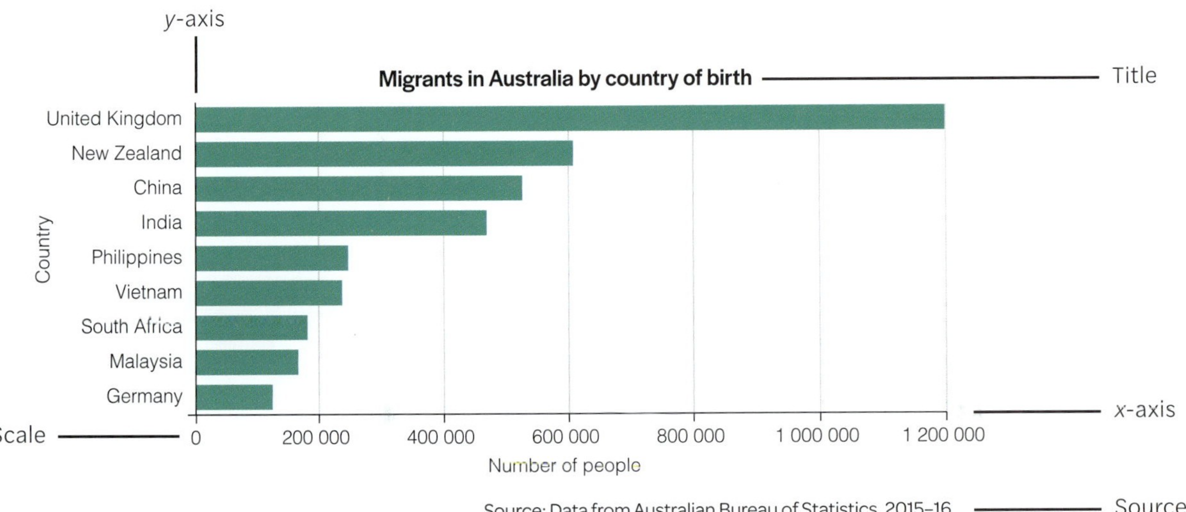

Source 18

Graph showing migrants in Australia by country of birth, 2015–16

Climate graphs

A **climate graph** is simply two graphs in one.

The graph below shows the temperature (degrees Celsius) on the left *y*-axis, which relates to the red average temperature line; and the amount of precipitation (mm) on the right y-axis, which relates to the blue columns. The horizontal axis or *x*-axis, shows us the months of the year. We notice that the red line, or average temperature, fluctuates throughout the year, peaking in December to March and decreasing from April to July. In a similar way, precipitation also fluctuates. However, it tends to be highest in the spring from August to November, with the exception of another peak in May. Using two PQE analyses (one for temperature and one for precipitation) is the best way to describe the patterns we observe in a climate graph.

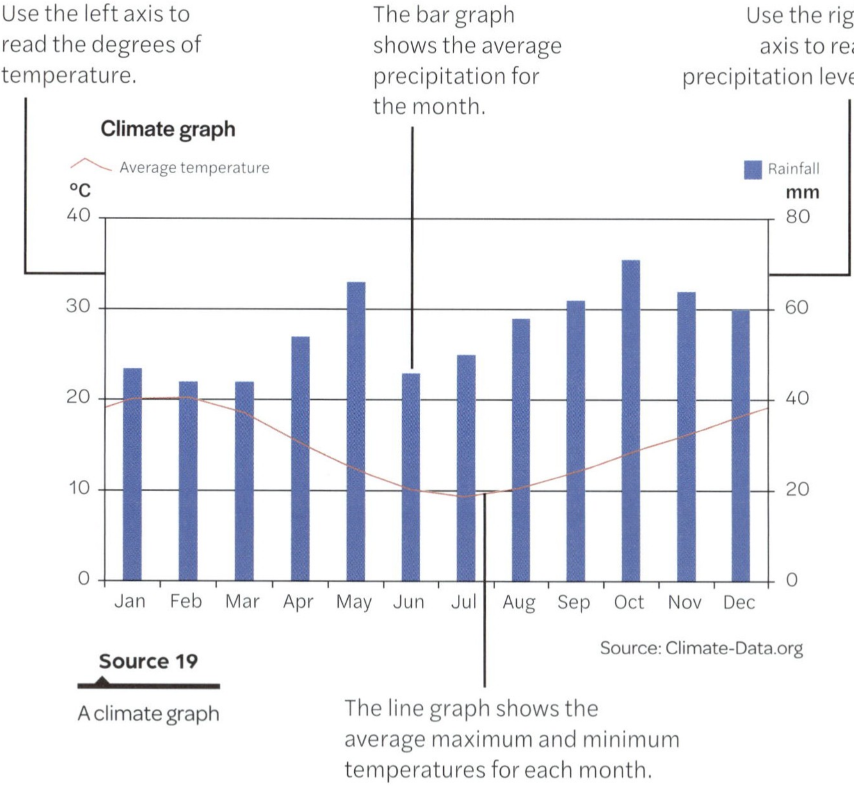

Source 19

A climate graph

Here we need to apply SALTS! Label the axes, provide a title and a source. The legend is part of the axis labels: red line denote average temperature in degrees Celsius; blue columns denote rainfall in mm.

What is the difference between weather and climate?

Look outside your window and describe today's temperature and the amount of rainfall. What you have just described is the weather. Weather changes daily but we can usually predict it up to 10 days in advance.

Now close your eyes and describe the 'climate' of Australia. Do you imagine Australia as mostly hot and dry? Just because we can describe Australia as hot and dry, does not mean it is like this everywhere, all year round. Unlike the weather, climate helps us describe the yearly (annual) average temperature and level of precipitation (rainfall, snow etc.) in a region or country. Climate graphs help us visualise the climate of a region or country. Climate change describes how the average temperature and precipitation levels of a location change over time.

Population pyramids

Population pyramids are graphs that show the number of females and males in particular age groups in a population; they are like bar graphs turned on one side.

Population pyramids can be made on a local, national or global scale. On a population pyramid, female data is normally shown on the right side and male data on the left. The length of each bar represents the number of males or females within that age group.

The shape of the pyramid tells us about the population:

- Triangular means the population is growing, because there are more young people than old
- Box-like means growth is slow or stable, as there are roughly equal numbers of old and young people.
- If the shape becomes wider towards the top, like a reduced pentagon, it represents an ageing or declining population, as there are more middle-aged people than young people.

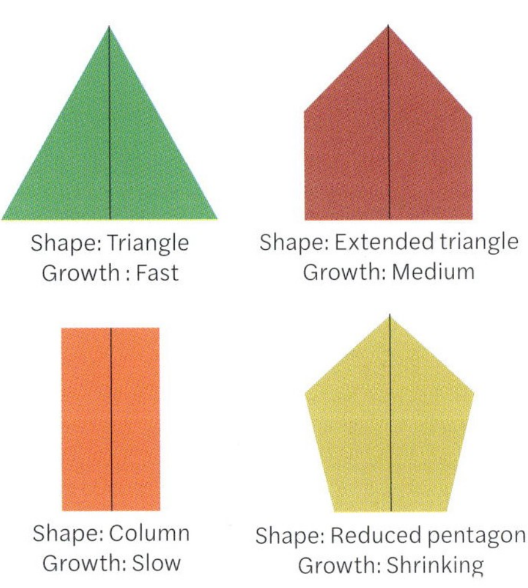

Source 20

Various population pyramids. The shape of the pyramid tells you about the population.

How do I interpret a population pyramid?

Source 21 shows population pyramids for Australia in 1955 and 2018. In 1955, there were many young people in the age range 0–14. Approximately 5.5 per cent of the population were 0–4-year-old males and 5.3 per cent were 0–4-year-old females. We also had many working-age people. For example, 4.1 per cent of the population were 30–34-year-old males and 3.8 per cent were 30–34-year-old females. In 2018, there are significantly fewer young people and more old people in our population. Our population also grew to 25 million people.

Source 21

Population pyramids showing Australia's population in 1955 (left) and 2018 (right)

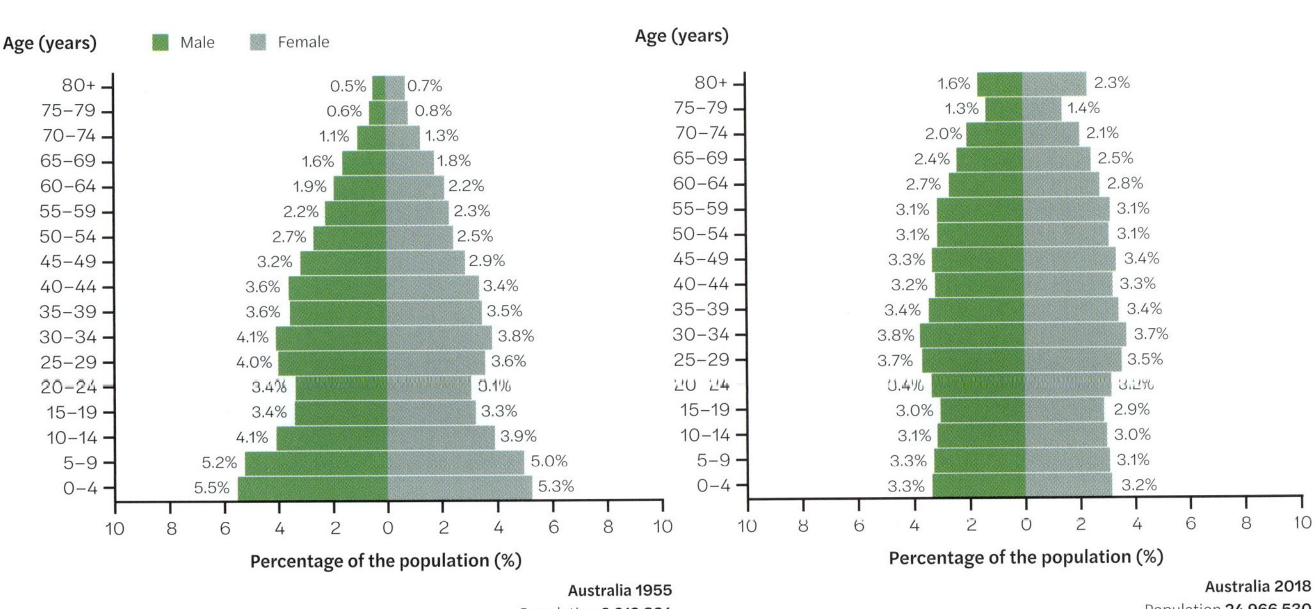

Source: PopulationPyramid

HOW TO

Satellite images

Satellite images are pictures of Earth taken from space. Satellite images give us the most information if taken during the day. Satellites orbit Earth and constantly take images and so we can use this data to see a change over time in land cover or other spatial patterns.

When trying to interpret a satellite image, it is best if you look for colour and shapes. Colour can give you an idea of land cover. For example, green usually indicates vegetation, blue is the colour of water and brown is desert or barren land. White can indicate either snow or clouds and so we use shapes to tell them apart. Clouds tend to look fluffy and can sometimes (depending when the images were taken) cast a shadow. Snow usually appears on the tops of mountains and you can observe where it is melting or following the slopes. Looking for shapes is also helpful when identifying rivers or reservoirs, which can be seen as meandering blue lines.

Seasonal variations can sometimes be very clear in satellite images – in spring and summer, green vegetation flourishes, while during winter, white snow may fall or vegetation may die, leaving brown bare ground.

Source 22

Satellite images taken at night provide different information than those taken during the day. This photo of the US shows the extent of its urban areas.

Source 23

This satellite image was taken at the same place over four seasons. Satellite images are useful to show change over time.

Source 24

Example of an image taken from a satellite that orbits around Earth. We can use this information to answer geographical questions and understand spatial patterns.

Infographics

Collecting and displaying data is an important skill for Geographers. As quantitative data tends to be quite complicated, we can sometimes use infographics to communicate these findings. **Infographics** are posters that use images, graphs, shapes and colour to illustrate to the audience key patterns and trends. Infographics are an engaging way to present information if used properly.

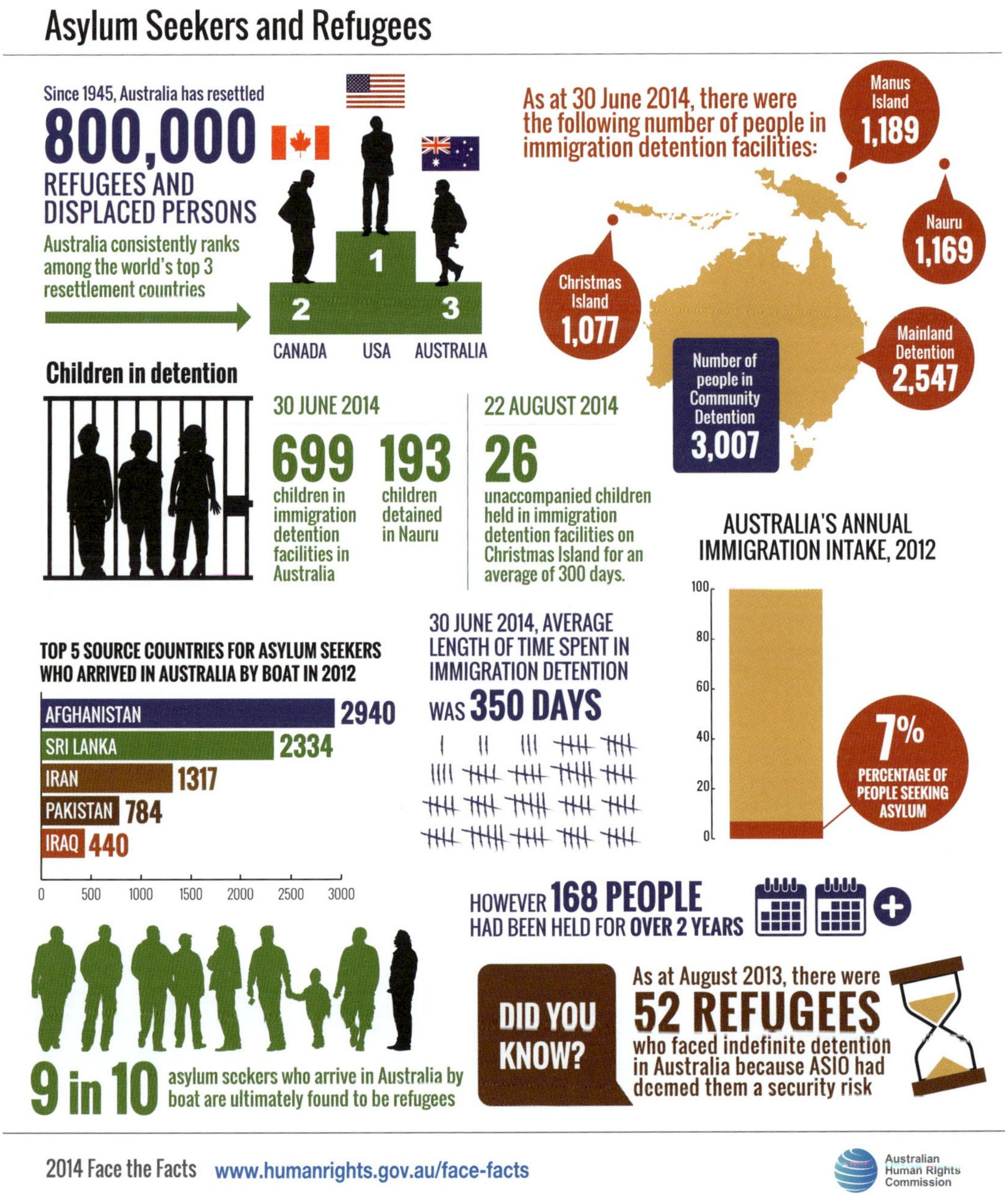

Asylum Seekers and Refugees

Since 1945, Australia has resettled **800,000** REFUGEES AND DISPLACED PERSONS

Australia consistently ranks among the world's top 3 resettlement countries

1 USA, 2 CANADA, 3 AUSTRALIA

As at 30 June 2014, there were the following number of people in immigration detention facilities:

Manus Island 1,189

Nauru 1,169

Christmas Island 1,077

Mainland Detention 2,547

Number of people in Community Detention 3,007

Children in detention

30 JUNE 2014

699 children in immigration detention facilities in Australia

193 children detained in Nauru

22 AUGUST 2014

26 unaccompanied children held in immigration detention facilities on Christmas Island for an average of 300 days.

AUSTRALIA'S ANNUAL IMMIGRATION INTAKE, 2012

100, 80, 60, 40, 20, 0

7% PERCENTAGE OF PEOPLE SEEKING ASYLUM

TOP 5 SOURCE COUNTRIES FOR ASYLUM SEEKERS WHO ARRIVED IN AUSTRALIA BY BOAT IN 2012

Country	Number
AFGHANISTAN	2940
SRI LANKA	2334
IRAN	1317
PAKISTAN	784
IRAQ	440

0, 500, 1000, 1500, 2000, 2500, 3000

30 JUNE 2014, AVERAGE LENGTH OF TIME SPENT IN IMMIGRATION DETENTION WAS **350 DAYS**

HOWEVER **168 PEOPLE** HAD BEEN HELD FOR **OVER 2 YEARS**

9 in 10 asylum seekers who arrive in Australia by boat are ultimately found to be refugees

DID YOU KNOW?

As at August 2013, there were **52 REFUGEES** who faced indefinite detention in Australia because ASIO had deemed them a security risk

2014 Face the Facts www.humanrights.gov.au/face-facts

Australian Human Rights Commission

Source 25

An infographic

Cross-sections

A cross-section shows the shape of a geographical landscape or landform from the side, as if it had been sliced by a knife. The cross-section in Source 26 shows Uluru as you would view it if you were on the ground.

Contour lines are used to construct a cross-section. The overlay on the aerial photograph of Uluru in Source 26 is a contour map. As discussed previously, the contour lines are the numbered lines that join places of equal height. Close contour lines indicate a steep slope and widely spaced contours mean a gentle slope.

Source 27 shows how to draw a cross section:

a Place a straight edge along the line between points X and Y. Mark each contour line at the point it touches the edge of the paper and record the height.

b Draw a vertical scale similar to the one in Source 27, with one centimetre representing 100 metres. Place the marked edge of the paper underneath and mark the appropriate height directly above with a dot. Join the dots with a smooth line and add labels.

Source 26

An aerial photo of Uluru with contour lines overlaid. (The ABCD labels are to help students answer question 6 on page 41.)

C
A
B
D
N
550 m
600 m
650 m
700 m
750 m
800 m
850 m
850 m
800 m
750 m
700 m
650 m
600 m
550 m
0 0.5 1 km

Source: Contour map from Matilda Education Australia, Geoscience Australia

Source 27

A simple cross-section of Uluru

Height above sea level in metres
800
700
600
500
X 600 700 800 800 700 600 Y

Glossary

absolute location a location's exact place on Earth, often described in terms of latitude and longitude

active (volcano) a volcano that is erupting or likely to erupt in the future

aesthetic value appreciation of how a place looks and its perceived beauty, uniqueness or connection to a person

backwash water returning to the sea from the beach

basalt dark-coloured volcanic (igneous) rock formed by the cooling of lava

beach nourishment replenishment of a beach by sand from another source

bias to be for or against an idea, a person or a group, especially in a way that could be thought of as unfair

block diagram a three-dimensional diagram that shows how a system or feature works

block mountain a mountain that forms when blocks of rocks are forced upward or downward along faults or cracks in Earth's crust; also known as a **horst**

blowout an eroded bare patch on a sand dune usually created by damage to vegetation that then allows the wind to carve out a crater in the dune

BOLTSS(NA) acronym for Border, Orientation, Legend, Titles, Scale and Source (Neatness and Accuracy); used when constructing maps

breaking wave a wave that is forced to spill forward as the bottom of the wave orbit comes into contact with the sea floor and slows down

breakwater a rock wall built parallel to the beach and located in the water, designed to direct the force of the waves at the wall and not at the beach and sand dunes behind it

butte an isolated hill with vertical sides and a small flat top; smaller than a mesa

chemical weathering when rocks are broken down by a chemical change

civil war war between citizens in the same country

climate the average weather in a certain place, over a period of time

climate graph a graph showing average rainfall and temperature for an area over a period of time

commuter town a town separated from suburbs by farmland and open space, but connected by roads and rail; people in these towns commute to the city for work

constructive wave a small and infrequent wave that deposits sand on the beach rather than scouring it away; responsible for creating depositional landforms such as beaches, sand dunes, sand bars, spits and tombolos

conurbation a process whereby whole cities merge together into a continuous urban area

converging margin an area on Earth where tectonic plates collide

counterurbanisation a process whereby people move from an urban area to a rural or coastal area to escape urban problems such as expensive housing, traffic and pollution

crest the top of a wave; where it peaks

cross-section an image that shows the shape of a geographical landscape or landform as though it has been sliced vertically by a knife

currency a system of money in general use in a country for the exchange of goods and services

decentralisation a process whereby governments encourage population growth and job creation in areas outside major cities

deforestation the clearing of a forest to make room for a different land use such as farming

democracy rule by the citizens, who elect officials and leaders

demonstration a protest by a group of people in support of a cause

desert dry environments that receive less than 250 millimetres of rain per year; they can be hot or cold and be rocky, sandy or even covered in ice

desertification the process whereby land along the edge of a desert becomes damaged by drought and through overuse by humans, and is replaced by desert

destructive wave a large and powerful wave that constantly crashes onto the coastline, wearing away cliffs and digging out areas of beach

direct action using protests, strikes, petitions and other forms of protest to achieve demands

diverging margin where tectonic plates move apart

dome mountain a mountain that forms when magma is injected between two layers of sedimentary rock, causing the top layer to bulge upwards

dormant (volcano) an inactive volcano that may erupt in the future

drainage basin an area of land where precipitation collects and drains into a river, bay or other body of water

dredge to clear or deepen an ocean or river bed by scooping up material with a machine known as a dredger

dune a mound or ridge of sand or loose sediment piled up by the wind, especially on beaches or in deserts

earthquake the sudden shaking of the ground as a result of movement in the Earth's crust

economy how money is made and used in a country or region, including the production and consumption of goods and services

effusive eruption a volcanic eruption where magma flows out of the volcano as a lava

epicentre the central point of a difficult or unpleasant situation, such as an earthquake or a pandemic

equator a geographic coordinate line through the middle of Earth dividing it into the northern and southern hemispheres; the line is at latitude 0°

erosion the process whereby rock particles are moved by flowing air (wind), water (rivers, sea and rain) or ice (glaciers)

exchange rate the value of one **currency** compared to the value of another currency

explosive eruption a volcanic eruption where magma full of bubbling gas explodes as it reaches the surface

extinct (volcano) a volcano that is unlikely to ever erupt again

flow diagram a diagram showing how one thing flows to another; arrows show the movement between stages, or show how different items are connected

fold mountain a mountain that forms when two tectonic plates collide and rock is folded upwards

geographic characteristics occurring feature of a place, such as its landforms and ecosystems

geomorphic hazard a danger that originates at or near Earth's surface, such as a volcanic eruption, an earthquake or a landslide

graben dropped blocks along a fault line in the formation of block mountains

Gross Domestic Product the total value of all goods produced and all services provided in a country over a year

groundswell wave a steep and fast wave formed by a strong wind storm in the open ocean that creates long wavelength waves that move quickly into shallow water before breaking

groyne a wall that extends out from a beach into the sea, designed to trap the sand moving along the beach and also direct waves away from the shore, reducing erosion

headland narrow, high land jutting out into the sea, formed of harder rock that resists erosion

horst uplifted blocks along a fault line in the formation of block mountains

hot spot a location on Earth where tectonic plates move over very hot parts of Earth's mantle and rocks melt to generate magma for volcanic activity

humanitarian aid donated goods such as food, blankets and medicine, and help such as doctors and transport, given to save lives and provide assistance to people following a natural or human-made disaster

hypothesis a proposed explanation used as the starting point for further investigation

inflation a general increase in prices in an economy, leading to an increase in the cost of living

infographic a poster that uses images, graphs, shapes and colour to illustrate key trends and patterns in data

infrastructure physical and organisational requirements needed for the operation of a society; for example, roads and water supplies

inselberg an isolated steep-sided mountain, uncovered as the softer land around it is eroded away

internal migration migration of people within the boundaries of a country or region

international migration migration of people from one country to another

lahar a fast-moving mudflow of volcanic ash and rock that flows down the slopes of a volcano either during an eruption or following heavy rains after an eruption

latitude a geographic coordinate line running east–west parallel to the equator; it measures the position of a place on Earth's surface north or south of the equator in degrees (°), minutes (") and seconds (')

landform a natural feature of Earth's surface such as a mountain, valley, sand dune or cliff

landscape all the visible features of an area of land, including landforms; different types of landscapes include coastal, desert, riverine, mountain and polar

landslide the mass movement of soil, rock and debris down a slope

lava the name given to magma when it reaches Earth's surface

less economically developed country (LEDC) a low-income country experiencing severe barriers to development; also referred to as a developing country

lobby an attempt by individuals or groups to influence the decisions of government and public opinion on issues such as deforestation, pollution and the impact of introduced species

longitude a geographic coordinate line running north–south perpendicular to the equator; it measures the position of a place on Earth's surface east or west of the Prime Meridian (i.e. 0° longitude) in degrees (°), minutes (") and seconds (')

longshore drift the movement of sand in a zigzag fashion along the beach until it meets an obstacle such as a headland or groyne

maar crater a crater formed by the explosive interaction between magma and groundwater; they are usually shallow, with steep sides and often filled with water

magma liquid rock from the mantle beneath Earth's crust that rises through volcanic vents to Earth's surface

mantle the solid layer under Earth's crust in which parts can melt into magma and move to the surface through volcanoes

marketing action to promote the selling of a product or service

mass wasting the movement of rocks downhill due to gravity, such as rock falls and landslides

mechanical weathering a process whereby rocks are broken into pieces by changes in temperature, being split open by expanding roots or ice

megacity a city with 10 million or more people

mesa a broad, isolated flat-topped landform found in desert areas; the eroded remains of a wide plateau

migration permanent or semi-permanent movement from one place to another

monsoon a seasonal wind, occurring in the Asia and Pacific regions, that brings large amounts of rain

more economically developed country (MEDC) a country with a strong economy, in which most people have access to good education, health care and employment opportunities

mountain a large **landform** that sits above the surrounding land, usually in the form of a peak; steeper than a hill

mountain range a series of peaks and high land linked in a line

northern hemisphere the half of Earth north of the Equator

pandemic a disease that spreads in multiple countries around the world at the same time, usually affecting a large number of people

petition signed support of a statement used to lobby national governments and the United Nations to support causes; used to persuade politicians that there is large support for a cause

phenomena things that are observed to exist or happen

photo essay a series of photographs used to visually present information about the characteristics of a place or process

plateau a large flat area of land that is high above sea level

population density a measure of how many people live in a particular region or area

population pyramid a graph showing the number of males and females living in age groups in a population

PQE an acronym for Pattern, Quantify and Exception; used to describe spatial patterns or graphs

primary dune a mound of sand at the rear of a beach

primary method a data-gathering activity undertaken in the field, such as field sketches

pyroclastic flow a huge avalanche of hot gas and tephra that can move at speeds of up to 700 kilometres per hour and heat to temperatures of 1000°C, destroying everything is its path

qualitative data non-numerical data based on qualities or characteristics

quantitative data numerical data based on measurements or counts

reef a long line of jagged rock, coral or sand sitting just above or below the surface of the ocean

refraction a change in the direction of a wave as it approaches land

refugee a person threatened by conflict or a natural disaster who is forced to flee their home to seek safety, food and shelter in another country

relative location the whereabouts of a location described by referring to another place, landmark or even time; for example, Melbourne is east of Geelong, or Sydney is 878 km from Melbourne

research question an idea to be investigated, or a problem to be solved through research

reurbanisation when people move back into inner city areas that have been improved through urban renewal

rift valley a valley with a wide, flat floor caused by dropped blocks known as graben from the formation of block mountains

rural based in or relating to living in the countryside rather than a city or large town

rural to urban migration a form of migration where large numbers of rural-dwellers move to urban areas with the promise of better job opportunities and services

SALTS an acronym for Scale, Axis, Legend, Title and Source; used in creating a graph

satellite image an image of Earth taken from space

scoria cone a small volcano with relatively steep sides, usually formed as the result of a single volcanic eruption

sea wall a wall that runs parallel to the shore at the back of a beach, designed to stop erosion by waves and storms and protect the paths, roads and houses built along the shore

secondary dune a mound of sand that is further inland behind a primary dune, with more soil to promote vegetation such as shrubs and trees

secondary method a data-gathering activity undertaken outside of field studies, such as research

sedimentary rock a rock that forms when layers of sediment are pressed together by the weight of overlying rock

seismic wave a wave generated from the sudden release of energy in or beneath Earth's crust during an earthquake

SHEEPT an acronym for Social, Historical, Environmental, Economic, Political and Technological factors; used to describe and analyse spatial patterns

sketch a simple drawing made to record data when in the field

skilled worker a worker with specialist training or qualifications

slum a highly populated residential area with densely packed, poor-quality housing with few services, often inhabited by people living in poverty and with few choices of where to live

soil liquefaction when wet soil turns to liquid mud during an earthquake, collapsing buildings

southern hemisphere the half of the Earth south of the Equator

spatial technology computer systems that interact with real-world locations in some way

spheroidal weathering the rounding of rocks caused by chemical weathering where rocks disintegrate along the edges and corners are rounded by water coming from more than one direction

spit a narrow bar of deposited sand linked to the coast

stack a pillar of rock separated from cliffs or created from a collapsed arch

subduction the process whereby a continental plate meets an oceanic plate, and the oceanic crust bends and moves down toward the upper mantle where the edge of the plate melts into magma that feeds volcanoes along the converging margin

suburb a medium-density region of housing that surrounds the central business district of a city

suburbanisation a trend in city development where land on a city's outer edge that was previously used for farming or being kept as a natural environment is made available for housing

sustainable able to be maintained at a certain level, such as energy or air quality

swash water and suspended materials such as sand flowing onto the beach

swell a collection of waves formed by winds that can travel for long distances through the ocean; measured from the crest to the trough

tectonic plate a slab of solid rock that comprises Earth's crust and upper mantle, that can collide with other plates to form mountains and volcanoes, and generate earthquakes, and move apart to form ocean trenches

tephra solid material ejected from a volcano during an eruption, such as volcanic bombs and ash

TOASTIE an acronym for Title, Orientation, Annotations, Scale, Time, Information and Edge

tombolo a landform that forms when waves deposit a bar of sand to link an island to the coast

tourism the business of providing services for tourists such as accommodation, transport, tours and other activities

training walls walls built on either side of a river mouth to stop sand from blocking it

transform margin where the edges of tectonic plates slide past one another; the plates can become stuck and then break free with a sudden movement, triggering an earthquake

trough the bottom of a wave

tsunami a huge ocean wave generated by an offshore earthquake, volcano or landslide

urban based in or relating to a city or large town

urban renewal the redevelopment of an existing urban area, such as the urban renewal of former docklands to create housing

urban sprawl the continual expansion of suburbs at the edges of large cities onto natural or farming land

urbanisation an increase in the proportion of people living in urban areas

volcanic bomb large rocks launched high into the air during an explosive volcanic eruption

volcano a vent in Earth's crust from which lava, gas and ash can pour during a volcanic eruption

wadi a dry watercourse in a deep gorge that divides a plateau

wave when the surface of water moves up and down as energy created by wind (or earthquakes and volcanic eruptions) moves through the water; the wind forces the water to rise and gravity pulls it back down again

wave-cut platform a flat eroded area of rock at the base of a coastal cliff

weathering a process where rock is worn away, broken down or dissolved into smaller and smaller pieces by water, heat and cold; chemical weathering is a chemical change in a rock, while mechanical weathering involves rocks being broken into pieces

Index

Acknowledgements

The author and publisher are grateful to the following for permission to reproduce copyright material:

PHOTOGRAPHS: AAP/Itsuo Inouye/AP, **32**, /Josh Jerga, **113** (top), /Bruce Omori/EPA, **30–1** (bottom), /Dan Peled, **46**, /Tourism Australia, **42**, /U.S. Geological Survey/AP, **31** (top); Airview Online, **120–1** (top), Alamy Stock Photo/4k-Clips, **84** (bottom), /David Ball, **4** (bottom left), **41**, /Christine Osborne Pictures, **90** (left), /Neale Clark/robertharding, **90–1** (bottom right), /Paul Franklin/Eye Ubiquitous, **23** (top), /Sean Gallagher/National Geographic Image Collection, **107** (top), /Hemis, **29** (top), /Jrstock, **137** (bottom left), /Khoon Lay Gan, **135** (top), /Makhnach_S, **134**, /Richard Milnes, **78** (bottom), **145** (bottom left & centre left), /Premraj K.P, **104–5** (top), /QAI Publishing/Universal Images Group North America LLC, **34–5** (centre), /Dmitry Rukhlenko - Travel Photos, **i** (centre left), **87**, /Sean Sprague, **108–9** (centre), **123**, /Simon Stone/Science Photo Library, **48** (top), /David Wall, **67** (top), **145** (bottom right); Bay Side Library Service/Rose Stereographs Co. Image 19875, **145** (top left); Fairfax/Edwina Pickles, **70–1** (bottom); Fotolibra/Miles Kelly, **33** (right); Gallery Stock/© Paola + Murray, i (bottom left), **133**; Getty Images/Atlantic Digital, **28–9** (bottom), /Debbie Barrow/EyeEm, **130**, /Felix Cesare, **162**, /Copyright Wacharachat Vaiyaboon hangingpixels, **27** (top), /DEA/D'ARCO EDITORI, **22** (left), /DigitalGlobe, **4** (bottom right), **151**, /Dorling Kindersley, **12–3**, **18** (left), **21** (top), /Jim Evans, **59** (left), /Michael Fay, **45** (bottom), /JIJI PRESS/AFP, **81**, /Nigel Killeen, **91** (top), /John S Lander, **53**, /MediaNews Group/Orange County Register, **128**, /Son Nguyen/AFP, **114–5** (centre), /Alok Ojha/EyeEm, **61**, /Hannah Peters, **49**, /QAI Publishing, **80**, /RAW.Exposed, **137** (bottom right), /Galen Rowell, **15**, /Jewel Samad, **82–3** (bottom), /Satellite Earth Art, **149** (bottom left), /Qilai Shen, **111** (bottom), /SOPA Images, **83** (top), /vladimir zakharov, **89**; iStockphoto/1001slide, **16–7**, /AlexanderYershov, **8**, **50**, **86**, **124** (arch icon), /Orbon Alija, **131**, /ampueroleonardo, **i** (background), /aquasolid, **i** (top right), **51**, /Christoph Burgstedt, **48** (bottom right), /calvindexter, **137** (top), /DarrenTierney, **161**, /duha127, **86**, **152–5**, /ewg3D, **129**, /FrankRamspott, **136** (bottom), /GlobalP, **140** (top), /Lemanieh, **156–60** (clouds), /leonello, **136** (top), /Rainer Lesniewski, **111** (top), /LUHUANFENG, **136** (centre), /MasterLu, **2–3**, /PeterHermesFurian, **103** (top), /Saturated, **132**, /SeanPavonePhoto, **98** (bottom), /Sebalos, **146** (top), /john shepherd, **143** (top), /simonbradfield, i (top left), **1**, /SolStock, **i** (bottom right), **125**, /tonda, **156–60** (desert), /Steven Tritton, **54–5**; Jason Juta, **24–5**; Image used by permission: ©2019 JL Darling, LLC, www.RiteintheRain.com, **127**; Mapcarta, © OpenStreetMap and Mapbox, **79** (bottom); Katy Murenu, **119**; Museums Victoria, Unknown photographer, collections.museumvictoria.com.au/items/1688875, **145** (top right); NASA Earth Observatory image by Robert Simmon, using Suomi NPP VIIRS data provided courtesy of Chris Elvidge (NOAA National Geophysical Data Center), **149** (top); NASA images courtesy Jeff Schmaltz LANCE/EOSDIS MODIS Rapid Response Team, GSFC, **149** (bottom left); National Geographic Creative/Pierre Mion, **45** (centre); Danielle O'Leary, **135** (bottom); Oz Aerial/Daryl Jones, **27** (centre); Erin Schubert, **142** (bottom); Science Photo Library/NASA, **108** (bottom); Shutterstock.com/Bjoern Alberts, **38** (top), /Galyna Andrushko, **38** (bottom centre), /arindambanerjee, **33** (left), /AsiaTravel, **109** (top), /David Bostock, **75** (top), /Dorothy Chiron, **11**, /Eugene Drobitko, **59** (top right), /evenfh, **35** (top), **50**, /fritz16, **55** (inset), /Aldona Griskeviciene, **144** (bottom left), /Alan Heartfield, **35** (bottom), /ibreakstock, **92**, /iofoto, **95** (bottom), /ixpert, **22–3**, /Andrea Izzotti, **36**, /Jackson Stock Photography, **47**, /katacarix, **70–1** (top), /Visun Khankasem, **69** (top), /kojihirano, **39** (bottom), /Steve Lovegrove, **45** (top), /Maurizio De Mattei, **38** (bottom right), /Doug Meek, **18–9**, /Dominic Meinardi, **14**, /Naeblys, **24** (left), /Layne V. Naylor, **39** (centre), /onemu, **102**, /pisaphotography, **62–3**, /Radiokafka, **100–1**, /Matyas Rehak, **103** (bottom), /Bruce Rolff, **93**, /saiko3p, **140–1** (bottom), /solarseven, i (top centre), **9**, /T photography, **96–7** (bottom), /Darren Tierney, **57** (top), **76–7** (bottom), /totajla, **37**, /Constantine Vladimirovich, **104** (bottom), /Martin Valigursky, **68**, /Taras Vyshnya, **116–7** (bottom), /www.sandatlas.org, **38** (bottom left); Southern Habitat, **75** (bottom); © The State of Queensland 1930, 1973, 1986, 2003, 2004, 2011, **77** (top left), **66** (bottom left), **66** (bottom right), **67** (bottom), **77** (top right), **76** (top), **84** (centre). Licensed under Creative Commons Attribution licence; State Library of Victoria [H38508], **79** (top), **145** (centre right); Unsplash/Johnny Bhalla, **43** (top right), /Bailey Mahon, **65**; U.S. Geological Survey, **122** (top); Peter Van Noorden, **57** (bottom), **66** (top).

OTHER MATERIAL: Map and graph, data from Australian Bureau of Statistics, 3412.0 - Migration, Australia, 2015–16. Licensed under CC BY 4.0 licence, **113** (bottom), **146** (bottom); Infographic, © Australian Human Rights Commission, 2014. Used with permission, **150**; Map, based on data from Beck, H.E., N.E. Zimmermann, T.R. McVicar, N. Vergopolan, A. Berg, E.F. Wood: Present and future Köppen-Geiger climate classification maps at 1-km resolution, Scientific Data 5:180214, doi:10.1038/sdata.2018.214 (2018), Licensed under CC BY 4.0 licence, **39** (top); Map, based on data from Bird, P. (2003), An updated digital model of plate boundaries, Geochem. Geophys. Geosyst., 4, 1027, doi:10.1029/2001GC000252, 3, and United Nations Office for Disaster Risk Reduction (UNDRR), Catalog of Earthquakes 1970–2014, updated November 24, 2015. Licensed under CC BY 4.0 licence, **20–1**; Map, based on data from Boyce, Julie, Catalogue of volcanoes and vents of the Newer Volcanics Province of southeast Australia (February 2015 update) and Victoria State Government, Seamless Geology 2014, **26–7**; Map, Bounford.com and UNEP/GRID-Arendal www.grida.no/resources/5563, **84** (top); Graph, based on data from Climate-Data.org, **147**; Map, data © Commonwealth of Australia 2017, Bureau of Meteorology. Licensed under CC BY 3.0 AU licence, **5** (bottom); Graph, based on data from Department of Home Affairs,

Historical Migration Statistics, gov.au/data/dataset/historical-migration-statistics and Department of Immigration and Border Protection, Permanent Additions to Australia's Resident Population, data.gov.au/dataset/ds-dga-e87976fd-c545-4ec0-ab5b-034080868624/details, **122** (bottom); Map screenshot images are the intellectual property of Esri and is used herein with permission. Copyright © 2020 Esri and its licensors. All rights reserved, **30** (top); Map, data from European Commission Emergency Response Coordination Centre, 25 October 2018 © European Union, 2018. Licensed under CC BY 4.0 licence, base satellite photograph courtesy of NASA Earth Observatory, **82** (left); Map, based on data from GEODATA TOPO 250K Series 2 - (Personal Geodatabase format). Geoscience Australia, Canberra, 2006, pid. geoscience.gov.au/dataset/ga/63999, Licensed under CC BY 4.0 licence, **4** (bottom right); Map, based on data from Hansen, M. C., P. V. Potapov, R. Moore, M. Hancher, S. A. Turubanova, A. Tyukavina, D. Thau, S. V. Stehman, S. J. Goetz, T. R. Loveland, A. Kommareddy, A. Egorov, L. Chini, C. O. Justice, and J. R. G. Townshend, 2013, *High-Resolution Global Maps of 21st-Century Forest Cover Change*, Science 342 (15 November): 850–53. Data available on-line from: earthenginepartners.appspot.com/science-2013-global-forest. Licensed under CC BY 4.0 licence, **97** (top); Map, based on data from Natural Earth and European Environment Agency, World digital elevation model. Licensed under CC BY 2.5 DK licence, **17** (top); Map, data © OpenStreetMap contributors and available under the Open Database License, **78** (top), **134–5**; Map, Parks Australia, **40** (bottom); Graphs, www.populationpyramid.net, **149**; Map, Bill Rankin, www.radicalcartography.net, **96** (left); Graph, Reserve Bank of Australia, 2018. Licensed under CC BY 4.0 licence, **122** (centre right); Map, Hannah Ritchie (2018) - "Urbanization". Published online at ourworldindata.org/urbanization. Data from World Bank – World Development Indicators and United Nations Population Division. World Urbanization Prospects: 2018 Revision, **99**; Map, Max Roser and Esteban Ortiz-Ospina (2016) - "Literacy". Published online at ourworldindata.org/literacy. Data from CIA Factbook (2016), **139**; Maps and graph, Max Roser, Hannah Ritchie, Esteban Ortiz-Ospina and Joe Hasell (2020) - "Coronavirus Pandemic (COVID-19)". Published online at ourworldindata.org/coronavirus. Based on data published by European Centre for Disease Prevention and Control (ECDC), **114** (left), **115** (top); Illustration, Erin Schubert, **142–3** (centre); Map, based on Statistics Indonesia. Population Density by Province, 2000–2015 (Indonesia). Jakarta, Indonesia: Statistics Indonesia, 2 Mar. 2017, **106**; Graph, based on Indonesia Statistical Bureau (BPS), Proyeksi Penduduk, 2005, **107** (bottom); Advertisements, Tourism Australia, **72**, **73** (top); Graph, Tourism Australia 2020 Report, **73** (bottom); Illustration, U.S. Department of Agriculture, **144** (bottom right); Map, based on data from U.S. Geological Survey Data Series 948, 33, Falcone, J.A., 2015, U.S. conterminous wall-to-wall anthropogenic land use trends (NWALT), 1974–2012, **97** (top); Map republished with permission from the Victorian Department of Jobs, Precincts and Regions, **120** (bottom); Map, Worldmapper, map no. 367, **112**.

While every care has been taken to trace and acknowledge copyright, the publisher tenders their apologies for any accidental infringement where copyright has proved untraceable. They would be pleased to come to a suitable arrangement with the rightful owner in each case .

FLIP BOOK

good humanities

Flip your book over and begin your journey through **History**